Encyclopedia of Soybean: Global Importance Volume IV

Edited by **Albert Marinelli and Kiara Woods**

New York

Published by Callisto Reference,
106 Park Avenue, Suite 200,
New York, NY 10016, USA
www.callistoreference.com

Encyclopedia of Soybean: Global Importance
Volume IV
Edited by Albert Marinelli and Kiara Woods

International Standard Book Number: 978-1-63239-299-2 (Hardback)

Contents

Preface

Legumes are an essential part of a healthy diet. They are an excellent source of carbohydrates, protein, vitamins and minerals. Soybean has its significance in the overall agriculture and trade as well as in its contribution to food supply. Soybean is the richest in terms of protein and has no cholesterol in comparison with animal food sources and conventional legume. In addition to this, soybean is a cheap source of food and has medicinal qualities because of its photochemical, genistein, isoflavones content. It has been found that Soybean is very helpful in fighting diabetes, heart disease and cancer. Presently, Soybean protein and calories are being employed for the prevention of body wasting which is mostly because of HIV. Soybean's nutritional value is of extreme importance in places where medication facilities are not available. This book covers significant topics such as Soybean in monogastric nutrition, nutrition and health, potential use of soybean flour and Brazilian soybean varieties.

After months of intensive research and writing, this book is the end result of all who devoted their time and efforts in the initiation and progress of this book. It will surely be a source of reference in enhancing the required knowledge of the new developments in the area. During the course of developing this book, certain measures such as accuracy, authenticity and research focused analytical studies were given preference in order to produce a comprehensive book in the area of study.

This book would not have been possible without the efforts of the authors and the publisher. I extend my sincere thanks to them. Secondly, I express my gratitude to my family and well-wishers. And most importantly, I thank my students for constantly expressing their willingness and curiosity in enhancing their knowledge in the field, which encourages me to take up further research projects for the advancement of the area.

Editor

Soybean in Monogastric Nutrition: Modifications to Add Value and Disease Prevention Properties

Samuel N. Nahashon and
Agnes K. Kilonzo-Nthenge

Additional information is available at the end of the chapter

1. Introduction

Soybean (*Glycine max*), a leguminous oilseed and one of the world's largest and most efficient sources of plant protein, has on average crude protein content of about 37-38% and 20% fat on a dry matter basis. The crude protein content of soybean varies with geographical region and damage to the soybean crop can cause a significant decrease in the crude protein content of the soybean. On the other hand, processed soybean meal which is commonly used in monogastric feeding contains about 44-48% crude protein (NRC, 1998). This high crude protein content of soybean and soybean meal in conjunction with high energy due to significant fat content and low fiber content make soybean an ideal source of protein for humans and also ideal feed ingredient in monogastric animals feeding (Table 1). The heat processed soybean is the form primarily used for human consumption and it contains lower crude protein concentration (37%) when compared to soybean meal which is produced from solvent extracted seeds and seeds without hulls (44% and 49% CP, respectively). The soybean meal is the common form of soybean utilized in animal feeding. While other nutrients such as calcium, potassium and zinc also tend to be lower in heat treated soybean than in the soybean meals, the energy and fat content is higher in the heated soybean than the soybean meals.

Previous reports have shown that soybean and soybean meal contains a balanced amino acid profile when compared with other oilseed meals, although it is deficient in methionine and lysine (Zhou et al., 2005). The comparisons of the amino acid composition of soybean and soybean meal which are routinely utilized in human and monogastric feeding are presented in Table 2.

Nutrient	Soybean seeds[2]	Soybean Meal[3]	Soybean Meal[4]
IFN[5]	5-04-597	5-04-604	5-04-612
Crude Protein, %	37	44	49
Energy, kcal/kg	3,300	2,230	2,440
Crude fat, %	18	0.8	1.0
Crude fiber, %	5.5	7.0	3.9
Calcium, %	0.25	0.29	0.27
Phosphorus[6], %	-	0.65	0.62
Phosphorus[7], %	0.53	0.27	0.24
Potassium, %	1.61	2.00	1.98
Iron, mg/kg	80	120	170
Zinc, mg/kg	25	40	55

[1]National Research Council 1994. [2]Heat processed. [3]Seeds, meal solvent extracted.

[4]Seeds without hulls, meal solvent extracted. [5]International feed number. [6]Total phosphorus.

[7]Non-phytate or available phosphorus.

Table 1. Comparison of selected nutrient composition of soybean and soybean meal[1]

Soybean boasts a well balanced amino acid profile with high digestibility when compared with other oilseeds. In soybean, the digestibility coefficients of lysine are estimated to be 91% (NRC, 1994) whereas that of cysteine and phenylalanine is estimated at 83-93 (Bandegan et al, 2010). Previous reports cite evidence that soya is a rich source of amino acids (Angkanaporn et al., 1996). Holle (1995) reported that soybean meal provides the best balance for amino acids when compared with other oilseeds and thus makes it a more suitable plant source protein for human and monogastric food animals. According to Kohl-Meier (1990), soybean accounts for more than 50% of the world's protein meal. The form in which soybean is utilized for human or monogastric feeding determines the nutritional value in terms of content and bioavailability of amino acids as described in the following section.

2. Anti-nutritional properties of soybean

In their natural form, soybeans contain anti-nutrients or phytochemicals which bear toxic effects when ingested by both humans and monogastric food animals. These anti-nutrients are nature's means of protection for the soybean plant from invasion by animals, bacteria, viruses and even fungi in the ecosystem. The major anti-nutrients in soybean are phytates, protease enzyme inhibitors, soyin, goitrogens, hemagglutinins or lectins, giotrogens, cyanogens, saponins, estrogens, antigens, non-starch polysaccharides and soy oligosaccharide. Although most of these anti-nutritional compounds in soybean were discussed in Nahashon and Kilonzo-Nthenge (2011), additional reviews of some of the major anti-nutritional factors are presented as follows:

Nutrient	Soybean seeds[2]	Soybean Meal[3]	Soybean Meal[4]
IFN[6]	5-04-597	5-04-604	5-04-612
	----------------------------------(%)----------------------------------		
Arginine	2.59	3.14	3.48
Lysine	2.25	2.69	2.96
Methionine	0.53	0.62	0.67
Cystine	0.54	0.66	0.72
Tryptophan	0.51	0.74	0.74
Histidine	0.90	1.17	1.28
Leucine	2.75	3.39	3.74
Isoleucine	1.56	1.96	2.12
Phenylalanine	1.78	2.16	2.34
Threonine	1.41	1.72	1.87
Valine	1.65	2.07	2.22
Glycine	1.55	1.90	2.05
Serine	1.87	2.29	2.48
Tyrosine	1.34	1.91	1.95

[1] National Research Council, 1994.

[2] Seeds, heat processed.

[3] Seeds, meal solvent extracted.

[4] Seeds without hulls, meal solvent extracted.

[5] International feed number.

Table 2. Comparison of selected amino acid composition of soybean and Soybean meals[1]

2.1. Phytates

Phytic acid (inositol hexakisphosphate), the storage form of phosphorus in seeds such as those of soybean is considered an anti-nutritional factor in monogastric nutrition. Raboy et al. (1984) cited evidence that phytic acid accounted for 67-78% of the total phosphorus in mature soybean seeds and these seed contain about 1.4-2.3% phytic acid which varies with soybean cultivars. In plants phytic acid is the principal store of phosphate and also serves as natural plant antioxidant. Earlier reports (Asada et al., 1969) suggested that phytic acid in soybean not only makes phosphorus unavailable, but also reduces the bioavailability of other trace elements such as zinc and calcium and the digestibility of amino acids (Ravindran, 1999). Ravindran et al., (1999) reported that in the presence of phytate, soybean protein forms complexes with the phytate. Heaney et al. (1991) reported that the absorption of calcium from soybean-based diets was higher in low-phytate soybean when compared with high phytate-soybean. This supports the assertion that soybean has the potential to form phytate-mineral-complex which inhibits the availability of the minerals to monogastric animals. In soybean, phytate is usually a mixture of calcium/magnesium/potassium salts of inositol hexaphosphoric acid which adversely affects mineral bioavailability and protein solubility when present in animal feeds (Liener, 1994). Reports of Vucenik and Shamsuddin (2003) point that inositol bears biological significance as antioxidant in mammalian cells. However, it interferes with mineral utilization and is the primary cause of low phosphorus utilization in soy-based poultry and swine diets. Phytin also chelates other minerals such as Calcium, Zinc, iron, Manganese and Copper, rendering them unavailable to the animals. Soybean has the highest amount of phytate when compared to all legumes and cereal grains. The phytates have been reported to be resistant to cooking temperatures.

2.2. Protease enzyme inhibitors

Proteases refer to a group of enzymes whose catalytic function is to hydrolyze, cleave or breakdown peptide bonds of proteins. They are also called proteolytic enzymes that include trypsin, chymotrypsin, elastase, carboxypeptidase, and aminopeptidase which convert protein (polypeptides, dipeptides, and tripeptides) into free amino acids which are readily absorbed through the small intestine into the blood stream. Protease or trypsin inhibitors of soybean have been reported to hinder the activity of the proteolytic enzymes trypsin and chymotrypsin in monogastric animals which in turn lowers protein digestibility (Liener and Kakade, 1980). Other reports (Liener and Kakade, 1969; Rackis, 1972) confirmed that trypsin inhibitors were key substances in soybean that adversely affected its utilization by chicks, rats and mice. Kunitz, (1946) isolated trypsin inhibitor from raw soybeans and demonstrated that it was associated with growth inhibition. These protease inhibitors were also reported to inhibit Vitamin B_{12} availability (Baliga et al., 1954). Later studies have also shown that the presence of dietary soybean trypsin inhibitors caused a significant increase in pancreatic proteases (Temler et al., 1984). To the benefit of the soybean plant, soybean protease inhibitors serve as storage proteins in seeds, regulate endogenous proteinases, and also protect the plant and seeds against insect and/or microbial proteinases (Hwang et al., 1978). These protease inhibitors contain about 20% of the sulfur-containing amino acids methionine and cysteine, which are also the most limiting essential amino acids in soybean seeds (Hwang et al., 1978).

Recent reports (Dilger et al., 2004; Opapeju et al. 2006; Coca-Sinova et al. 2008) show that the nutritional value of soybean meal for monogastric animals is significantly hindered by these protease inhibitors which interfere with feed intake and nutrient metabolism. They reported that soybeans with high content protease inhibitors, especially trypsin inhibitors adversely affect protein digestibility and amino acid availability. In earlier reports, Birth et al. (1993) cited evidence that ingestion of food containing trypsin inhibitor by pigs increased endogenous nitrogen losses hence the effect of the trypsin inhibitors affected nitrogen balance more by losses of amino acids of endogenous secretion than by losses of dietary amino acids. This may be due to compromised integrity of the gastrointestinal lining leading to reduction of absorptive surface. However, Gertler et al. (1967) attributed the depression of protein digestibility to reduced proteolysis and absorption of the exogenous or dietary protein which was caused by inhibition of pancreatic proteases.

2.3. Hemagglutinins or lectins

Soybean hemagglutinins or lectins are glycoproteins that resemble some animal glycoproteins, such as ovalbumin and are rich in the acidic amino acids while being low in the sulfur-containing amino acids methionine and cysteine. According to Lis *et al.* (1966), the only carbohydrates serving as constituents of soybean hemagglutinins are mannose and glucosamine. Soybean hemaglutinins are a component of soybeans that were characterized by Schulze *et al.* (1995) as being anti-nutritional. Oliveira *et al.* (1989) reported that these glycoproteins bind to cellular surfaces via specific oligosaccharides or glycopeptides. They exhibit high binding affinity to small intestinal epithelium (Pusztai, 1991) which impairs the brush border and interfere with nutrient absorption. Hemaglutinins have also been implicated in producing structural changes in the intestinal epithelium and resisting gut proteolysis (Pusztai *et al.*, 1990), changes which in most cases result in impairment of the brush border and ulceration of villi (Oliveira *et al.*, 1989). This occurrence result in significant decrease in the absorptive surface and increase endogenous nitrogen losses as reported by Oliveira and Sgarbierri (1986) and Schulze *et al.* (1995). Consequently, Pusztai *et al.* (1990) observed that hemagglutinins depressed growth rate in young animals. Hemagglutinins are known to promote blood clotting or facilitating clumping together of red blood cells. It has however been concluded that soybean hemagglutinins play a minor role in the deleterious effect contributed by anti-nutritional factors in raw soybean.

2.4. Giotrogens and estrogens

Soybean is known to produce estrogenic isoflavones which bind to the estrogen receptors. According to Doerge and Sheehan (2002), such estrogenicity was implicated in toxicity and estrogen-mediated carcinogenesis in rats. Genistein is the major soy isoflavones of great concern in conferring estrogenic effect especially in women. Although the possible gitrogenic effect of the soybean isoflavones has not been researched extensively, certain soy components may present some antithyroid actions, endocrine disruption, and carcinogenesis in animals and humans as well. Messina, (2006) reported that Soybean contains flavonoids that may impair the activity of the enzymes thyroperoxidase. Earlier reports (Divi and Doerge, 1996) indicate that plant-derived foods such as soybean contain flavonoids which are widely distributed, possessing numerous biological activities that include antithyroidism in experimental animals and humans. A study was conducted to evaluate inhibition of thyroid peroxidase (TPO), the enzyme that catalyzes thyroid hormone biosynthesis, by 13 commonly consumed flavonoids (Divi and Doerge, 1996). Consequently, most flavonoids tested including genistein and daidzein were potent inhibitors of TPO (Figure 1). They suggested that chronic consumption of flavonoids, especially suicide substrates, could play a role in the etiology of thyroid cancer. More recent reports (Messina, 2006; Xiao, 2008; Zimmermann, 2009) have also shown that use of soy-based formula without added iodine can produce goiter and hypothyroidism in infants, but in healthy adults soy-based products appear to have negligible adverse effects on thyroid function.

Other reports (Fort, 1990) have also shown that concentrations of soy isoflavones resulting from increased consumption of soy-based formulas inhibited thyroxine synthesis inducing goiter and hypothyroidism and autoimmune thyroid disease in infants. Still many questions linger on the full Impacts of soy products on thyroid function, reproduction and carcinogenesis, hence the need for further research in this context. According to Divi et al. (1997), the IC50 values for inhibition of TPO-catalyzed reactions by genistein and daidzein were ca. 1-10 microM, concentrations that approach the total isoflavone levels (ca. 1 microM). More recent finding using normal rats (Chang and Doerge, 2000) however suggest that, even though substantial amounts of TPO activity are lost concomitant to consumption of soy isoflavone, the remaining enzymatic activity is sufficient to maintain thyroid homeostasis in the absence of additional perturbations. On the other hand, additional factors other than the soybean isoflavones can also cause overt thyroid toxicity. These may include other soybean fractions, iodine deficiency, defects of hormone synthesis that may be caused by gene mutations or environmental and random factors including dietary factors that may be goitrogenic.

Environmental estrogens, on the other hand, are classified into two main categories namely phytoestrogens which are of plant origin and xenoestrogens which are synthetic (Dubey et al., 2000). Soybeans contain phytoestrogens which can cause enlargement of the reproductive tract disrupting reproductive efficiency in various species, including humans (Rosselli et al., 2000), and rats (Medlock et al., 1995). In some cases these estrogens are hydrolyzed in the digestive tract to form poisonous compounds such as hydrogen cyanide. Woclawek-potocka et al. (2004) reported that phytoestrogens acting as endocrine disruptors may induce various pathologies in the female reproductive tract. Studies have shown that soy-derived phytoestrogens and their metabolites disrupt reproductive efficiency and uterus function by modulating the ratio of PGF2a to PGE2. Because of the structural and functional similarities of phytoestrogens and endogenous estrogens, there is the likelihood that the plant-derived substances modulate prostaglandin synthesis in the bovine endometrium, impairing reproduction. Previous research has shown that phytoestrogens may act like antagonists or agonists of endogenous estrogens (Rosselli et al., 2000; Nejaty and Lacey, 2001).

2.5. Allergens and antigens

Today, food allergies are a common serious health threat and food safety concern around the world. The increasing use of soybean (*Glycine max*) products in processed foods owing to its nutritional and health promoting properties, poses a potential threat to individuals who allergic to foods and especially sinsitive to soybean. According to Cordle (2004), there are about 16 potential soy protein allergens that have been identified and the Food and Agriculture Organization of the United Nations includes soy in its list of the 8 most significant food allergens. While many of these soy allergenic proteins have not been fully characterized, their allergenicity can be mild to life-threatening anaphylaxis. Consequently, consumers who have allergies to soybean and soybean products are often at a risk of serious or life threatening allergic reaction if they consume these products.

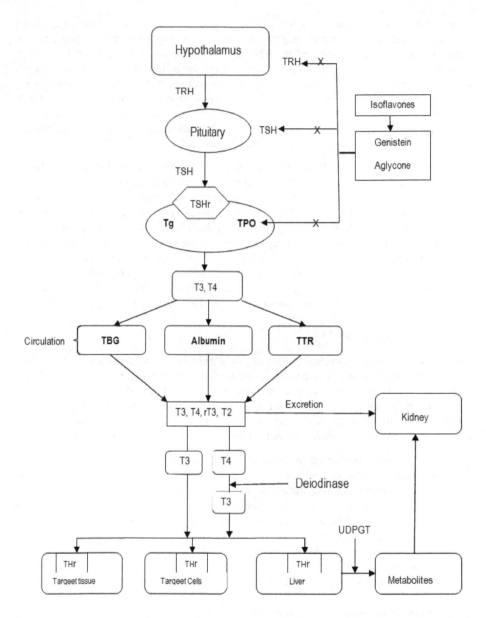

Figure 1. Schematic presentation of inhibition of TPO-catalyzed reactions by genistein and daidzein. TPO:thyroid peroxidase; UDPGT:Uridine diphosphate glucuronyltransferase; TBG:thyroxine binding globulin; TTR: Transthyretin; THr: thyroxine receptor; T:thyroxine; TSH:Thyroid stimulating hormone; TRH:Thyrotropin-releasing hormone; Tg:Thyroglobulin.

One of these low abundance proteins, Gly m Bd 30 K, also referred to as P34, is considered to be a dominant immunodominant soybean allergen. It is a member of the papain protease superfamily with a glycine in the conserved catalytic cysteine position found in all other cysteine proteases (Herman et al., 2003). The P34 protein is thought to trigger most allergic reactions to soy and it accumulates during seed maturation localizing in the protein storage vacuoles but not in the oil bodies (Kalinski et al., 1992). Several attempts have been made to understand and remove the allergenic proteins in soybean. Using a "gene silencing" technique, researchers were able to "knock out" a gene that makes Antigenic factors glycinin and β-conglycinin removal increases animal performance.

The prevalence of soybean allergy is estimated at 0.4% in children and 0.3%in adults North America (Sicherer and Sampson, 2010). A study was conducted to determine the relationship between adverse health outcomes and occupational risk factors among workers at a soy processing plant (Cummings et al, 2010). They reported that asthma and symptoms of asthma were associated with immune reactivity to soy dust. Such occurences are a occupational hazard and have led to strict regulatory oversight in soybean plants and food manufacturing plants that process food products containing soybean. There have been numerous recalls by the FDA (FDA Enforcement Reports) of several products containing soy proteins, paste, oils and flour for reasons ranging due to improper labeling. Unlisted soy protein on the product label is considered a potential hazard for people who may be allergic to soy.

2.6. Cyanogens and saponins

Soybean and other closely related legumes which are common food ingredients for human and monogastric animals have been recognized to contain cyanogenic compounds (Montgomery, 1980). The content of cyanide was reported at 0.07-0.3 pg of hydrogen cyanide/g of sample in soy protein products and 1.24 pg/g in soybean hulls when browning was kept to a minimum (Honig et al.,1983). Cyanide is considered toxic even in small amounts, hence where soy are a major constituent of a diet, there are concerns of cyanide toxicity.

On the other hand, saponins are unabsorbable glucosides of steroids, steroid alkaloids or triterpenes found in soybeans germ and cotyledons and other leguminous plants. They form lather in aqueous solutions and impart a bitter test or flavor in feed, resulting in reduction of feed consumption. In severe cases they cause haemolysis of red blood cells and diarrhea (Oakwndull, 1981). Raw soybeans have been reported to contain 2-5 g of saponins/100 g soy products. Although soybean saponins possess anti-nutritional properties, some are edible and have been reported to possess some health benefits. They have been shown to stimulate the immune system, bind to cholesterol and make it unavailable for absorption and allowing its clearance into the colon and eventual excretion (Elias et al., 1990).

2.7. Non-starch polyssacharides and soy oligosaccharides

Soybean oligosaccharides (OS) such as raffinose and stachyose are carbohydrates consisting of relatively small number of monosaccharides and they have been reported to influence ileal nutrient digestibility and fecal consistency in monogastric animals (Smiricky et al.,

2002). According to Leske et al., (1993), raffinose and stachyose represent about 4-6% of soybean dry matter. The digestion of OS in the small intestine is limited because mammals lack α-galactosidase necessary to hydrolyze the α 1,6 linkages present in OS (Slominski, 1994). Previous research has demonstrated that soy OS are responsible for increasing intestinal viscosity of digesta and as a result interfere with digestion of nutrients by decreasing their interaction with digestive enzymes (Smits and Annison, 1996). Irish and Balnave (1993) demonstrated that stachyose derived from the oligosaccharides of soyabean meals exert anti-nutritive effects in broilers fed high concentrations soyabean meal as the sole protein concentrate. The OS in soybean, raffinose and stachyose, are not eliminated by heat treatment during processing (Leske et al., 1993). In earlier reports (Coon et al.,1990) observed that removal of the OS from SBM in poultry diets increased the true metabolizable energy value of the diet by 20 percent.

The digestion of oligosaccharides in the small and large intestine is aided by beneficial microbial fermentation (Hayakawa et al., 1990). Certain oligosaccharides, however, are considered to be prebiotic compounds because they are not hydrolyzed in the upper gastrointestinal tract and are able to favorably alter the colonic microflora, conferring beneficial effects of digestion and fermentation of carbohydrates to the host. More recent studies have demonstrated that feeding a higher level of an oligosaccharide (8 g/kg) to chicks depressed metabolizable energy and amino acid digestibility (Biggs et al., 2007). Smiricky-Tjardes et al. (2003) reported the presence of significant quantities of galactooligosaccharides in soy-based swine diets. These soy oligosaccharides are partially fermented by gut microflora functioning as prebiotics which promote selective growth for beneficial bacteria. The high content of enzyme inhibitors in unfermented soybeans interferes with thecomplete digestion of carbohydrates and proteins from soybeans. When foods are not completely digested because of enzyme inhibitors, bacteria in the large intestine try to do the job, and this can cause discomfort, bloating, and embarrassment. The enzyme inhibiting properties of soybean compounds the low levels of digestive enzymes, a common phenomenon especially in elderly people.

3. Modifications of soybean to enhance nutritional value and health benefits

3.1. Genetic modifications (GMO's)

About 99 percent of the soybean that we consume is genetically modified (GMO) and referred to as GMO soybean. The genetic modifications in soybean are primarily meant to improve the yield and nutritional value of soybean, reduce allergenicity, create resistance to certain diseases or disease causing pathogens and/or confer tolerance to herbicides or adverse climatic or environmental conditions. For example, transgene-induced gene silencing was used to prevent the accumulation of Gly m Bd 30 K protein in soybean seeds. The Gly m Bd 30 K-silenced plants and their seeds lacked any compositional, developmental, structural, or ultrastructural phenotypic differences when compared with control plants (Herman et al., 2003). Current GMO crops, including soybean, have not been shown to add any additional allergenic risk beyond the intrinsic risks already present (Herman 2003).

The enhancement of the nutritional value of soybean and soybean products through genetic engineering of soybeans has been reported. Through genetic engineering completely new fatty acid biosynthetic pathways have been introduced into soybeans from exotic plants and various microorganisms (Cahoon et al., 1999; Wallis, Watts, and Browse, 2002). Soybean oil is used in many food applications and therefore alterring its composition, especially the fatty acid composition would benefit the consumer. Several fatty acids especially the omega-3 have been reported to possess health benefit to the consumer. Engineered soybean lines that are rich in oleic acid (producing stable oil that does not need to be partially hydrogenated and is thus free of trans fatty acids) and lines lower in saturated fatty acids have been produced (Kinney & Knowlton, 1998; Buhr et al., 2002). These high-oleic and low-saturated soybean oils provide the potential benefits to human health and point to the positive impact of the achievements in biotechnology that promote human health.

Bioactive polyunsaturated fatty acids are also known to confer beneficial and positive effects in humans' health (Knapp et al., 2003). These bioactive fatty acids can also be found in oils other than soybean can also be produced in soybean through genetic engineering. Thus, introducing these into soybean can enhance the existing health benefits of soybean with the complementary benefits of bioactive lipids and other compounds.These polyunsaturated fatty acids are known to mediate their heart-healthy effects by mechanisms independent of those of soy protein and they have been previously researched extensively (Kelley and Erikson, 2003; Knapp et al., 2003).

Previously attempts were made to increase the oxidative stability of soybean oil by increasing the composition in soybean of the fatty acids oleic and stearic and decrease linoleic acid content of the soybean without creating *trans* or polyunsaturated fatty acids (Clemente, 2009). According to Clemente, (2009) and Clemente and Cahoon, (2009), DuPont announced the creation of a high oleic fatty acid soybean, with levels of oleic acid greater than 80%. This product was due for release into the market in 2010. Soybean mutants with elevated and reduced palmitate have been developed (Rahman et al., 1999). While the palmitate content of commercial soybean cultivars is approximately 11%, elevated palmitate content in soybean oil may be important for the production of some food and industrial products.

Soy foods have also been reported (Sirtori and Lovati, 2001) to have the potential for reducing blood low-density lipoprotein (LDL) cholesterol concentrations in humans. According to Weggemans and Trautwein, (2003), this positive health effect appears to be directly related to the soy storage proteins rather than other components. The bulk of soy protein (more than 80%) is contributed by two major classes of storage protein, conglycinin (11S globulins) and beta-conglycinin (7S globulins). It has been possible to produce soybean transgenic lines with either 7S or no 11S protein using gene-silencing techniques (Kinney & Fader, 2002).

For many years soybean has been defined as a crop with the best amino acid composition within all cultivated protein crops and is most widely utilized for human and monogastric foods as a primary source of protein (Wenzel, 2008). Since the amino acids are directly used in the genetic formation of proteins and fatty acids, this makes the soybean invaluable in oil production and primary protein source of choice to many. There has been attempt through genetic engineering to modifiy the soybean to enhance its oxidative stability by changing the proportion of certain fatty acids, which would provide a more useful and abundant oil supply with health benefits to the consumer. The enhancement of soybean oil content, Clemente et al. (2009) achieved this goal by introducing a seed-specific transgene for a DGAT2-type enzyme from the oil-accumulating fungus *Umbelopsis ramanniana* into soybean. Without disrupting the protein content, the oil content was increased from approximately 20% of the seed weight to approximately 21.5%.

There has also been an attempt to genetically modify soybean to enhance flavors. These compounds are associated with the oxidation of the polyunsaturated fatty acids linoleic and linolenic acids (Frankel, 1987). There are hundreds of volatile compounds associated with bad flavors in soy preparations (Stephan and Steinhart, 1999), and these compounds are the predominant fatty acids in soybean oil whose oxidation during bean storage and processing results in the formation of secondary products of lipid oxidation that impart off-flavors to soy protein products.

Drought tolerant varieties of soybean have also been developed through genetic engineering. The Roundup Ready soybean, also known as soybean 40-3-2, is a transgenic soybean that has been immunized to the Roundup herbicide. Although soybean's natural trypsin inhibitors provide protection against pests, weeds still remain a major challenge in soy farming (Wenzel, 2008). An herbicide used to control weeds in soybean farming contained glyphosate which inhibited the expression of the soybean plant's ESPSP gene. According to Wenzel, (2008), the gene is involved in the maintenance of the "biosynthesis of aromatic metabolites," and killed the plant along with the weeds for which the herbicide was meant. Consequently, the soybean was genetically engineered by transferring a plasmid which provided immunity to glyphosate-containing herbicides was transferred to the soybean cells through the cauliflower mosaic virus, perfecting the Roundup Ready soybean.

Since drought stress is a major constraint to the production and yield stability of soybean, integrated approaches using molecular breeding and genetic engineering have also provided new opportunities for developing high yield and drought resistance in soybeans (Manavalan et al., 2009). Recently, Yang et al. (2010) pointed out that genetic engineering must be employed to exploit yield potential and maintaining yield stability of soybean production in water-limited environments in order to guarantee the supply of food for the growing human population and for food animals. On the other hand, new soybean varieties that are resistant to diseases and pests are being developed. Hoffman et al. (1999) observed that plants commonly respond to pathogen infection by increasing ethylene production. They suggested that soybean can be altered by genetic manipulation using mutagenesis to generate soybean lines with reduced sensitivity to ethylene. Two new genetic loci were identified, Etr1 and Etr2 and Plant lines with reduced ethylene sensitivity developed similar or less-severe disease symptoms in response to virulent Pseudomonas syringae. Other reports (Yi et al., 2004) indicate that *CaPF1*, a ERF/AP2 transcription factor in hot pepper plants may play dual roles in response to biotic and abiotic stress in plants and that through genetic engineering this factor could be modified to improve soybean disease resistance as well.

3.2. Enzyme supplementation

The diets of monogastric animals are primarily composed of feed ingredients of plant origin such as soybean. Soybean contains a variety of antinutritional factors such as phytin, non-starch polysaccharides, and protease inhibitors which limit the availability of nutrients that are essential for normal growth and performance, production or otherwise. Enzyme supplementation and soybean fermentation products have been used for a long time to improve the nutritional value and health-promoting properties of soybean (Kim et al., 1999). Recent reports (Kim et al., 2010) cite evidence that fermented soybean meal can effectively serve as an alternative protein source for nursery pigs at 3-7 weeks of age, possibly replacing the use of dried skim milk which tends to be more digestible than soybean. Feng et al. (2007) evaluated the effect of soybean meal fermented with *Aspergillus oryzae* on the activity of digestive enzymes and intestinal morphology of broilers. The fermentation had no significant influence on the activity of lipase, amylase and protease enzymes. However, they observed a significant increase in duodenal villus height and a decrease in crypts depth a sign of improved morphology of the absorptive surface. Earlier studies by Kiers et al. (2003) indicated that fermentation of soybean resulted in an increase nutrient solubility and digestibility in broilers.

The availability of phosphorus in soybean is about 30 percent; hence diets of monogastric animals must be supplemented with inorganic phosphorus or supplemented with the phytase enzyme to improve the utilization of phytate phosphorus (NRC, 1994; Richter, 1994). Phytase (myo-inositol-hexakisphosphate phosphohydrolase) is an enzyme that catalyzes the hydrolysis of phytic acid, an indigestible inorganic form of phosphorus in oil seeds such as soybean. As a result, phytases increase the digestion of phosphorus, consequently increasing its utilization and reducing its excretion by monogastric animals. The phytase enzymes commonly used in monogastric animal feeding are derived from yeast or fungi and bacteria. Nahashon et al. (1993) reported that phosphorus retention was improved in layers when the diet was supplemented with Lactobacillus bearing phytase activity. The use of phytase to hydrolyze phosphorus and possibly other mineral elements that may be bound onto phytate such as calcium, zinc, copper, manganese and ion, has been reported (Selle and Ravindran, 2007; Powell et al., 2011).

Recently, Liu et al. (2007) demonstrated that phytase supplementation in soybean-based diets significantly improved the digestibility of phosphorus and Calcium by 11.08 and 9.81%, respectively. A 2- 8% improvement of the digestibility of amino acids was also noted since phytate also binds to protein forming phytate-protein complex. This complex is less soluble resulting in decreased digestibility of soybean protein (Carnovale et al., 1988). Earlier, Singh and Kirkorian (1982) reported that phytate also inhibits trypsin and pepsin activities. These findings suggest that phytase supplementation in soybean-based diets can improve the digestibility of calcium, phosphorus and proteins and indeed amino acids. Augspurger and Baker, (2004) observed that high dietary levels of efficacious phytase enzymes can release most of the P from phytate, but they do not necessarily improve protein utilization (Boling-

Frankenbach et al., 2001). Supplemental phytase has also been reported to improving dietary phosphorus utilization by pigs (Sands et al., 2001; Traylor et al., 2001). Other reports (Pillai et al., 2006) demonstrated that addition of E. coli phytase to P-deficient broiler diets improved growth, bone, and carcass characteristics. Most recently Rutherford et al. (2012) demonstrated that addition of microbial phytase to diets of broiler chickens improved significantly the availability of phytate phosphorus, total phosphorus, other minerals such as calcium, zinc, manganese etc. and amino acids.

Protease enzymes on the other hand break down long protein chains into short peptides which can be readily absorbed. These proteolytic enzymes whose catalytic function is to hydrolyze or breakdown peptide bonds of proteins include enzymes such as trypsin, chymotrypsin, pepsin, papain, elastase, plasmin, thrombin, and proteinase K. These enzymes can also be supplemented in feed or indirectly by feeding microbials that have the potential to produce these enzymes in the gastrointestinal tract of the host animals. On the other hand, carbohydrases such as xylanase and amylase are enzymes that catalyze the hydrolysis of carbohydrates into sugars which are readily available or metabolizable by monogastric animals. Soybean meal contains approximately 3% of soluble non-starch polysaccharides (NSP) and 16% of insoluble NSP (Irish and Balnave, 1993). The NSP in soybean is thus of negligible amounts to yield digesta viscosity problems. Therefore, diets that comprise soybean are considered to be highly digestible, hence requiring less use of carbohydrases. Previous reports have, however, pointed out that since these cereal grains contain some soluble NSP, there is still the need to supplement soybean based diets with these enzymes to further improve their nutritional value (Maisonnier-Grenier et al., 2004).

Supplementing soybean based diets with multicarbohydrase enzymes, a preparation containing nonstarch polysaccharide-degrading enzymes, phytases and proteases reveled that these enzymes improved nutrient utilization and growth performance of broiler chickens (Woyengo et al., 2010). Cowieson and Ravindran (2008) reported that when these enzyme combinations were fed in broler diets with both adequate and reduced energy and amino acid content, a 3% and 11% increase in apparent metabolizable energy and nitrogen retention, respectively, were observed. Also feeding other multicarbohydrase combinations containing xylanase, protease, and amylase resulted in significant improvements in feed conversion and body weight gain of broilers (Cowieson, 2005). In their recent review, Adeola and Cowieson (2011) suggested that when used together with phytase, nonstarch polysaccharide-hydrolyzing enzymes may increase the accessibility of phytase to phytin encapsulated in plant cell walls.

Although the enhancement of monogastric animal performance using enzyme supplements in feed have been extensively researched and documented, the benefits of enzymes such as the phytases and multicarbohydrase in soybean-based diets of monogastric animals have not been fully explored and require further research. There is still a great deal of uncertainty regarding the mode of action of these enzymes including the phytases, carbohydrases and proteases and their combination thereof in soybean based diets of monogastric animals. It is just fair to not that the future of enzymes in nonruminant animal production is promising and will require further research to elucidate the role of enzyme supplementation in promoting health and provide an understanding of the modes of action of these enzymes in modulating gene functions and their interactions thereof.

3.3. Probiotics and prebiotics supplementation

Probiotics, also known as direct-fed microbials, are live microbial feed supplements which beneficially affect the host animal by improving its intestinal microbial balance (Fuller, 1989). They have been reported to improve feed consumption, feed efficiency, health and metabolism of the host animal (Cheeke, 1991). The total collection of these probiotics, other gut microflora, their genetic elements or genomic materials and their interactive environment or the gastrointestinal tract of the host is termed the "microbiome". Currently efforts are underway to understand the microbiome and elucidate the mode of action of both probiotics and prebiotics due to a great interest in these gut microbiota and their health promoting properties and enhancement of performance of humans and monogastric animals. The scientific basis for the modes of action of probiotics and prebiotics is, therefore, beginning to emerge. According to report of Quigley (2012) a number of human disease states may benefit from the use of probiotics; these include diarrheal illnesses, inflammatory bowel diseases, certain infectious disorders, and irritable bowel syndrome. Prebiotics promote the growth of "good" bacteria, primarily through competitive exclusion resulting in a variety of health benefits. Probiotics have also been reported to: (1) improve feed intake and digestion and production performance (Nahashon et al., 1994a, 1994b, 1994c, 1996), (2) maintain a beneficial microbial population by competitive exclusion and antagonism (Fuller, 1989), and (3) alter bacterial metabolism (Cole et al., 1987; Jin et al., 1997). Nahashon et al (1994a) evaluated the phytase activity in lactobacilli probiotics and the role in the retention of phosphorus and calcium as well as egg production performance of Single Comb White Leghorn laying chickens. They reported phytase activity in the direct-fed microbial and that supplementation of the corn-soy based diets with the probiotics to a 0.25% available phosphorus diet improved phosphorus retention and layer performance.

Prebiotics are defined as non-digestible food ingredients that beneficially affect the host, selectively stimulating their growth or activity, or both of one or a limited number of bacteria in the colon and thus improve gut health (Gibson and Roberfroid, 1995). They are short-chain-fructo-oligosaccharides (sc-FOS) which consist of glucose linked to two, three or four fructose units. They are not absorbed in the small intestine but they undergo complete fermentation in the colon by colonic flora (Gibson and Roberfroid, 1995). They benefit humans and monogastric animals by: (1) releasing volatile fatty acids which are absorbed in the large intestine and contribute to the animal's energy supply; (2) enhance intestinal absorption of nitrogen, calcium, magnesium, iron, zinc and copper in rats (Ducros et al., 2005); (3) increase the number and/or activity of bifidobacteria and lactic acid bacteria (Hedin et al., 2007); and (4) since they are non-digestible, they provide surface for attachment by pathogenic bacteria and therefore facilitate the excretion of these pathogenic microorganisms.

According to Bouhnik et al. (1994) and Gibson and Roberfroid, (1995), fructooligosaccharides such as inulin, oligofructose, and other short-chain fructooligosaccharides can be fermented by beneficial bacteria such as bifidobacteria and lactobacilli and in turn control or reduce the growth of harmful bacteria such as *Clostridium perfringens* through competitive exclusion. The bifidobacteria and lactobacilli are generally classified as beneficial bacteria (Gibson and Wang, 1994; Flickinger et al., 2003). Although the mode of action of several of these oligosaccharides are still obscure, Biggs et al. (2007, pointed out that even low concentrations (4 g/kg) of an indigestible oligosaccharide can be fed to monogastric animals with no deleterious effects on metabolizable energy and amino acid digestibility. The benefit of utilizing oligosaccharides in soy-based diets of monogastric animals are due to the ability of these oligosaccharide to pass through to the hindgut of the monogastric animals intact and to be fermented by beneficial bacteria that are stimulated to grow and produce compounds that are beneficial to the host. These beneficial bacteria are also able to prevent the growth of bacteria such as *Escherichia coli* and *Clostridium perfringens* that can be harmful to the host through competitive exclusion (Gibson and Roberfroid, 1995). Biggs and Parsons (2007) reported an increase in the digestibility of a few amino acids was by some oligosaccharides in cecectomized roosters.

3.4. Amino acids and vitamin supplementation

Soybean is an excellent source of protein and vegetable oil for human and animal nutrition due to its balanced amino acid profile. According to Berry et al. (1962), methionine is the most limiting amino acid followed by lysine, and threonine. The level of supplementation of the amino acids lysine, methionine, threonine and glycine was evaluated (Waguespack et al., 2009). Feed efficiency decreased significantly in broilers fed diets supplemented with more than 0.3% L-Lysine but not in birds fed diets containing 0.25% L-Lysine. It was also observed that up to 0.25% L-Lysine could be added to corn-soy diets of broilers supplemented with methionine, threonine and glycine. Waguespack et al. (2009) also reported that arginine and valine were equally limiting after methionine, threonine and glycine in the diets containing 0.25% L-Lysine.

Earlier reports (Douglass and Persons, 2000) demonstrated a significant improvement in feed efficiency by methionine and Lysine supplementation in broilers diets. Studies have shown that excess heating by extrusion cooking or autoclaving of soybean during oil extraction can decrease lysine availability, hence requiring supplementation of this and other amino acids in soybean meal-based diets (Persons et al., 1992). It has also been further established that supplementation of raw soybean meal with methionine is an effective way of eliminating the potential nutritional deficiencies in both the raw and heated soybean meals. Due to the fact that raw soybean contains protein of low quality, supplementation of pig raw soy-based diets with cysteine and the B-complex vitamins exhibited significant improvement in performance (Peterson et al., 1941). Evaluating the effect of supplementation of turkey diets with dL-tocopheryl acetate, Sell et al. (1997) reported that feeding soybean based diets containing 6-20 IU tocopheryl acetate/kg improved performance of male turkeys from 1-day of age to market age.

3.5. Heat treatment and autoclaving

Heat treatment is a common procedure in soybean processing during extraction of oil and the inactivation of antinutritional factors such as trypsin inhibitors. The processing inactivates these protease inhibitors, although there has to be a balance in conditions of heat inactivation since excessive heating could also destroy other essential nutrients. Kwok et al. (1993) demonstrated that excess heat in the inactivation of protease inhibitors of soybean may increases Maillard reactions between the amino group of amino acids and reducing sugars and as a result decrease the digestibility of energy and amino acids by monogastric animals. Comparing the nutritive value of different heat treated commercial soybean meals, Veltmann et al. (1986) reported that compared to the normal meal, excessively heat-treated soybean meal had lower crude protein which also reflected lower essential and non-essential amino acids, and less trypsin inhibitors. Herkelman et al. (1991) evaluated the effect of heating time and sodium metabisulfite (SMBS) on the nutritional value of full fat soybeans for chicks. They observed that chicks fed the full-fat soybean achieved maximum performance when the soybeans were heated at 121 ° C for 40 min, and the SMBS decreased by one half the heating times required inactivating trypsin inhibitors.

4. Health benefits of soybean in human nutrition

4.1. General overview

Soybeans which boast rich content of protein (38-40%) of high quality and with a balanced amino acids profile are widely grown around the world and are the most important world source of edible oil and protein. Besides it's use in livestock feeds and to some extent biofuels, soybean is processed into products that are utilized for human consumption such as soybean oil and fermented soybean products which have long been utilized to prepare healthy human foods worldwide (Kim et al., 1999). Highly purified and oil-free food grade proteins isolates containing as high as 95% crude protein are commonly utilized in human foods. In the United States, 90% of the soybean is used for food especially as soybean oil (Smith and Wolf, 1961). In Asia and other parts of the world, soybean and soybean products are routinely utilized in large quantities in various forms of foods such as mature soybean, soybean flour, soybean meal, soybean milk, and also as oriental soybean products such as tofu, natto, miso, shoyu, and sprouts. In the recent past there has been increased focus on soybean as human food because of its health benefits. As a result, considerable research effort has been directed to evaluating the health benefits and increasing the uses of soybean in human foods. Abundant supplies of high protein soybean products and the rapid development of the soybean processing industry has also contributed significantly to the increased use of soybean as human food.

4.2. Selected components of soybean that confer health benefits

Soybean and soybean products have been acclaimed to confer health benefits to consumers because they contain substances that have been confirmed to bear health conferring properties. These substances include Iron, isoflavones, high content of protein rich in balanced amino acids, the sulfur containing amino acids methionine and cysteine, saponins, phytoestrogens, and the omega-3 fatty acids present in soybean oil. Soybeans are a major source of nonheme iron in diets of humans. Although some of the iron is unavailable for it is in the form of ferritin complexed with phytate, calcium and proteins, iron in soybean is a bioavailable source for human consumption. On the other hand, the benefits of omega-3 long chain fatty acids to heart health are well established (Lemke et al., 2010) and enrichment of soybean oil with these fatty acids has been a sustainable way of increasing tissue concentration of these omega-3 fatty acids and in reducing the risk of cardiovascular disease.

Soybean also contains nonstarch polysaccharide (NSP) hydrolysis products of soybean meal. These nonstarch polysaccharide hydrolysis products of soybean meal are beneficial in maintaining fluid balance during *Enterotoxigeic Escherichia Coli* (ETEC) infection and controlling ETEC-induced diarrhea in piglets. Soybean fermented with *R. oligosporus produce* antibacterial compounds that are active against some gram-positive microorganisms. The material can be extracted with water from soybeans fermented by *R. oligosporus*. Genistein and other soybean isoflavones slow the growth of blood vessels to tumors, another action that makes it popular as a cancer fighter.

4.3. Disease prevention properties of soybean

4.3.1. Cholesterol

Soybean, a popular source of protein for both humans and other monogastric animals due to its protein content and quality, especially the balance of amino acids, is an invaluable source of oil which contains fatty acids known to be effective in prevention of cardiovascular disease. The US Food and Drug Administration (1999) indicated that soybean proteins were responsible for prevention of cardiovascular disease. According to Lovatti et al. (1996) the soybean 7S or β-conglycinin has also been implicated in the upregulation of liver high-affinity LDL receptors. This protein was also shown to reduce plasma triglycerines in humans and rats (Aoyama et al., 2001). Later studies (Duranti et al., 2004) evaluated the effect of soybean 7S globulin subunits on the upregulation of LDL receptors. They reported that it lowered the expression of β-VLDL receptors induced by soybeam subunit. The oral admoinistration of soybean 7S globulin and the α-subunit significantly reduced plasma cholesterol and tryglycerides of hypercholesterolemic rats (Duranti et al., 2004). On the other hand, feeding soybean (25 g/day) was associated with lower total cholesterol concentrations in individuals with initial cholesterol concentratons of greater than 5.7 mmol/L (Bakhit et al., 1994). Later studies (Carroll, 1991) demonstrated that soybean protein lowered blood lipids in humans and experimental animals. Sirtori et al. (1985) also demonstrated that a 50% substitution of animal protein with soybean protein significantly reduced blood cholesterol concentrations of humans with type II familial hypercholesterolemia.

Soybean fiber has also been reported to reduce blood lipids whereas comsumption of cookies containing 25 g soybean cotyledon fiber was associated with a significant reduction in total plasma LDL cholesterol in hypercholesterolemic patients (Lo et al., 1986). Various mechanisms of reduction of cholesterol in humans and other monogastric animals have been proposed. These include the direct effect of soybean peptides which may modulate the endocrine regulation of catabolism and/or reduction in cholesterol biosynthesis (Bakhit et al., 1994). Most recently, Cho et al. (2007) suggest that soy peptides can effectively stimulate LDL-R transcription in the human liver cell line and reduce blood cholesterol level. They proposed several mechanism and component of the cholesterol lowering activity of soybean which include blockage of bile acid and/or cholesterol absorption, inhibition of cholesterol synthesis, and stimulation of low-density lipoprotein receptor (LDL-R) transcription. Similar observations were reported earlier by Beynen et al. (1986) that hypocholesterolemic effect of dietary soybean protein was caused primarily by its influence on the heterohepatic circulation of bile acids and cholesterol.

4.3.2. Cancer

Soybean is a rich protein source for humans and monogastric animals and contains about 0.2-1.5 mg/g of the isoflavones daidzein and genistin, and their glycones daidzein and genistkein (Wang and Murphy, 1994). These isoflavones have been proposed to possess anticarsinogenic properties which may be associated with their ability to serve as antioxidants which prevent fat rancidity, β-carotene bleaching and glutathione peroxidase activity (Hendrich et al., 1994), antiestrogens, and inhibiting the estrogen synthetase preadipocyte aromatase in humans (Aldercreutz et al., 1993). Adlercreutz et al. (1991) suggested that the low breast cancer incidence in Japanese women may be attributed to their consumption of feeds rich in soybean, a source of isoflavones. These isoflavones in soybean such as genistein confer anticarcinogenic effect primarily by inhibiting estrogen binding to the estrogen receptors; the soy isoflavones compete for estrogen receptors.

Xu et al. (1995) also hypothesized that soybean isoflavones possess anticarsinogenic properties. They anaerobically incubated soybean isoflavones with human feces and observed that intestinal half life daidzein and genistein were as little as 7.5 and 3.3 hrs, respectively. Hence the bioavailability of these isoflavones was depended on the ability of gut microflora to degrade these compounds. They attributed the cancer protective effects of the isoflavones to also the isoflavone metabolites such as methyl p-hydroxyphenylacetate, a monophenolic compound of both exogenous flavonoids and tyrosine which are inhibitors of hormone-dependent neoplastic cell proliferation. This compound has high affinity for nuclear type II binding site which is involved in cell growth regulation by estrogenic hormones (Xu et al., 1995).

In his report, Messina (1999) stated that soybean isoflavones may reduce the risk of prostrate cancer in men and breast cancer in women. The anti-cancer properties in soybean are attributed to the isoflavone genistein which influence signal transduction and the potential role in preventing and treating cancer. McMichael-Phillips et al. (1998) observed a significant enhancement of DNA synthesis by breast cells taken from biopsies of normal breast tissue from women with benign and malignant breast disease when these women were fed soybean for about two weeks. On other studies (Jing et al., 1993) reported that daidzein, one of the two primary isoflavones in soybean exhibited anti-cancer effects by inhibiting the growth of HL-60 cells implanted into the subrenal capsules of mice. The anticancer effects of the isoflavone genistein may be attributed to its antioxidant properties and its ability to inhibit several enzymes that are involved in signal transduction (Wei et al., 1993) including tyrosine protein kinase (Akiyama et al., 1987), ribosomal S6 kinase (Linassier et al., 1990), Map kinase (Thoburn et al., 1994), the inhibition of the activity of DNA topoisomerase and increasing the concentration of transforming growth factor β (TGFβ) as reported by Benson and Colletta (1995).

The interest in soybean and soybean products has been driven by its potential health benefits, especially in prevention of various forms of cancer by the soybean isoflavones genistein, deidzein and glycitein. In more recent studies, Su et al. (2000) reported that isoflavones played a protective role against bladder cancer cells. They also observed that both genistein and combined isoflavones exhibited significant tumor seppressing effects. According to Messina and Barnes (1991), increased soybean consumption reduced the risk of breast, colon and breast cancer for people living in Asia as opposed to people living in the United States and Western Europe. A comprehensive review of the interelationship between diet and cancer by The World Cancer Research Fund (1997) revealed that vegetable intake decreased the risk of colon cancer. The increased consumption of soybean and soybean products have also been reported to reduce the risk of colon cancer in some human and animal populations.

While examining the ability of dietary soybean components to inhibit the growth of prostate cancer, Zhou et al. (1999) reported that dietary soybean products inhibited experimental prostrate tumor growth through direct effects in the tumor cells and indirectly through the effect on tumor neurovesculature. Earlier reports (Herbert et al., 1998) showed that increased consumptio of soybean products contributed to reduction in prostrate cancer risk. Phytochemicals in soybean have been reported to posses anticarcinogenic properties (Messina et al., 1994). Zhou et al. (1999) further observed that soybean isoflavones and phytochemicals inhibited LNCaP cell proliferation, blocked cell cycle progression and enhanced DNA fragmentation which is a marker for opoptosis or programmed cell death. Datta et al., (1997) reported that soybean is capable of oxidizing benzo (a) pyrene-7, 8-dihydrodiol and 2-aminofluorine which are known to cause developmental toxicity or transplacental carcinogenicity in mammals. In other studies, Wei et al., (1995) cited evidence that genistein's antioxidant properties and antiproliferative effects may be responsible for its anticarcinogenic effects. Therefore, high content of genistein in soybean and its high bioavailability increases soybean's potential for prevention of various forms of cancer.

Soybean saponins have also been cited as potential contributors to the health promoting properties of soybean and soybean products. Saponins are chemical structures consisting of triterpenoidal or steroidal aglycones with various carbohydrates moieties in plants. Saponins are excellent emulsifyers since they bear both hydrophilic and hydrophobic regions and they tend to inhibit colon tumor cell proliferation in vitro. Various saponins have demonstrated antimutagenic and anticarcinogenic effecta against cell lines. More recent studies (Ellington et al., 2005) suggested that the B-group soyasaponins may be colon cancer suppressive component os soybean serving as potential chemopreventative phytochemical. Therefore, soybean and soybean products should be explored further for their potential in prevention and treatment of the various forms of cancer.

4.3.3. Osteoporosis

Osteoporosis is a degenerative thinning of the bones that is associated with decreasing estrogen levels which is a common problem with aging, especially in women. According to Ikenda et al. (2006), soybeans and soybean products which contain large amounts of menaquinone-7 (vitamin K2) may help prevent the development of osteoporosis. Soybeans have also generated interest in connection with osteoporosis because they contain a phytoestrogens called isoflavones, which are believed to have potential as substitute for estrogen without its adverse side effects. Intake of Natto, an ancient Japanese food of fermented soybeans, was reported to bear properties that were preventative of postmenopausal bone loss through the effects of menaquinone-7 or bioavailable isoflavones which were more abundant in natto than in other soybean products (Ikenda et al., 2006). Heaney (1996) described vitamin K functionally as a cofactor of γ-carboxylase enzyme which mediates the conversion of undercarboxylated osteocalcin to carboxylated osteocalcin by transforming the glutamyl residue of osteocalcin into carboxyglutamic acid residue. The carboxyglutamic acids have high affinity for calcium ions in hydroxyapatite and regulate the growth of these crystals in bone formation. Therefore there is sufficient evidence to suggest that fermented soybean products can effectively maintain bone stiffness (Katsuyama et al., 2002) by increasing serum levels of menaquinone-7 and γ-carboxylated osteocalcin (Kaneki et al., 2001) as well as maintaining bone mineral density.

Soybean and soybean-based diets for human contain naturally occurring bioactive compounds known as phytochemicals that have been cited to confer long-term health benefits (Setchell, 1998). These phytoestrogens primarily occur as glycosides bearing a weak estrogen-like activity which allows them to bind to the estrogen receptor (Miksicek, 1994) and therefore are of great significance as remedy where estrogen levels decline due to old age. These isoflavones can serve as alternative to estrogen therapy in the treatment of existing low bone mass or osteoporosis. They present potential naturally occuring alternative to hormone or drug therapy (estrogen) that would minimize bone loss in menopausal women.

In other studies, Picherit et al. (2001) assessed the dose-dependent effect of daily soybean isoflavones consumption in reversing bone loss in adult ovariectomized rats. They reported that in adult ovariectomized rats, daily soybean isoflavone consumption decreased bone turnover but did not reverse established osteopenia. In earlier studies using a rat model, Arjmandi et al. (1996) evaluated the potential for soybean protein isolate to prevent bone loss induced by ovarian hormone deficiency. They reported an increase in femoral and vertebral bone densities in rats that were fed soybean diets possibly due to the presence of isoflavones in soybean.

5. Nutritional benefits of soybean in other monogastric animals

5.1. Poultry

Soybean meal is the primary protein source in corn-soy based poultry rations. It is fed to poultry as soybean meal and is primarily the by-product of soybean oil extraction; it's the ground defatted flakes. Various studies have been conducted to evaluate methods of enhancing the acceptability of soybean and the enhancement of its nutritional value in poultry feeding. For instance, a study was conducted to evaluate the effect of extruding or expander processing prior to solvent extraction on the nutritional value of soybean meal (SBM) for broiler chicks. The results of this study indicate that pre-solvent processing method (expander or non-expander) had no significant effect on the nutritional value of SBM for broiler chicks. Both Methionine and Lysine supplementation increased feed efficiency (Douglas and Persons, 2000). Several other studies (Coca-Sinova et al. 2008; Dilger et al. 2004; Opapeju et al. 2006) have evaluated various methods of enhancing the digestibility of individual amino acids and protein of soybean meal.

The guinea fowl is classified as poultry and although its production is not popular as chickens, it is gradually gaining popularity and acceptance as alternative meat to chicken. It is also gradually finding its share of the global market for poultry and poultry products. Lacking however, is estimates for nutrient requirements of the guinea fowl. Recent efforts have focused on evaluating the growth pattern of the guinea fowl (Nahashon et al. 2010) and their nutrient requirements (Nahashon et al. 2009, 2010, 2011). The soybean meal has been utilized extensively as the sole protein source for the guinea fowl providing accurate estimate for the nutrient requirements for both the Pearl Grey and French varieties of the guinea fowl.

5.2. Swine

Soybean meal and soybean products have also been used extensively in swine production because of its relatively high concentration of protein (44 to 48%) and its excellent profile of highly digestible amino acids. Soy protein provides most amino acids that are deficient in most cereal grains commonly fed as energy sources in swine production. However, as opposed to feeding animal source proteins, when raw soybean is fed to young pigs as the primary protein source, dramatic slowdown in body weight gains were reported even with supplementation of amino acids such as methionine and cysteine (Peterson et al., 1942). The animal source proteins tend to exhibit higher digestibility than plant source proteins such as soybean and therefore better suited for nursery pigs (Kim et al., 2009; Gottlob et al., 2006). The low digestibility of raw soybean by young pigs is therefore attributed to the low nutritive value of the raw soybean protein. Due to the high cost of feeding and also the antinutritional factors in raw soybean, there have been attempts to minimize the amount of soybean in swine rations and also to improve its digestibility. The digestibility of the amino acids of soybean by swine has also been researched quite extensively (Smiricky-Tjardes et al. 2002; Grala et al., 1998; NRC 1998).

The supplementations of raw soybean with the amino acids threonine and the B- complex vitamins have been reported to enhance growth in young pigs. Also fermented soybean meals, soybean meals with enzyme supplements and extruded soybean meals have been used extensively in swine diets and they tend to improve performance especially of young pigs (Kim et al., 2006). Kim et al. (2009) reported an increase in crude protein concentration from 50.3 to 55.3% by fermentation of soybean meal with *A. oryzae* without affecting the balance of limiting amino acids for pigs. Bruce et al. (2006) evaluated the inclusion of soybean

(SB) processing byproducts such as gums, oil, and soapstock into soybean meal. Addition of these processing by-products significantly reduced the nutritive value of the resultant meal. Smiricky-Tjardes et al. (2003) evaluated other approaches such as the addition of galactooligosaccharides on ileal nutrient digestibility to enhance and expand the utilization of soybean in swine production.

5.3. Aquatic life

The feeding value of soybean as a rich protein source has also been extended to aquaculture. The high protein level makes soybean meal a key ingredient for aquaculture feeds since soybean meal is considerably less expensive than traditionally used marine animal meals. However, soybeans not contain complete amino acid profiles and usually are deficient in the essential amino acids lysine and methionine. Therefore, other protein sources should be used in combination with soybean to overcome the deficiencies. Soybean meal and genetically modified soybean products have also been employed in aquaculture (Hammond et al. 1995). Naylora et al. (2009) points to the importance of fish oils and fishmeal as a protein source in food animal production and also the extensive use of soybean and soybean products as protein supplements in aquaculture feeds.

5.4. Companion animals

The term "companion animals" refers to the entire spectrum of animal species which are considered as 'pets' such as cats, dogs, fish, rabbits, rodents, cage birds, and even non-indigenous species. Large animals such as horses, as well as small ruminants such as the goats and sheep have also been classified as companion animals as well because they contribute to human companionship; they have an important role to play in our society. The companion animal industry is rapidly growing sector of the global economy and so is the need for provision of adequate nutritional regimens for optimum growth, production and reproduction. Relatively few data are available on the nutrient digestibilities of plant-based protein sources by companion animals.

Plant source proteins such as soybean and soybean products are predominantly used in diets of companion animals. Soybeans are an essential part of Plant-based protein sources and are generally less variable in chemical composition than animal-based protein sources especially in nutrients such as calcium and phosphorus. The effects of including selected soybean protein sources in dog diets on nutrient digestion at the ileum and in the total tract, as well as on fecal characteristics, were evaluated (Clapper et al. 2001). Apparent amino acid digestibility at the terminal ileum, excluding methionine, threonine, alanine, and glycine, were higher ($P < 0.01$) for soybean protein-containing diets when compared with diets containing other sources of protein.

The effects of soybean hulls containing varying ratios of insoluble: soluble (I: S) fiber on nutrient digestibility and fecal characteristics of dogs were evaluated (Burkhalter et al., 2001). Ileal digestibility of dry matter by dogs fed the soybean hulls treatments responded quadratically ($P < 0.05$) to I: S fiber diets, with digestibility coefficients decreasing as the I:S approached 3.2. In other studies (Tso and Ling, 1930) reported that the blood cholesterol value of rabbits is slightly higher in animals fed the soybeans diets than in controls. However, differences in cholesterol levels between rabbits fed on cooked and raw soybeans were not statistically significant. Also the blood of rabbits fed exclusively on water-soaked raw soybeans showed an increase in uric acid, urea nitrogen, inorganic phosphorus and cholesterol. After extending this study to feeding cooked soybeans, there are no demonstrable changes in the blood composition of rabbits whether they were fed cooked or raw soybeans.

The optimum concentration of a mixture of soybean hulls and defatted grape seed meal (SHDG) for rabbits was evaluated (Necodemus et al. 2007). They observed that SHDG could be included up to 26.7% in diets for fattening rabbits and lactating does that meet ADL and particle size requirements. In another study, Angora goat doelings (average BW 22.1 kg) were used to examine the effects of dietary crude protein level and degradability on mohair fiber production (Sahlu et al. 1992). They reported that plasma glucose was elevated 2 hours after feeding in the goats fed conventional, solvent-extracted soybean meal, whereas glucagon concentrations were greater at 0 and 4 h in the group fed expelled, heat-treated soybean meal.

Author details

Samuel N. Nahashon[1*] and Agnes K. Kilonzo-Nthenge[2*]

*Address all correspondence to: snahashon@tnstate.edu

1 Department of Agricultural Sciences, Tennessee State University, Nashville, TN, USA

2 Department of Family and Consumer Sciences, Tennessee State University, Nashville, TN, USA

References

[1] Aljmandi, B. H., Elekel, L., Hollis, B. W., Amin, D., Stacewicz-Sapuntzakis, M., Guo, P., & Kukreja, S. C. (1996). Dietary soybean protein prevents bone loss in an ovarectomized rat model of osteoperosos. *J. Nutr.*, 126, 161-167.

[2] Adeola, O., & Cowieson, A. J. (2011). Board-invited review: Opportunities and challenges in using exogenous enzymes to improve nonruminant animal production. *J. Anim. Sci,*, 89, 3189-3218.

[3] Adachi, M., Kanamori, J., Masuda, T., Yagasaki, K., Kitamura, K., Mikami, B., & Utsumi, S. (2003). Crystal structure of soybean 11S globulin: Glycinin A3B4 homohexamer. *Proceedings of the National Academy of Sciences, USA,*, 100, 7395-7400.

[4]   Adler-Nissen, J. (1978). Enzymatic hydrolysis of soy protein for nutritional fortification of low pH food. *Annals of Nutrition and Alimentation,*, 32, 205-216.

[5]   Adlercreutz, H., Bannwart, C., Wahala, K., Makela, T., Brunow, G., Hase, T., Arosemena, P. J., Kellis, J. T., Jr , , & Vickey, E. L. (1993). Inhibition of human alromatase by mammalian lignans and isoflavonoids phytoestrogens. *J. Steroid Biochem. Mol. Biol.,* 44, 147-153.

[6]   Adlercreutz, H., Honjo, S., Higashi, A., Fotsis, T., Hamalainem, E., Hasegawa, T., & Okada, H. (1991). Urinary excretion of lignans and isoflavonoid phytoestrogens in Japanese men and women consuming a traditional Japanese diet. *Am. J. Clin. Nutr.,* 54, 1093-1100.

[7]   Akiyama, T., Ishida, J., Nakagawa, S., et al. (1987). . Genistein, a specific inhibitor of tyrosine specific protein kinases. J. Biol. Chem. ., 262, 5592-5595.

[8]   Asada, K., Tanaka, K., & Kasai, Z. (1969). Formation of phytic acid in cereal grains. Ann NY Acad Sci , 165, 801-814.

[9]   Applegate, T. J., Webel, D. M., & Lei, X. G. (2003). Efficacy of a phytase derived from Escherichia coli and expressed in yeast on phosphorus utilization and bone mineralization in turkey poults. *Poult. Sci.,* 82, 1726-1732.

[10] Angkanaporn, K., Ravindran, V., & Bryden, W. L. (1996). Additivity of apparent and true ileal amino acid digestibilities in soybean meal, sunflower meal, and meat and bone meal for broilers. *Poultry Science,* 75(9), 1098-1103.

[11] Augspurger, N. R., & Baker, D. H. (2004). High dietary phytase levels maximize phytate-phosphorus utilization but do not affect protein utilization in chicks fed phosphorusor amino acid-deficient diets. *J. Anim. Sci.,* 82, 1100-1107.

[12] Barbazan, M. (2004). Liquid swine manure as a phosphorous source for corn-soybean rotation. Thesis. Iowa State University, Ames, Iowa.

[13] Beilinson, V., Chen, Z., Shoemaker, C., Fischer, L., Goldberg, B., & Nielsen, C. (2002). Genomic organization of glycinin genes in soybean. *Theoretical and Applied Genetics,* 104, 1132-1140.

[14] Buhr, T., Sato, S., Ebrahim, F., Xing, A., Zhou, Y., Mathiesen, M., et al. (2002). Ribozyme termination of RNA transcripts down-regulate seed fatty acid genes in transgenic soybean. *The Plant Journal,*, 3, 155-63.

[15] Beynen, A. C., Van Der Meer, R., West, C. E., Cugano, M., & Kritchevsky, D. (1986). Possible mechanisms underlying he differential cholesterolemic effects of dietary casein and soy protein. In:, *Nutritional effects on Cholesterol Metabolism,* (Beynen, A. C, ed),, 29-45, Transmondial, Voorthuizen, The Netherlands.

[16] Benson, J. R., & Colletta, A. A. (1995). Transforming growth factor β. prospects for cancer prevention and treatment. *Clin. Immunother,* 4, 249-258.

[17] Biggs, P. & Pearsons, C. M. (2007). The Effects of Several Oligosaccharides on True Amino Acid Digestibility and True Metabolizable Energy in Cecectomized and Conventional Roosters. *Poultry Science*, 86, 1161-1165.

[18] Braden, C. R. (2006). Salmonella enterica serotype enteritidis and eggs: a national epidemic in the United States. *Clin. Infect. Dis.*, 43, 512-517.

[19] Barth, Christian A., Bruta Lãoehding, Martin Schmitz & Hans Hagemeister. (1993). Soybean trypsin inhibitor(s) reduce absorption of exogenous and increase loss of endogenous protein. *J. Nutr.*, 123, 2195-2200.

[20] Beynen, A. C., Winnubst, E. N. W., & West, C. E. (1983). The effect of replacement of dietary soybean protein by casein on the fecal excretion of neutral steroids in rabbits. *Z. Theraphysiol. Tierrer. naehr. Futtermittelkd.*, 49, 43-49.

[21] Berry, T. H., Becker, D. E., Rasmussen, O. G., Jensen, A. H., & Norton, H. W. (1962). The Limiting Amino Acids in Soybean Protein. *J ANIM SCI*, 21(3).

[22] Burkhalter, T. M., Merchen, N. R., Bauer, L. L., Murray, S. M., Patil, A. R., Brent, J. L., Jr , , Fahey, G. C., & Jr 200, . (2001). The Ratio of Insoluble to Soluble Fiber Components in Soybean Hulls Affects Ileal and Total-Tract Nutrient Digestibilities and Fecal Characteristics of Dogs. *J. Nutr.*, 131, 1978-1985.

[23] Biggs, P., Parsons, C. M., & Fahey, G. C. (2007). The Effects of Several Oligosaccharides on Growth Performance, Nutrient Digestibilities, and Cecal Microbial Populations in Young Chicks. *Poultry Science*, 86, 2327-2336.

[24] Bouhnik, Y., Flourie', B., Ouarne, F., Riottot, M., Bisetti, N., Bornet, F., & Rambaud, J. (1994). Effects of prolonged ingestion of fructo-oligosaccharides on colonic bifidobacteria, fecal enzymes and bile acids in humans. *Gastroenterology*, 106, A598, (Abstr.).

[25] Balloun, S. L., & 198, . (1980). Soybean meal processing. In: K.C. Lepley (Ed) Soybean Meal in Poultry. The Ovid Bell press, Inc., Fulton, Missouri. , 6-12.

[26] Bhakit, R., Klein, B. P., Essex-Sorlie, D., Ham, J. O., Erdman, J. W., & Porter, S. M. (1994). Intake of 25 g of soybean protein with or wothout soybean fiber alters plasma lipids in men with elevated cholesterol concentrations. *J. Nutr.*, 124, 213-222.

[27] Carrol, K. K. (1991). Review of clinical studies on cholesterol lowering response to soy protein. *J. Am. Diet. Assoc.*, 91, 820-827.

[28] Cahoon, E. B., Carlson, T. J., Ripp, K. G., Schweiger, B. J., Cook, G. A., Hall, S. E., & Kinney, A. J. (1999). Biosynthetic origin of comnjugated double bonds: production of fatty acids components of high-value drying oils in transgenic soybean embryos. *Proceedings of the National Academy of Sciences, USA,*, 22, 12935-12940.

[29] Clemente, T. E., & Cahoon, E. B. (2009). Soybean Oil: Genetic Approaches for Modification of Functionality and Total Content. *Plant Physiology*, 151, 1030-1040.

[30] Cook, D. R. (1998). The effect of dietary soybean isoflavones on the rate and efficiency of growth and carcass muscle content in pigs and rats. Ph.D. Dissertation. Iowa State Univ.,.

[31] Cowieson, A. J. (2005). Factors that affect the nutritional value of maize for broilers. *Anim. Feed Sci. Technol.*, 119, 293-305.

[32] Protease inhibitors and carcinogenesis: a review. Cancer Invest , (1996). , 14, 597-608.

[33] Clapper, G. M. C. M., Grieshop, N. R., Merchen, J. C., Russett, J. L., Brent Jr, , Fahey, G. C., & Jr , . (2001). Ileal and total tract nutrient digestibilities and fecal characteristics of dogs as affected by soybean protein inclusion in dry, extruded diets. *J ANIM SCI.*, 79, 1523-1532.

[34] Clemente, Tom E., & Edgar, B. Cahoon. (2009). Soybean Oil: Genetic Approaches for Modification of Functionality and Total Content. *Plant Physiology.*, 151(3), 1030-40.

[35] Coca-Sinova, A., Valencia,† D. G., Jimenez-Moreno,† E., Lazaro,† R., & Mateos, G. G. (2008). Apparent Ileal Digestibility of Energy, Nitrogen, and Amino Acids of Soybean Meals of Different Origin in Broilers. *Poultry Science*, 87, 2613-2623.

[36] Cheryan, M. (1980). Phytic acid interactions in food systems. *CRC Crit. Rev. Food Sci. Nutr.*, 13, 297-335.

[37] Cowieson, A. J., & Ravindran, V. (2008). Effect of exogenous enzymes in maize-based diets varying in nutrient density for young broilers: Growth performance and digestibility of energy, minerals and amino acids. *Br. Poult. Sci.*, 49, 37-44.

[38] Coon, C. N., Leske, K. L., Akavanichan, O., & Cheng, T. K. (1990). Effect of oligosaccharide-free soybean meal on true metabolizable energy and fiber digestion in adult roosters. *Poult. Sci.*, 69, 787-793.

[39] Cordle, C. T. . (2004). Soy Protein Allergy: Incidence and Relative Severity· 1213S EOF-1219S EOF.

[40] Crump J.A, Griffin P.M, Angulo FJ. (2002). Bacterial contamination of animal feed and itsrelationship to human foodborne illness. *Clin Infect Dis*, 35, 859-865.

[41] Cheeke, P. R. (1991). Applied Animal Nutrition. AVI Publishing Company, Inc., Wesport, CT.

[42] Cole, C. B., Fuller, R., & Newport, M. J. (1987). The effect of diluted yogurt on the gut microbiology and growth of piglets. *Food Microbiol.*, 4, 83-85.

[43] Choct, M. (2006). Enzymes for the feed industry: Past, present and future. *World's Poult. Sci. J.*, 62, 5-16.

[44] Cummings, K. J., Gaughan, D. M., Kullman, G. J., Beezhold, D. H., Green, B. J., Blachere, F. M., Bledsoe, T., Kreiss, K., & Cox-Ganser, J. (2010). Adverse respiratory outcomes associated with occupational exposures at a soy processing plant. *ERJ*, 36, 1007-1015.

[45] Datta, K., Sherblom, P. M., & Kulkarni, A. P. (1997). Co-oxidative metabolism of 4-aminobiphenyl by lipoxygenase from soybean and human term placenta. *Drug Metab. Desp.*, 25, 196-205.

[46] Dalluge, J. J., Eliason, E., & Frazer, S. (2003). Simultaneous identification of soyasaponins and isoflavones and quantification of soyasaponin Bb in soy products using liquid chromatography/electrospray ionization-mass spectrometry. *Journal of Agricultural and Food Chemistry*, 51, 3520-3524.

[47] Drewnowski, A. (2001). The science and complexity of bitter taste. *Nutrition Review,*, 59, 163-169.

[48] Duranti, M., Lovati, M. R., Dani, V., Barbiroli, A., Scarafoni, A., Castinglion, S., Ponzone, C., & Morazzoni, P. (2004). The α-subunit fro soybean 7S globulin lowers plasma lipids and upregulates liver β-VLDL receptors in rats fed a hypercholesterolemic diet. *J. Nutr.*, 134, 1334-1339.

[49] Dauglas, M. W., & Parsons, C. M. (2000). Effect of presolvent extraction processing method on the nutritional value of soybean meal for chicks. *Poult. Sci.*, 79, 1623-1626.

[50] Dilger, R. N., Sanders, J. S., Ragland, D., & Adiola, O. (2004). Digestibility of nitrogen and amino acids in soybean meal with added soyhulls. *J. Anim. Sci.*, 82, 715-724.

[51] Dixon, R. A. (2004). Phytoestrogens. Annu. Rev. Plant Biol. , 55, 225-261.

[52] Davis, Hancock. D. D., Rice, D. H., Call, D. R., Di Giacomo, R., Samadpour, M., et al. (2003). Feedstuffs as a vehicle of cattle exposure to Escherichia coli O157:H7 and Salmonella enterica. *Vet Microbiol*, 95, 199-210.

[53] De Rham, O., & Jost, T. (1979). Phytate-protein interactions in soybean extracts and low-phytate soy protein products. *J. Food Sci.*, 44, 596-600.

[54] Doerge, DR, & Sheehan, DM. (2002). Goitrogenic and estrogenic activity of soy isoflavones. *Environ Health Perspect.*, 110(3), 349-53.

[55] Divi, RL, Chang, HC, & Doerge, DR. (1997). Anti-thyroid isoflavones from soybean: isolation, characterization, and mechanisms of action. *Biochem Pharmacol.*, 54(10), 1087-96.

[56] Dubey, R. K., Rosselli, M., Imthurn, B., Keller, P. J., & Jackson, E. K. (2000). Vascular effects of environmental oestrogens: implications for reproductive and vascular health. *Hum Reprod Update*, 4, 351-363.

[57] Du, Pont. H. L. (2007). The growing threat of foodborne bacterial enteropathogens of animal origin. *Clinical Infectious Diseases*, 45, 1353-1361.

[58] Ducros, V., Arnaud, J., Tahiri, M., Coudray, C., Bornet, F., Bouteloup-Demange, C., Brouns, F., Rayssiguier, Y., & Roussel, A. M. (2005). Influence of short-chain fructo-oligosaccharides (sc-FOS) on absorption of Cu, Zn, and Se in healthy postmenopausal women. *J. American College of Nutr.*, 24(1), 30-37.

[59] DeRouchey J.M, Tokach, M.D., Nelsen J. L, Goodband, R. D., Dritz, S. S., Wood-worth, J. C., & James B. W. (2002). Comparison of spray-dried blood meal and blood cells in diets for nursery pigs. *J. Anim. Sci.*, 80, 2879 EOF-86 EOF.

[60] Elias, R., De Meo, M., Vidal-Ollivier, E., et al. (1990). Antimutagenic activity of some saponins isolated from Calendula officinalis L., C arvensis L., and Hedra helix L.". *Mutagenesis*, 5, 327-331.

[61] Ellington, A. A., Barlow, M., & Singletary, K. W. (2005). Induction of macroautopha-gy in human colon cancer cells by soybean B-group triterpenoid saponins. *Carcinio-genesis*, 26, 159-167.

[62] Eliot, M., Herman, Ricki. M., Helm, Rudolf., Jung, , & Anthony, J. Kinney. (2003). Ge-netic Modification Removes an Immunodominant Allergen from Soybean. (1), 1-36.

[63] Fort, P., Moses, N., & Fasano, M. (1990). Breast and soy-formula feedings in early in-fancy and the prevalence of autoimmune thyroid disease in children. *J Am Coll Nutr*, 9, 164-167.

[64] Frankel, E. N. (1987). Secondary products of lipid oxidation. *Chemistry and Physics of Lipids*, 44, 73-85.

[65] Friedman, M., & Brandon, D. L. (2001). Nutritional and health benefits of soy pro-teins. *Journal of Agricultural and Food Chemistry*, 49, 1069-1086.

[66] Flickinger, E. A., Schreijen, E. M. W. C., Patil, A. R., Hussein, H. S., Grieshop, C. M., Merchen, N. R., Fahey, G. C., & Jr , . (2003). Nutrient digestibilities, microbial popula-tions, and protein catabolites as affected by fructan supplementation of dog diets. *J. Anim. Sci.*, 81, 2008-2018.

[67] Fuller, R. (1989). Probiotics in man and animals. A review. *J. Appl. Bacteriol.*, 66, 365-378.

[68] Gudbrandsen, O. A., Wergedahl, H., Mork, S., Liaset, B., Espe, M., & Berge, R. K. (2006). Dietary soya protein concentrate enriched with isoflavones reduced fatty liv-er, increased hepatic fatty acid oxidation and decreased the hepatic mRNA level of VLDL receptor in obese Zucker rats. *Br. J. Nutr.*, 96, 249-257.

[69] Green, S., Bertrand, S. L., Duron, M. J., & Maillard, R. (1987). Digestibility of amino acids in soyabean, sunflower and groundnut meals, determined with intact and cae-cectomised cockerels. *British Poultry Science*, 28(4), 643-652.

[70] Garcia, M. C., Torre, M., Marina, M. L., & Laborda, F. (1997). Composition and char-acterization of soyabean and related products. *Critical Reviews in Food Science and Nu-trition*, 37, 361-391.

[71] Grala, W., Verstegen, M. A., Jansman, A. J. M., Huisman, J., & Van Leeusen, P. (1998). Ileal apparent protein and amino acid digestibilitiesand endogenous nitrogenlosses in pigs fedsoybean and rapeseed products. *J. Anim. Sci.*, 76, 557-568.

[72] Gertler, A., Birk, Y., & Bondi, A. (1967). A comparative study of the nutritional and physiological significance of pure soybean trypsin inhibitors and of ethanol-extracted soybean meals in chicks and rats. *J. Nutr.*, 91, 358-370.

[73] Goldflus, F., Ceccantini, M., & Santos, W. (2006). Amino acid content of soybean samples collected in different brazilian states-Harvest 2003/2004. *Braz. J. Poult. Sci.*, 8, 105-111.

[74] Gibson, G. R., & Wang, X. (1994). Inhibitory effects of bifidobacteria on other colonic bacteria. *J. Appl. Bacteriol.*, 77, 412-420.

[75] Gibson, G. R., & Roberfroid, M. B. (1995). Dietary modulationof the human colonic microbiota-Introducing the concept of prebiotics. *J. Nutr.*, 125, 1401-1412.

[76] Ghadge, V. N., Upase, B. T., & Patil, P. V. (2009). Effect of replacing groundnut cake by soybean meal on performance of broilers. *Veterinary World,*, 2(5), 183-184.

[77] Gaylord, B. S., Heeger, A. J., & Bazan G. C. (2003). DNA hybridization detection with water-soluble conjugated polymers and chromophore-labeled single-stranded DNA. *J. Am. Chem. Soc.*, 125, 896 EOF-900 EOF.

[78] Grieshop, C. M., Catzere, C. T., Clapper, G. M., Flickinger, E. A., Bauer, L. L., Frazier, R. L., & Fahey, G. C. Jr. (2003). Chemical and nutritional characteristics of United States soybeans and soybean meals. *J. Agric. Food Chem.*, 51, 7684-7691.

[79] Grieshop, C. M., & Fahey, G. R. (2001). Comparison of quality characteristics of soybeans from Brazil, China and the United States. *J. Agric Food Chem*, 49, 2669-2673.

[80] Halina, I., Nathan, S., & Ephrain, K. (1966). Soybean hemagglutinin, a plant glycoprotein. I. Isolation of a glycopeptide.

[81] Hong Y. H., Wang, T. C., Huang, C. J., Cheng, W. Y., & Lin, B. F. (2008). Soy isoflavones supplementation alleviates disease severity in autoimmune-prone MRL-lpr/lpr mice. *Lupus*, 17, 814-821.

[82] Heaney, R. P., Weaver, C. M., & Fitzsimmons, M. L. (1991). Soybean phytate content: effect on calciumabsorption. *Am. J. Clin Nutr*, 53, 745-747.

[83] Herman, E. M., Ricki, M., Helm, R., Anthony, J., & Kinney, J. (2003). Genetic Modification Removes an Immunodominant Allergen from Soybean. *Plant Physiology*, 132(1).

[84] Harris, L. J., Farber, J. N., Beuchat, L. R., Parish, M. E., Suslow, T. V., Garrett, E. H., & Busta, F. F. (2003). Outbreaks associated with fresh produce: incidence, growth, and survival of pathogens in fresh and fresh-cut produce. *Comprehensive Reviews in Food Science and Food Safety Supplement*, 79-141.

[85] Herman, E. M. (2003). Genetically modified soybeans and food allergies. *Journal of Experimental Botany*, 54(386), 1317-1319.

[86]   Hwang, David. l., Wen-kuang, Yang., Donald, E., & Foard, . (1978). Rapid Release of Protease Inhibitors from Soybeans. *Plant Physiol.*, 61, 30-34.

[87]   Hayakawa, K., Mitzutani, J., Wada, K., Masai, T., Yoshihara, I., & Mitsuoka, T. (1990). Effects of soybean oligosaccharides on human fecal flora. *Microbiol. Ecol. Health Dis.*, 3, 293-303.

[88]   Harada, J. J., Barker, S. J., & Goldberg, R. B. (1989). Soybean beta-conglycinin genes are clustered in several DNA regions and are regulated by transcriptional and post-transcriptional processes. *The Plant Cell*, 1, 415-425.

[89]   Herman, E. M., Helm, R. M., Jung, R., & Kinney, A. J. (2003). Genetic modification removes an immunodominant allergen from soybean. *Plant Physiology,*, 132, 36-43.

[90]   Hildebrand, D. F. (1989). Lipoxygenase. *Physiologia Plantarum,*, 76, 249-253.

[91]   Hitz, W. D., Carlson, T. J., Kerr, P. S., & Sebastian, S. A. (2002). Biochemical and molecular characterization of a mutation that confers a decreased raffinosaccharide and phytic acid phenotype on soybean seeds. *Plant Physiology,*, 128, 650-660.

[92]   Hedin, C., et al. (2007). Evidence for the use of probiotics and prebiotics in inflammatory bowel disease: a review of clinical trials. *Proc Nutr Soc.*, 66(3), 307-15.

[93]   Hendrich, S., Lee, K. W., Xu, X., Wang, H. J., & Murphy, P. A. (1994). Defining food components as new nutrients. *J. Nutr.*, 124, 1789S EOF-1792S EOF.

[94]   Herkelman, K. L., Cromwel, G. L., & Stahly, T. (1991). Effect of heating time and sodium metabisulfite on the nutritional value of full-fat soybeans for chicks. *J. Anim. Sci.*, 69, 4477-4486.

[95]   Hoaglund CM, VM Thomas, MK Peterson, and RW Knott. (1992). Effects of supplemental protein source and metabolizable energy intake on nutritional status in pregnant ewes. *J. Anim. Sci.*, 70, 273 EOF-80 EOF.

[96]   Hammond, B. C. J. L., Vicini, C. F., Hartnell, M. W., Naylor, C. D., Knight, E. H., Robinson, R. L., Fuchs, , & Padgette, S. R. (1995). The feeding value of soybeans fed to rats, chickens, catfish and dairy cattle is not altered by genetic incorporation of glyphosate tolerance. *J. Nutr.*, 126, 3-717.

[97]   Heaney, R. P. (1996). Nutrition and the risk of osteoporosis. In: Marcus R Fieldman D, Kelsey J. L., editors., *Osteoperosis*, Academic press: San Diego, CA;, 483-509.

[98]   Holle, D. G. (1995). Amino Acids. *Ratite Feeds and Feeding*, 1-2, 58-59.

[99]   Herkelman, K. L., Cromwell, G. L., & Stahly, T. S. (1991). Effect of heating time and sodium metabisulfite on the nutritional value of full-fat soybean for chicks. *J. Anim. Sci.*, 69, 4477-4486.

[100] Honig, D. H., Elaine, M., Hockridge, Robert. M., Gould, , Joseph, J., & Rackis, . (1983). Determination of Cyanide in Soybeans and Soybean Products. *Journal of Agricultural & Food Chemistry,*, 272-275.

[101] Irish, G. G., & Balnave, D. (1993). Non-starch polysaccharides and broiler performance on diets containing soyabean meal as the sole protein concentrate. *Australian Journal of Agricultural Research*, 44(7), 1483-1499.

[102] Ikuomola, D. S., & Eniola, K. I. T. (2010). Microbiological Quality and Public Health Risk Associated with Beske: Fried Soybean Snack, Retailed in Ikeji-Arakeji, Osun State, Nigeria. *Nigeria Journal of Microbiology,*, 24(1), 2114-2118.

[103] Irish, G. G., & Balnave, D. (1993). Non-starch polysaccharides and broiler performance on diets containing soyabean meal as the sole protein concentrate. *Aust. J. Agric. Res.*, 44, 1483-1499.

[104] Izabelawoclawek-potocka, Mamadou. M., Bah, Anna., Korzekwa, Mariusz. K., Piskula, Wiesławwiczkowski., Andrzej, Depta., & Skarzynski, D. J. (2005). Soybean-Derived Phytoestrogens Regulate Prostaglandin Secretion in Endometrium During Cattle Estrous Cycle and Early Pregnancy. *Experimental biology and Med*, 230, 189-199.

[105] Irish, G. G., & Balnave, D. (1993). Non-starch polysaccharides and broiler performance on diets containing soyabean meal as the sole protein concentrate. *Australian Journal of Agricultural Research*, 44(7), 1483-1499.

[106] Jin, L. Z., Ho, Y. W., Abdullah, N., & , S. Jalaludin(1997). Probiotics in poultry: modes of action. World's Poultry Sci. J. , 53, 353-368.

[107] Jung, W., Yu, O., Lau, S. M., O'Keefe, D. P., Odell, J., Fader, G., & Mc Gonigle, B. (2000). Identification and expression of isoflavone synthase, the key enzyme for biosynthesis of isoflavones in legumes. *Nature Biotechnology*, 18, 208-212.

[108] Sherill, J. D., Morgan, Sparks., M.D., John, M. M., Kemppainen, B. W., Bartol, F. F., Morrison, E. E., & Akingbemi, B. T. (2010). Developmental Exposures of Male Rats to Soy Isoflavones Impact Leydig Cell Differentiation. *Biology of Reproduction*, 83, 488-501.

[109] Jing, Y., Nakaya, K., & Han, R. (1993). Differentiation of promyelocytic leukemia cells HL-60 induced by daidzein in vitro and in vivo. *Anti Cancer Res.*, 13, 1049S-1054S.

[110] Kaneki, M., Hedjes, S. J., Hosol, T., Fijiwara, S., Lyons, A., Crean, S. J., Ishida, N., Nakagawa, M., & Takechi, M. (2001). Japanese fermented soybean food as the major determinant of the largegeographic difference in circulating levels of vitamin k2: possible implications of hip fracture risk. *Nutrition*, 17, 315-321.

[111] Kunihiko, U., Chieko, T. & Isao, K. (2010). Inactivation of Bacillus subtilis spores in soybean milk by radio-frequencyflash heating. *Journal of Food Engineering*, 100, 622-626.

[112] Katsuki, Y., Yasuda, K., Ueda, K, & Naoi, Y. (1978). *Annu. Rep. Tokyo Metrop. Res. Lab. Public Health*, 29, 26l.

[113] Knaus, W. F, Beermann, D. H., Robinson, T. F., Fox, D. G., & Finnerty, K. D. (1998). Effects of a dietary mixture of meat and bone meal, feather meal, blood meal, and

fish meal on nitrogen utilization in finishing Holstein steers. *J. Anim. Sci.*, 76, 1481 EOF-7 EOF.

[114] Kats, L. J., Nelssen J. L., Tokach, M. D., Goodband, R. D., Hansen, J. A., & Laurin, J. L. (1994). The effect of spray-dried porcine plasma on growth performance in the early-weaned pig. *J. Anim. Sci.*, 72, 2075 EOF-81 EOF.

[115] Karr-Lilienthal, L. K., Mershem, N. R., Grieshop, C. M., Flahaven, M. A., Mahan, D. C., Watts, N. D., Fahey, G. C., & Jr , . (2004). Ileal amino acid digestibilities by pigs fed soybean meals from five major soybean producing countries. *J. Anim Sci*, 82, 3198-3209.

[116] Karr-Lilienthal, L. K., Grienshop, C. M., Spears, J. K., Fahey, G. C., & Jr , . (2005). Amino acid, carbohydrate, and fat composition of soybean meals prepared at 55 commercial U.S. soybean processing plants. *J. Agric. Food Chem.*, 53, 2146-2150.

[117] Karr-Lilienthal, L. K., Grieshop, C. N., Merchen, N. R., Mahan, D. C., Fahey, G. C., & Jr , . (2004). Chemical composition and protein quality comparisons of soybeans and soybean meals from five leading soybean-producing countries. *J. Agric. Food Chem.*, 52, 6193-6199.

[118] Katsuyama, H., Ideguchi, S., Fukunaga, M., Saijoh, K., & Sunami, S. (2002). Usual dietary intake of fermented soybeans (natto) is associated with bone mineral density in premenopausal women. *J. Nutr. Sci. Vitaminol (Tokyo).*, 48, 207-215.

[119] Kalinski, A., Melroy, D. L., Dwivedi, R. S., & Herman, E. M. (1992). A soybean vacuolar protein (p34) related to thiol proteases is synthesized as a glycoprotein precursor during seed maturation. *The Journal of Biological Chemistry*, 267, 12068-12076.

[120] Kelley, D. S., & Erikson, K. L. (2003). Modulation of body composition and immune cell functions by conjugated linoleic acid in humans and animal models:benefits vs. risks. *Lipids*, 38, 377-390.

[121] Kinney, A. J. (1996). Development of genetically engineered soybean oils for food applications. *Journal of Food Lipids*, 3, 273-292.

[122] Kinney, A. J., & Knowlton, S. (1998). Designer oils: the high oleic acid soybean. In S. Roller & S. Harlander (Eds.),, *Genetic modification in the food industry: A strategy for food quality improvement*, 193-213, London: Blackie.

[123] Kinney, A. J., & Fader, G. M. (2002). *U.S. Patent* [6], Washington, DC: US Patent and Trademark Office.

[124] Kinsella, J. E. (1979). Functional properties of soy proteins. *Journal of the American Oil Chemists Society,,* 56, 242-258.

[125] Kitamura, K. (1995). Genetic improvement of nutrional and food processing quality in soybean. *Japan Agricultural Research Quarterly,,* 29, 1-8.

[126] Kitts, D. D., & Weiler, K. (2003). Bioactive proteins and peptides from food sources. Applications of bioprocesses used in isolation and recovery. *Current Pharmaceutical Design*, 9, 1309-1323.

[127] Knapp, H. R., Salem, N., Jr , , & Cunnane, S. (2003). Dietary fats and health. *Proceedings of the 5th Congress of the International Society for the Study of Fatty Acids and Lipids*,, 38, 297-496.

[128] Kunitz, M. (1946). Crystalline soybean trypsin inhibitor. *J. Gen. Phygiol.*,, 9, 149-154.

[129] Kwok, K. C., Qin, W. H., Tsang, J. C. 1993. Heat Inactivation of Trypsin Inhibitors in Soymilk at Ultra-High Temperatures. J. Food Science 58:859–862.

[130] Liu, N., Liu, G. H., Li, F. D., Sands, J. S., Zhang, S., Zheng, A. J., & Ru, Y. J. (2007). Efficacy of Phytases on Egg Produ. ction and Nutrient Digestibility in Layers Fed Reduced Phosphorus Diets. *Poultry Science*, 86, 2337-2342.

[131] Lim, H. S., Namkung, H., & Paik, I. K. (2003). Effects of phytase supplementation on the performance, egg quality, and phosphorous excretion of laying hens fed different levels of dietary calcium and nonphytate phosphorous. *Poult. Sci.*, 82, 92-99.

[132] Lenehan, N. A., De Rouchey, J. M., Goodband, R. D., Tokach, M. D., Dritz, S. S., Nelssen, J. L., Groesbeck, C. N., & Lawrence, K. R. (2007). Evaluation of soy protein concentrates in nursery pig diets. *J. Anim. Sci.*, 85, 3013-3021.

[133] Lemke, S. L., Vicini, J. L., Su, H., Goldtein, D. A., Nemeth, M. A., Krul, E. S., & Harris, W. S. (2010). Dietary intake of steriodonic acid-enriched soybean oil increases the omega-3 index:randomized, double-blind clinical study of afficacy and safety. *J. Clin. Nutr.*, 92, 766-775.

[134] Liener, I. E. (1953). Soyin, a toxic protein from the soybean. *J. Nutrition*,, 49, 527.

[135] Liener, I. E. (1994). Implications of antinutritional components in soybean foods. *Crit. Rev. Food Sci. Nutr.*, 34, 31-67.

[136] Liu, Jiang-Gong and Lin Tser-KeShun,. (2008). Survival of Listeria monocytogenes inoculated in retail soymilk products. *Food Control*,, 19(2008), 862-867.

[137] Linassier, C., Pierr, M., Le Pacco-B, J., & Pierre, J. (1990). Mechanisms of action in NIH-3T3 cells of genistein, an inhibitor of EGF receptor tyrosine kinase activity. *Biochem Pharmacol*, 39, 187-193.

[138] Leske, K. L., Jevne, C. J., & Coon, C. N. (1993). Effect of oligosaccharide additions on nitrogen-corrected truemetabolizable energy of soy protein concentrate. *Poult. Sci.*, 72, 664-668.

[139] Lienei, I. E., & Kakade, M. L. (1969). Protease inhibitors. In I. E. Liener (Ed.), *Toxic Constituents of Plant Foodstuffs.*, Academic Press, New York.

[140] Liener, I. E. (1981). Factors affecting the nutritional quality of soy products. *J. Am. Oil. Chem. Soc.*, 58, 406-415.

[141] Liener, L. E., & Kakade, M. L. (1980). Protease Inhibitors. In: I. E. Liener (Ed), *Toxic Constituents of Plant Foodstuffs.*, 7-71, Academic Press, New York.

[142] Linz, A. L., Xiao, R. J., Ferguson, M., Badger, T. M., & Simmen, F. A. (2004). Developmental feeding of soy protein isolate to male rats inhibits formation of colonic aberrant crypt foci: Major response differences to azoxymethane (AOM) in Sprague-Dawley rats from two colonies. *FASEB J.*, 18, A127.

[143] Lovati, M. R., Manzoni, C., Corsini, A., Granata, A., Fumagalli, R., & Sirtori, C. R. (1996). 7S globulin from soybean is metabolized in human cell cultures by a specific uptake and degradation system. *Journal of Nutrition,*, 126, 2831-2842.

[144] Lovati, M. R., Manzoni, C., Corsini, A., Granata, A., Fumagalli, R., Sirtori, C., & , R. (1992). Low density lipoprotein (LDL) receptor activity.

[145] Madden, R. H., Murray, K. A., & Gilmore, A. (2004). Determination of principal points of product contamination during beef carcass dressing process in Northern Ireland. J. of. *Food Prot.*, 67, 1494-1496.

[146] Foley, S.L., White, D. G., McDermott, P.F., Walker, R.D., & Messina, M. J. (1999). Legumes and soybean: Overview of their nutritional profiles and health effect. *Am. J. Clin. Nutr.*, 70, 439S-450S.

[147] Mac, Donald. R. S., Guo, J. Y., Copeland, J., Browning, J. D., Jr , , Sleper, D., Rottinghous, G. E., & Berhow, M. A. (2005). Environmental influences on isoflavones and saponins in soybean and their role in colon cancer. *J. Nutr.*, 135, 1239-1242.

[148] Mc Andrews, G. M., Liebman, M., Cambardella, C. A., & Richard, T. L. (2006). Residual effects of composted and fresh solid swine (Sus scrofa L.) manure on soybean [Glycine max (L.) Merr.] growth and yield. *Agron. J.*, 98, 873-882.

[149] Montgomery, R. D. (1980). In, *Toxic Constituents of Plant Foodstuffs*, , 2nd ed.; Liener, I. E., Ed.; Academic Press: New York,.

[150] Maciorowski, K. G., Jones, F. T., Pillai, S. D., & Ricke, S. C. (2004). Incidence, sources, and control of foodborne Salmonella spp. in poultry feed. World's Poultry Sct. J. , 60, 446-457.

[151] Montazer-Sadegh, R., Ebrahim-Nezhad, Y., & Maheri-Sis, N. (2008). Replacement of different levels of rapeseed meal with soubean meal on broiler performance. *Asian Journal of Animal and Vet. Advances*, 3(5), 278-285.

[152] Mc Cue, P., & Shetty, K. (2004). Health benefits of soy isoflavonoids and strategies for enhancement: A review. *Crit. Rev.Food Sci. Nutr.*, 44, 361-367.

[153] Zimmermann, Michael B. (2009). *Iodine Deficiency Endocrine Reviews,*, 30(4), 376-408.

[154] Messina, M., & Redmond, G. (2006). Effects of soy protein and soybean isoflavones on thyroid function in healthy adults and hypothyroid patients: a review of the relevant literature. *Thyroid*, 16, 249-258.

[155] Manavalan, L. P., Satish, K., Guttikonda-Son, Lam., Phan, Tran., Henry, T., & Nguyen, . (2009). Physiological and Molecular Approaches to Improve Drought Resistance in Soybean. *Plant Cell Physiol*, 50(7), 1260-1276.

[156] Medlock K. L, Branham W. S, & Sheehan D. M. (1995). Effects of coumestrol and equol on the developing reproductive tract of the rat. *Proc Soc Exp Biol Med*, 208, 67-71.

[157] Moizzudin, S. (2003). Soybean meal quality in US and World markets. MS. Thesis, IA Iowa State University.

[158] Marty, B. J., Chavez, E. R., & De Lange, C. F. M. (1994). Recovery of amino acids at the distal ileum for determining apparent and true ileal amino acid digestibilities in growing pigs fed various heat-processed full-fat soybean products. *J. Anim. Sci.*, 72, 2029-2037.

[159] Martin-C, J., & Valeille, K. (2002). Conjugated linoleic acids: All the same or to everyone its own function? *Reproduction Nutrition Development,*, 42, 525-536.

[160] Maruyama, N., Adachi, M., Takahashi, K., Yagasaki, K., Kohno, M., Takenaka, et., & al, . (2001). Crystal structures of recombinant and native soybean beta-conglycinin beta homotrimers. *European Journal of Biochemistry,*, 268, 3595-3604.

[161] Maruyama, N., Mohamed, Salleh. M. R., Takahashi, K., Yagasaki, K., Goto, H., Hontani, N., et al. (2002). Structure-physicochemical function relationships of soybean beta-conglyxcinin heterotrimers. *Journal of Agricultural and Food Chemistry,*, 50, 4323-4326.

[162] Messina, M., Gardner, C., & Barnes, S. (2002). Gaining insight into the health effects of soy but a long way still to go. *Journal of Nutrition,*, 132, 547S EOF-551S EOF.

[163] Mohamed, Salleh. M. R., Maruyama, N., Adachi, M., Hontani, N., Saka, S., Kato, N., et al. (2002). Comparison of protein chemical and physicochemical properties of rapeseed cruciferin with those of soybean glycinin. *Journal of Agricultural and Food Chemistry*, 42, 525-536.

[164] Messina, M. J., Persky, V., Setchell, K. D., & Barnes, S. (1994). Soy intake and cancer risk: A review of the in-vitro and vivo data. *Nutr. Cancer*, 21, 113-131.

[165] Messina, M. J., & Loprinzi, C. L. (2001). Soy for breast cancer survivors: A critical review of the literature. *J. Nutr.*, 131, 3095S EOF-108S EOF.

[166] Messina, M. J., & Barnes, S. (1991). The role of soy products in reducing risk of cancer. *J. Nutr. Cancer Instit.*, 83, 541-546.

[167] Messina, M. J. (1999). Legumes and soybeans: Overview of their nutritional profiles and health effects. *Am. J. Clin. Nutr.*, 70, 439S EOF-450S EOF.

[168] Mc Michael-Phillips, D. F., Hardingm, C., & Morton, M. (1998). Effect of soy protein supplementation on epithelial proliferation in histologically normal human breasts. Am. J. Clin. Nutr. S., 68, 1431S EOF-1435S EOF.

[169] Milsicek, R. J. (1994). Interaction of naturally occuring nonsteroidal estrogens with expressed recombinant human estrogen receptor. *J. Steroid Biochem. Mol. Biol.*, 49, 153-160.

[170] Nahashon, S. N., & Kilonzo-Nthenge, A. (2011). Advances in soybean and soybean by-products in monogastric nutrition and health. In, *Soybean and Nutrition*, Ed. H. A. El-shemy, Intech, Rijeka, Croatia.

[171] Nahashon, S. N., Nakaue, Harry S., & Mirosh, Larry W. (1996). Performance of Single Comb White Leghorns given a diet supplemented with a live microbial during the growing and egg laying phases. *Animal Feed Science and Technology*, 57, 25-38.

[172] Nahashon, S. N., Nakaue, H. S., & Mirosh, L. W. (1994a). Phytase activity, phosphorus and calcium retention and performance of Single Comb White Leghorn layers fed diets containing two levels of available phosphorus and supplemented with direct-fed microbials. *Poultry Science*, 73, 1552-1562.

[173] Nahashon, S. N., Nakaue, H. S., & Mirosh, L. W. (1994b). Production variables and nutrient retention in Single Comb White Leghorn laying pullets fed diets supplemented with direct-fed microbials. *Poultry Science*, 73, 1699-1711.

[174] Nahashon, S. N., Nakaue, H. S., Snyder, S. P., & Mirosh, L. W. (1994c). Performance of Single Comb White Leghorn layers fed corn-soybean meal and barley-corn-soybean meal diets supplemented with a direct-fed microbial. *Poultry Science*, 73, 1712-1723.

[175] Nahashon, S. N., Adefope, N., & Wright, D. (2011). Effect of floor density on growth performance of Pearl Grey guinea fowl replacement pullets. *Poultry Science*, 90, 1371-1378.

[176] Nahashon, S. N., Nahashon, A., Akuley, , & Adefope, N. (2010). Genetic relatedness of Pearl Grey guinea fowl and Single Comb White Leghorn chickens. *Journal of Poultry Science*, 47, 280-287.

[177] Nahashon, S. N., Aggrey, S. E., Adefope, N. A., Amenyenu, A., & Wright, D. (2010). Gomperts-Laird model prediction of optimum utilization of crude protein and metabolizable energy by French guinea fowl broilers. *Poultry Science*, 89, 52-57.

[178] Nahashon, S. N., Adefope, N., Amenyenu, A., & Wright, D. (2009). The effect of floor density on growth performance and carcass characteristics of French guinea broilers. *Poultry Science*, 88, 2461-2467.

[179] Tyus, I. I. J., Nahashon, S. N., Adefope, N., & Wright, D. (2009). Production performance of single comb white leghorn chickens fed growing diets containing blood meal and supplemental blood meal. *Journal of Poultry Science*, 46, 313-321.

[180] Nagano, T., Fukuda, Y., & Akasaka, T. (1996). Dynamic viscoelastic study on the gelation properties of beta-conglycinin-rich and glycinin-rich soybean protein isolates. *Journal of Agricultural and Food Chemistry,*, 44, 3484-3488.

[181] Noordermeer, M. A., Veldink, G. A., & Vliegenthart, J. F. (2001). Fatty acid hydroperoxide lyase: a plant cytochrome p450 enzyme involved in wound healing and pest resistance. *ChemBioChem,*, 2, 494-504.

[182] Naylora, R. L., Ronald, W., Hardyb, Dominique. P., Bureauc, Alice., Chiua, Matthew., Elliottd, Anthony. P., Farrelle, Ian., Forstere, Delbert. M., Gatlinf,g, Rebecca. J., Goldburgh, Katheline., Huac, , Peter, D., & Nicholsi, . (2009). Feeding aquaculture in an era of finite resources. *PNAS*, 106, 36-15103.

[183] Nelson, A. I., Wijeratne, W. B., Yeh, S. W., Wei, T. M., & Wei, L. S. (1987). Dry extrusion as an aid to mechanically expelling of oil from soybean. *J. Am. Oil Chem. Soc.*, 64, 1341-1347.

[184] NRC. (1998). Nutrient requirements of swine. 10th ed. National Academy Press, Washington, DC.

[185] Nejaty, H., Lacey, M., & Whitehead, S. A. (2001). Differing effects of endocrine disrupting chemicals on basal and FSH-stimulated progesterone production in rat granulose-luteal cells. *Exp Biol Med*, 226, 570-576.

[186] Nicodemus, N. J., García, R., Carabaño, , & De Blas, J. C. (2007). Effect of substitution of a soybean hull and grape seed meal mixture for traditional fiber sources on digestion and performance of growing rabbits and lactating does *J. Anim Sci.*, 85, 181-187.

[187] Opapeju, F. O., , A., Golian, A., Nyachoti, C. M., & Campbell, L. D. (2006). Amino acid digestibility in dry extruded-expelled soybean meal fed to pigs and poultry. *J. Anim. Sci.*, 84, 1130-1137.

[188] Oakwndull, O. C. (1981). Saponins in Food- A Review. *Food Chem*, 6, 19-40.

[189] Odunsi, A. A. (2003). Blend of bovine blood and rumen digesta as a replacement for fishmeal and groundnut cake in layer diets. *Inter. J. Poult. Sci.*, 2, 58.

[190] Onwudike, OC. (1981). Effect of various protein sources on egg production in a tropical environment. *Trop. Anim. Prod.*, 6, 249.

[191] Fletciier, a. P., Marks, g. S., Marshall, r. D., & Neuberger, A. (1963). *Biochem J.*, 8.

[192] Plummer, T. H., , J. R., & , C. H. W. Hirs.(1963). J. Biol. Chem., , 238, 1396.

[193] Panda, A. K., Rao, S. V. R., Raju, M. V. L. N., & Bhanja, S. K. (2005). Effect of microbial phytase on production performance of White Leghorn layers fed on a diet low in non-phytate phosphorus. *Br. Poult. Sci.*, 46, 464-469.

[194] Payne, R. L., Bidner, T. D., Southern, L. L., & Mc Millin, K. W. (2001). Dietary Effects of Soy Isoflavones on Growth and Carcass Traits of Commercial Broilers. *Poultry Science*, 80, 1201-1207.

[195] Peganova, S., & Eder, K. (2002). Studies on requirement and excess of isoleucine in laying hens. *Poult. Sci.*, 81, 1714 EOF-21 EOF.

[196] Powell, S., Bidner, T. D., & Southern, L. L. (2011). Phytase supplementation improved growth performance and bone characteristics in broilers fed varying levels of dietary calcium 1. *Poultry Science*, 90, 604-608.

[197] Potter, S. M., Baum, J. A., Teng, H., Stillman, R. J., Shay, N. F., Erdman Jr, , & , J. W. (1998). Soy protein and isoflavones: Their effects on blood lipids and bone density in postmenopausal women. *American Journal of Clinical Nutrition,*, 68, 1375S EOF-1379S EOF.

[198] Persons, C. M., Hashimoto, K., Wedekind, K. J., Han, Y., & Baker, D. H. (1992). Effect of overprocessing on availability of amino acids and energy in soybean meal. *Poult. Sci.*, 71, 133-140.

[199] Paterson, W. J., Hostetler, E. H., & Shaw, A. O. (1942). Studies in feeding soybean to pigs. *J. Anim. Sci.*, 1, 360.

[200] Palacios, M. F., Easter, R. A., Soltwedel, K. T., Parsons, C. M., Douglas, M. W., Hymo-witz, T., & Pettigrew, J. E. (2004). Effect of soybean variety and processing on growth performance of young chicks and pigs. *J. Anim. Sci.*, 82, 1108-1114.

[201] Peisker, M. (2001). Manufacturing of soy protein concentrate for animal nutrition. 103-107, *Feed Manufacturing in the Mediterranean Region*, Improving Safety: From Feed to Food. J.Brufau, ed. CIHEAM-IAMZ, Reus, Spain.

[202] Pillai, P. B., O'Connor-Dennie, T., Owens, C. M., & Emmert, J. L. (2006). Efficacy of an Escherichia coli Phytase in Broilers Fed Adequate or Reduced Phosphorus Diets and Its Effect on Carcass Characteristics. *Poultry Science*, 85, 1737-1745.

[203] Perilla, N. S., Cruz, M. P., de Belalcazar, F., & Diaz, G. J. (1997). Effect of temperature of wet extrusion on the nutritional value of full-fat soyabeans for broiler chickens. *British Poultry Science*, 38(4), 412-416.

[204] Quigley, E. M. M. (2012). Prebiotics and Probiotics Their Role in the Management of Gastrointestinal Disorders in Adults. *Nutr. Clin. Pract.*, 27(2), 195-200.

[205] Rackis, J. J. (1975). Oligosaccharides of food legumes: Alpha-galactosidase activity and the flatus problem. In: A. Jeanes and J.Hodges (ed.) Physiological Effects of Food Carbohydrates., *Am. Chem. Soc.,*, Washington, DC.

[206] Ravindran, V., Cabahug, S., Ravindran, G., & Bryden, W. L. (1999). Effects of phytase supplementation, individually and in combination with glycanase, on the nutritive value of wheat and barley. *Poult. Sci.*, 78, 1588-1595.

[207] Rahman, S. M. T., Kinoshita, T., Anai, , & Takagi, Y. (1999). Genetic relationships between loci for palmitate contents in soybean mutants. *J Hered*, 90(3), 423 EOF-427 EOF.

[208] Ravindran, V., Morel, P. C., Partridge, G. G., Hruby, M., & Sands, J. S. (2006). Influence of an Escherichia coli-derived phytase on nutrient utilization in broiler starters fed diets containing varying concentrations of phytic acid. *Poult. Sci.*, 85, 82-89.

[209] Rhodes, P. J., Fedorka-Cray, S., Simjee, , & Zhao, S. (2006). Comparison of subtyping methods for differentiating Salmonella enterica serova Typhimurium isolates obtained from animal food sources. *J. Clin. Microbiol.*, 44, 3569-3577.

[210] Roman, S., Temler, Charles. A., Dormono, Eliane., Simon, Brigette., Morel, , & Christine, Mettraux. (1984). Proteins, Their Hydrolysates and SoybeanTrypsin Inhibitor. *J. Nutr.*, 114, 270-278.

[211] Rangngang M. D, Nelson M. L and S. M Parish. (1997). Ruminal undegradability of blood meal and effects of blood meal on ruminal and postruminal digestion in steers consuming vegetative orchard grass hay. *J. Anim. Sci.*, 75, 2788.

[212] Rosselli, M., Reinhard, K., Imthurn, B., Keller, P. J., & Dubey, R. K. (2000). Cellular and biochemical mechanisms by which environmental estrogens influence reproductive function. *Hum Reprod. Update*, 6, 332-350.

[213] Riblett, A. L., Herald, T. J., Schmidt, K. A., & Tilley, K. A. (2001). Characterization of beta-conglycinin and glycinin soy protein fractions from four selected soybean genotypes. *Journal of Agricultural and Food Chemistry*, 49, 4983-4989.

[214] Raboy, V., & Dickinson, D. B. (1984). Effect of phosphorus and zinc nutrition on soybean seed phytic acid and zinc. *Plant Physiol.*, 75, 1094-1098.

[215] Raboy, V., Dickinson, D. B., & Below, F. E. (1984). Variation in seed total phosphorus, phytic acid, zinc, calcium, magnesium, and protein among lines of Glycine Max and G. Soja. *Crop Sci.*, 24, 431-434.

[216] Rackis, J. J., Wolf, W. J., & Baker, E. C. (1986). Protease inhibitors in plant foods: Content and inactivation. In: M. Friedman (Ed), *Nutritional and Toxicological Significance of Enzyme Inhibitors in Foods*, 299-347, Plenum Publishing, New York.

[217] Rackis, J. J. (1972). Biologically active components. In Smith and Circle (Ed.), *Soybeans: Chemistry and Technology*, Vol. I. The Avi Publ. Co., Inc.,Westport, Conn.

[218] Savage, J. H., et al. (2010). The natural history of soy allergy. *J. Allergy Clin. Immunol.*, 125, 683-686.

[219] Sahlu, T. J. M., Fernandez, C. D., Lu, , & Manning, R. (1992). Dietary protein level and ruminal degradability for mohair production in Angora goats. *J ANIM SCI*, 70, 1526-1533.

[220] Sicherer, S. H., & Sampson, H. A. (2010). Food allergy. J. Allergy Clin. Immunol. S125., 125, S116.

[221] Smiricky, M. R., Grieshop, C. M., Albin, D. M., & Wubben, J. E. (2002). The influence of soy oligosaccharides on apparent and true ileal amino acid digestibilities and fecal consistency in growing pigs. *J. Anim. Sci.*, 80, 2433-2441.

[222] Sampson, H.A. (2004). Update on food allergy. *J Allergy Clin Immunol;*, 113, 805-19.

[223] Sicherer, SH. (2002). Food allergy. *Lancet*, 360, 701-10.

[224] Sell, J. L., Soto-Salanova, M., Pierre, P., & Jeffrey, M. (1997). Influence of supplementing corn-soybean diets with vitamin E on performance and selected physiological traits of male turkeys. *Poult. Sci.*, 76, 1405-1417.

[225] Setchell, K. D. (1998). Phytoestrogens: The niochemistry, physiology, and implications for human health of soy isoflavones. *Am. J. Cclin Nutr.*, 68:, S1333-S1346.

[226] Scheiber, M. D., & Reber, R. W. (1999). Isoflavones and postmenopausal bone health:A variable alternative to estrogen therapy? *Menopause*, 6, 233-241.

[227] Slominski, B. A. (1994). Hydrolysis of galactooligosaccharides by commercial preparations of α-galactosidase and β-fructofuranose:Potential for use as dietary additives. *J. Sci. Food Agric.*, 65, 323-330.

[228] Smiricky-Tjardes, M. R., Flickinger, E. A., Grieshop, C. M., Bauer, L. L., Murphy, M. R., & Fahey, G. C. Jr. (2003). In vitro fermentation characteristics of selected oligosaccharides by swine fecal microflora. *J. Anim. Sci.*, 81, 2505-2514.

[229] Sahn, K. S., Maxwell, C. V., Southern, L. L., & Buchanan, D. S. (1994). Improved soybean protein sources for early weaned pigs.:Ii. Effect on ileal amino acid digestibility. *J. Anim. Sci.*, 72, 631-637.

[230] Smiricky-Tjardes, M. R., Grieshop, C. M., Flickinger, E. A., Bauer, L. L., & Fahey, G. C. Jr. (2003). Dietary galactooligosaccharides affect ileal and total0-tract nutrient digestibility, ileal and fecal bacteria concentrations, and ileal fermentative characteristics of growing pigs. *J. Anim. Sci.*, 81, 2535-2545.

[231] Sirtori, C. R., Zucchi-Dentone, C., Sirtori, M., Gatti, E., Descovich, G. C., Gaddi, A., Cattin, L., Da, Col. P. G., Senin, U., Mannarino, E., Avellone, G., Colombo, L., Fragiacomo, C., Noseda, G., & Lenzi, S. (1985). Cholesterol-lowering and HDL-raising properties of lecithinated soy proteins in tyoe II hyperlipidemia patients. *Ann. Nutr. and Metab.*, 29, 348-357.

[232] Smiricky-Tjardes, M. R., Grieshop, C. M., Albin, D. M., Wubben, J. E., Gabert, V. M., & Fahey, G. C. Jr. (2002). The influence of soy oligosaccharides on apparent and true ileal amino acid digestibilities and fecal consistency in growing pigs. *J. Anim. Sci.*, 80, 2433-2441.

[233] Smith, A. K., & Wolf, W. J. (1961). Food uses and properties of soybean protein. I. *Food Technology.*, 15(4).

[234] Schor, A., & Gagliostro, G. A. (2001). Undegradable protein supplementation to early-lactation dairy cows in grazing conditions. *J. Dairy Sci.*, 84, 1597 EOF-606 EOF.

[235] Seong-Jun, Cho, Juillerat, Marcel A., & Lee, Cherl-Ho. (2007). Cholesterol Lowering Mechanism of Soybean Protein Hydrolysate. *J. Agric. Food Chem.*, 5, 10599-10604.

[236] Sirtori, C. R., & Lovati, M. R. (2001). Soy proteins and cardiovascular disease. *Current Atherosclerosis Reports*, 3, 47-53.

[237] Soyatech. (2003). New soyfoods market study shows Americans love their soy. Available on the World Wide Web:, http://www.soyatech.com.

[238] Stephan, A., & Steinhart, H. (1999). Identification of character impact odorants of different soybean lecithins. *Journal of Agricultural and Food Chemistry*, 47, 2854-2859.

[239] Sindt, M. H., Stock, Klop, R. A., Klopfenstein T. J. & Shain D. H. (1993). Effect of protein source and grain type on finishing calf performance and ruminal metabolism. *J. Anim. Sci.*, 71, 1047 EOF-56 EOF.

[240] Selle, P. H., & Ravindran, V. (2007). Microbial phytase in poultry nutrition. A review. *Anim. Feed Sci. Technol.*, 135, 1-41.

[241] Smits, C. H. M., & Annison, G. (1996). Non-starch plant polysaccharides in broiler nutrition-towards a physiologically valid Soy oligosaccharides and pigs 2441proach to their determination. *World's Poult. Sci. J.*, 52, 204-221.

[242] Stott, J. A., Hodgson, J. E., & Chaney, J. C. (1975). Incidence of Salmonella in animal feed and the effct of pelleting on content of Enterobacteriaceae. *J Appl. Bact.*, 39, 41-46.

[243] Sands, J. S., Ragland, D., Baxter, C., Joern, B. C., Sauber, T. E., & Adeola, O. (2001). Phosphorus bioavailability, growth performance, and nutrient balance in pigs fed high available phosphorus corn and phytase. *J. Anim. Sci.*, 79, 2134-2142.

[244] Subuh, A. M. H., Moti, M. A., Fritts, C. A., & Waldroup, P. W. (2002). Use of various ratios of extruded full fat soybean meal and dehulled solvent extracted ;soybean meal in broiler diets. *Int. J. Poult. Sci.*, 1, 9-12.

[245] Southern, L. L., Ponte, J. E., Watkins, K. L., & Coombs, D. F. (1990). amino acid-supplemented raw soybean diets for finishing swine. *J. Anim. Sci.*, 68, 2387-2393.

[246] Su, S., Yeh, T., Lei, H., & Chow, N. (2000). The potential for soybean foods as a chemoprevention approach for human urinary tract cancer. *Clin. Can. Res.*, 6, 230-236.

[247] Tyus, J., II., Nahashon S. N., Adefope, N. & Wright, D. (2008). Growth Performance of Single Comb White Leghorn Chicks Fed Diets Containing Blood Meal Supplemented with Isoleucine. *J. Poult. Sci.*, 45, 31-38.

[248] Tyus, J., II., Nahashon S. N., Adefope, N. & Wright, D. (2009). Production Performance of Single Comb White Leghorn Chickens Fed Growing Diets Containing Blood Meal and Supplemental Isoleucine. *J. Poult. Sci.*, 46, 313-321.

[249] Tharkur, M., & Hurburgh, C. R. (2007). Quality of US soybean meal compared to the quality of soybean meal from other origins. *J. Am Oil Chem Soc*, 84, 835-843.

[250] Thornburn, J., & Thornburn, T. (1994). The tyrosine kinase inhibitor, geistein, prevents α-adrenergic-induced cardiac muscle cell hypertrophy by inhibiting activation of the Ras-MAP kinase signaling pathway. *Biochem Biophys Res Commun*, 202, 1586-1591.

[251] Traylor, S. L., Cromwell, G. L., Lindermann, M. D., & Knabe, D. A. (2001). Effects of level of supplemental phytase on ileal digestibility of amino acids, calcium, and phosphorus in dehulled soybean meal for growing pigs. *J. Anim. Sci.*, 79, 2634-2642.

[252] Tso, E., & Ling, S. M. (1930). Changes in the Composition of Blood in Rabbits Fed on Raw and Cooked Soybeans. 28(3).

[253] U.S. Soybean Industry: Background Statistics and Information. (2010). United States Department of Agriculture, Economic Research Service. March 28,.

[254] U.S. Food and Drug Administration. (1999). Food labeling health claims: Soybean protein and coronary heart disease. Final rule. Fed. Regist., 645, 57699-57733.

[255] Utsumi, S., Katsube, T., Ishige, T., & Takaiwa, F. (1997). Molecular design of soybean glycinins with enhanced food qualities and development of crops producing such glycinins. *Advances in Experimental Medicine and Biology*, 415, 1-15.

[256] Vucenik, I., & Shamsuddin, A. M. (2003). Cancer Inhibition by Inositol Hexaphosphate (IP6) and Inositol:From Laboratory to Clinic. *J. Nutr.*, 133, 3778S-3784S.

[257] Veltmann, Jr., Hansen, B. C., Tanksley, T. D. Jr., Knabe, D., & Linton, S. L. (1986). Comparison of the nutritive value of different heat-treated commercial soybean meals: utilization by chicks in practical type rations. *Poult. Sci.*, 65, 1561-1570.

[258] Van Kempen, T. A. T. G., Kim, I. B., Jansman, A. J. M., Verstegen, M. W. A., Hancock, J. D., Lee, D. J., Gabert, V. M., Albin, D. M., Fahey, G. C., Jr , , Grieshop, C. M., & Mahan, D. (2002). Regional and processor variation in the ileal digestible amino acid content of soybean meals measured in growing swine. *J. Anim. Sci.*, 80, 429-439.

[259] Wang, Hwa. L., Doris, I., Ruttle, , & Hesseltine, C. W. (1969). Antibacterial Compound from a Soybean Product Fermented by Rhizopus oligosporus. *Exp Biol Med*, 131(2), 579-583.

[260] Wang, H. J., & Murphy, P. A. (1994). Isoflavon content in commercial soybean foods. *J. Agric. Food Chem.*, 42, 1666-1673.

[261] Wallis, J. G., Watts, J. L., & Browse, J. (2002). Polyunsaturated fatty acid synthesis: What will they think of next? *Trends in Biochemical Sciences*, 27, 467-473.

[262] Wansink, B., & Chan, N. (2001). Relation of soy consumption to nutritional knowledge. *Journal of Medical Food,*, 4, 145-150.

[263] Weggemans, R. M., & Trautwein, E. A. (2003). Relation between soy-associated isoflavones and LDL and HDL cholesterol concentrations in humans: A meta-analysis. *European Journal of Clinical Nutrition*, 57, 940-946.

[264] Waguespack, A. M., Powell, S., Bidner, T. D., Payne, R. L., & Southern, L. L. (2009). Effect of incremental levels of L-lysine and determination of the limiting amino acids in low crude protein corm-soybean meal diets for broilers. *Poult. Sci.*, 88, 1216-1226.

[265] Williams, J. E. (1981). Salmonella in poultry feeds a world-widereview. *World's Poult. Sci. J.*, 37, 97-105.

[266] Wu, G., Liu, Z., Bryant, M. M., Roland, D. A., & Sr , . (2006). Comparison of Natuphos and Phyzyme as phytase sources for commercial layers fed corn-soy diet. *Poult. Sci.*, 85, 64-69.

[267] Wilhelms, K. W., Scanes, C. G., & Anderson, L. L. (2006). Lack of Estrogenic or Antiestrogenic Actions of Soy Isoflavones in an Avian Model: The Japanese Quail. *Poultry Science*, 85, 1885-1889.

[268] Wenzel, G. (2008). The Experience of the ZKBS for Risk Assessment of Soybean. *Journal fuer Verbraucherschutz und Lebensmittelsicherheit/Journal of Consumer Protection and Food Safety.*, 3(2), 55-59.

[269] Wei, H., Bowen, R., Cai, Q., Barnes, S., & Wang, Y. (1995). Antioxidant and antipromotional effects of the soybean isoflavone genistein. *Exp. Biol. Med.*, 208, 124-130.

[270] Wei, H., Wei, L., Frenkel, K., Bowen, R., & Barnes, S. (1993). Inhibition of tumor promoter-induced hydrogen peroxide formation in vitro and in vivo by genistein. *Nutr. Cancer*, 20, 1-12.

[271] Woyengo, T. A., Slominski, B. A., & Jones, R. O. (2010). Growth performance and nutrient utilization of broiler chickens fed diets supplemented with phytase alone or in combination with citric acid and multicarbohydrase. *Poultry Science*, 89, 2221-2229.

[272] World production of soybean. (2009). Source: FAS/USDA, United States Department of Agriculture, Foreign Agricultural Service-, http://www.fas.usda.gov/psdonline/psdReport.aspx? hidReportRetrievalName=Table+07%3a+Soybeans%3a+World+Supply+and+Distribution&hidReportRetrievalID=706&hidReportRetrievalTemplateID=8November2009.

[273] Woodworth, J. C., Tokash, M. D., Goodband, R. D., Nelssen, J. L., O'Quinn, P. R., Knabe, D. A., & Said, N. W. (2001). Apparent ileal digestability of amino acids and the digestible and metabolizable energy content of dry extruded-expelled soybean meal and its effect on growth performance. *J. Anim. Sci.*, 79, 1280-1287.

[274] World Cancer Research Fund and American Institute for Cancer Research. (1997). Food, nutritionand the prevention of cancer. A global perspective. Banta BNook Group, Menasha, WI.

[275] Xiao, C. W. (2008). Health Effects of Soy Protein and Isoflavones in Humans. *Journal of Nutrition,*, 138, 12445-12495.

[276] Xiurong Wang, X., Xiaolong Yan and Hong Liao. (2010). Genetic improvement for phosphorus efficiency in soybean: a radical approach. *Ann Bot*, 106(1), 215-222.

[277] Xu, Xia. K. S., Harris, H., Wang, P. A., & Murphy, . (1995). Bioavailability of soybean isoflavones depends upon gut microflora in women. *J. Nutr.*, 125, 2307-2315.

[278] Wang, X., Nahashon, S. N., Tromondae, K. F., Bohannon-Stewart, A., & Adefope, A. (2010). An initial map of chromosomal segmental copy number variations in the chicken. *BMC Genomics*, 11, 351, doi:10.1186/1471-2164-11-351.

[279] Yi, S. Y., Jee-Hyub, Kim., Young-Hee, Joung., Sanghyeob, Lee., Woo-Taek, Kim., Seung, Hun., Yu, , & Doil, Choi. (2004). The Pepper Transcription Factor CaPF1 Confers Pathogen and Freezing Tolerance in Arabidopsis. *Plant Physiology*, 136, 2862-2874.

[280] Barbara V., Jiangxin W., & Huang, Y. (2010). Narrowing Down the Targets: Towards Successful Genetic Engineering of Drought-Tolerant Crops. *Mol. Plant*, 3(3), 469 EOF-490 EOF.

[281] Yagasaki, K., Takagi, T., Sakai, M., & Kitamura, K. (1997). Biochemical characterization of soybean protein consisting of different subunits of glycinin. *Journal of Agricultural and Food Chemistry*, 45, 656-660.

[282] Zimmermann, M. B. (2009). *Iodine Deficiency Endocrine Reviews*, 30(4), 376-408.

[283] Zhou, J., Gogger, E. T., Tanaka, T., Guo, Y., Blackburn, G. L., & Clinton, S. K. (1999). Soybean phytochemicals inhibit the growth of transplantable human prostrate carcinoma and tumor angiogenesis in mice. *J, Nutr.*, 129, 1628-1635.

Chapter 2

Soybean Meal and The Potential for Upgrading Its Feeding Value by Enzyme Supplementation

D. Pettersson and K. Pontoppidan

Additional information is available at the end of the chapter

1. Introduction

Soy is a crop of tremendous importance for the food industry but also in the animal feed industry. Soybeans, as well as other oilseed crops such as rape seed (canola), sunflower and palm kernel, are grown primarily for the production of vegetable oil for human consumption but the by-products after oil extraction are of similar importance as feed ingredients. Meals from these crops are obtained after the extraction of the vegetable oil from the seed. In addition, considerable amounts of cottonseed meal (a by-product of the cotton fibre production) are also available for animal feed. Soybean meal (SBM), the by-product after oil extraction of soybeans has become increasingly important as a feed component and is used in variable amounts in the feeding of all species in animal production, even to some extent in the feeding of farmed fish. On a global scale SBM is dominating the market for protein meals primarily due to the high content of good quality protein, making SBM an excellent ingredient in feed formulations. SBM is particularly important for poultry production, constituting approximately 30-40 % of a standard soy/maize diet, since broilers and layers require a high proportion of protein in their feed. It is generally estimated that approximately 46 % of all SBM produced for animal feed is used in poultry diets (broilers, layers and turkeys), while another 25 % is used for feeding pigs. In the US approximately 50 % of the SBM is used for poultry, 25 % for swine and 12 % for beef cattle. Although US is the largest producer of soybeans, Argentina is by far the largest exporter of SBM followed by Brazil and US. With very limited own production of soy the EU is one of the leading markets for import of SBM. [1]

Over the last 5 years the price for SBM has been increasing, and this trend is expected to continue in the future. Hence, with protein already being the second most expensive ingredient in e.g. poultry diets, there is a need to either replace SBM with other and cheaper protein ingredients in the diets or to increase the utilization of SBM nutrients e.g. by the use of enzymes.

2. Composition and nutritive value of soybean meal

SBM consists primarily of protein and carbohydrates in the form of indigestible neutral and acidic non-starch polysaccharides (NSP) as well as low molecular weight sugars (Table 1).

Due to the low content of easily available carbohydrates, and a high content of NSP belonging to the indigestible dietary fibre fraction [2] the apparent metabolisable energy value (AME) of SBM is low for broiler chickens and estimated to only about 9.5 MJ/kg fresh weight, while in growing pigs that, as opposed to poultry, have a high capacity for hind gut fermentation of the indigestible carbohydrate fraction, the AME value may be around 14.5 MJ/kg fresh weight [3].

	Range (g/kg dry matter)	
Component	Low	High
Crude protein	490	540
Starch	0	27
Crude fat	17	21
Low molecular weight sugars		
Mono saccharides	5	8
Sucrose	55	81
Raffinose series	53	67
Neutral non starch polysaccharides		
Rhamnose	3.7	5
Fucose	2.9	3.1
Arabinose	22	25
Xylose	15	18
Mannose	9	13
Galactose	37	40
Glucose	50	59
Acidic non starch polysaccharides		
Uronic acids	39	41

Table 1. Typical soybean meal gross chemical composition. (g/kg dry matter), compiled from literature, for crude protein, starch, crude fat and low molecular weight sugars, while the non-starch polysaccharide composition of a selection (n=6) of soybean meals were analysed at Novozymes A/S according to Theander *et al.* [4]. Data compiled from [2,3] as well as from internal analysis of neutral and acidic non starch polysaccharide constituents.

The composition of SBM may vary depending on the country of origin of the soybean, the cultivar, the processing and the inevitable year to year variation in growing conditions. Still most feeding tables contains very little, if any, information about the variability of e.g. amino acids (AA) and digestibility that may be expected although the data presented typically is based on several hundred samples collected over several years. Neither is the indigestible dietary fibre fraction well described in feed tables and at best values for neutral detergent and acid detergent fibre are provided.

2.1. Protein

Plants cultivated for their protein content are typically classified as angiosperms and belong to a number of different botanical families. Beans, peas, lupins and soybeans are all members of the leguminosae family while rape seed belong to the cruciferae family, sunflower and safflower are members of the compositae family and cotton belongs to the malvaceae [5]. On average the content of CP in the common raw oilseeds such as soybean, sunflower and rape seed ranges from 20-40 %. Due to the various processing steps and the subsequent concentration of the protein-containing fraction by solvent extraction, the average CP content of oilseed meals varies from 32 % in sunflower meal to over 50 % in some SBM [6]. SBM is used in feed rations for monogastric animals mainly due to the high protein content and also because of the superior AA profile compared to other plant protein products used as diet ingredients [3]. Poultry and swine diets are generally formulated based on AME and the level of CP. SBM has a high content of lysine, which makes it a good ingredient in poultry and swine diets as both of these species has a high requirement for this essential amino acid.

The CP fraction of SBM is made up of around 80 % storage proteins in the form of glycenin and β-conglycenin, approximately 5 % is represented as various anti nutritional factors (ANFs) and the remaining 15 % consists of other proteins. Most tables on nutrient composition of feedstuffs such as e.g. the NRC [7] operates with two types of SBM based on the crude protein CP content. One is the regular SBM with approximately 44 % CP, where a fraction of the hulls has been added back into the meal, and the other is dehulled SBM with approximately 48 % CP. The feed compound industry principally assumes that the digestible AA content of SBM per unit of protein is constant; disrespecting that variability may occur due to e.g. genotype, origin, processing and storage conditions [8-10]. The variability in SBMs has been nicely demonstrated by de Coca-Sinova et al. [10] who evaluated six SBMs from different origins and found considerable variation in the chemical composition and protein quality which translated into differences in AA digestibility in broilers, so that SBM with higher levels of CP and lower levels of trypsin inhibitor activity showed higher AA digestibility.

2.2. Carbohydrates

For broiler chickens vegetable protein sources typically have a low metabolisable energy content compared to cereals, due to a lower starch content and a higher content of indigestible NSP, which are part of the dietary fibre fraction (Table 1).

The structural features of the common vegetable protein sources are more complex than that of cereals but their cell walls still contains cellulose and hemicelluloses but in addition also high amounts of rhamnogalacturonans, assigned to the pectic polysaccharides. Quantification of the pectic fraction is not straightforward due to a more complex structure and since extraction procedures used for the analysis often causes an overestimation and the literature data may differ.

It is notable that AME table values and AME calculated on the content of CP, crude fat, starch and sucrose (Table 2) are quite similar for meals from soybean and sunflower while in meals from rape seed and cotton seed the discrepancy is 1.7 and 2.3 MJ/kg, respectively. The amount of pectin (61 g uronic acids/kg dry matter) and fibre matrix structure of rape seed most likely increases the water holding capacity of this raw material resulting in a poor nutrient availability for monogastric animals and in addition ANFs may reduce the energetic value. In cotton seed similar effects may be at play although the pectin content is quite similar to that of SBM (about 45 g/kg dry matter), while the ANFs are different and the cellulose content is considerably higher.

	Soybean	Rape seed	Sunflower	Cotton seed
Crude protein	450	337	426	426
Crude fat	10	23	29	29
Starch	5	0	0	0
Sucrose	70	58	33	16
Oligo, di and mono –saccharides, except sucrose	67	24	23	56
Cellulose	62	52	89	92
Total dietary fibre	233	354	326	340
Apparent metabolisable energy (AME) in broilers				
Table values AME (MJ/kg)[a]	9.5	5.9	6.2	6.3
Calculated AME (MJ/kg)[b]	9.3	7.6	6.9	8.6

Table 2. Average chemical composition. (g/kg fresh weight) of solvent extracted meal from soybean, rape seed, sunflower and cotton seed. Table compiled from [2,3]. [a]Table values from [3]. [b]Values based on crude protein, crude fat, starch and sucrose.

Plant cell walls are divided in primary and secondary walls and their composition will differ according to their stage of development (maturity). Primary cell walls are flexible and surround cells in growth and elongation whereas secondary walls surround cells in which growth has ceased. The secondary walls are lignified and thereby rigid. The primary cell wall is synthesized during cell expansion at the first stages of development and is composed of cellulose, hemicelluloses, pectic polysaccharides and many proteins (Table 3).

Cell walls are classified as Type I cell walls, which are generally the most common in the plant kingdom, or as Type II which is typical for grasses. The non-cellulosic polysaccharides of Type I cell walls are xyloglucans and about 35 % of the cell wall mass are pectins. Type I cell walls are found in all dicotyledons, the non-graminaceous monocotyledons, and gymnosperms. Type II walls have a low pectin and xyloglucan content and a high arabinoxylan content. Type II walls also contain mixed linked β-D-glucan and possess ester linked ferulic bridges in the xylan, which have not as yet been found in Type I walls [11].

In soybeans and other dicotelydons the pectin fraction consists of rhamnogalacturonans. The rhamnogalacturonan I consists of a main chain of galacturonic acid and rhamnose. Attached to this structure there are side chains of galactose and arabinose residues. There are also xylose and rhamnose side chains present as well as xylo-galacturonans. In addition traces of mannose polysaccharides may be found that possibly origin from an incomplete removal of the hull fraction from the meal [12].

In the cotelydons approximately 30 % of the NSP belong to the pectin fraction while in the hulls about 80-90 % of the non-starch polysaccharides are of pectic origin. The galactose content of SBM is generally higher than in other oilseed meals and is highly associated with the rhamnogalacturonans. This is not the case for rape seed meal, sunflower meal and cotton seed meal [13] where arabinans and arabinogalactans constitute the most important side chains. Since the NSP are indigestible by the endogenous enzymatic systems in the small intestine they can only be utilised through hind-gut fermentation, thereby providing short chain fatty acids that may be absorbed in the hind gut and utilised as an energy source by the animal. As a consequence the AME content of all oilseed meals is low for broiler chickens that have a limited capacity for hind gut fermentation, while it is higher for pigs [14].

The dietary fibre fraction is composed of different polysaccharide structures and their molecular structure and incorporation into the cell wall matrix determines their solubility characteristics. A high solubility favours fermentation and even poultry may to some extent ferment soluble NSP [15,16].

The principal hemicelluloses found in dicotelydons including soybeans are xyloglucans (Table 3) consisting of a glucose back bone with xylose side chains linked to the carbon 6 of the glucose residues in the back bone chain. It is a well acknowledged hypothesis that primary and secondary cell walls in dicotelydons are constructed in different ways. In the primary cell wall cellulose and xyloglucans interact in a network consisting of cellulose which is coated with a monolayer of xyloglucans. The secondary cell wall is composed of pectins, but not of the homogalacturonan type that is found in fruit and berries. In soybean the pectic polysaccharides are xylogalacturonans with a backbone of α-(1-4)-galacturonan residues and to this xylose residues are linked in β-(1-3) position. The xylogalacturonans are associated with regions that consist of rhamnogalacturonans type I and II, with a higher degree of branching in the type II [18,19]. Together with the xylogalacturonans the rhamnogalacturonan I and its side chains consisting of arabinans and arabinogalactans makes up the main part of the pectic substances [20].

Component		Approximate composition (%)	
		Primary cell walls	Secondary cell walls
Cellulose		30	45-50
Hemicellulose	Xyloglucan	25	25
	Xylans	5	
	ß-D-glucans	nd	nd
	Glucomannan	nd	nd
Pectins	Homogalacturonans	15	
	Rhamnogalacturonan I	15	0.1
	Rhamnogalacturonan II	5	
Glycoproteins	Arabinogalactanprotein	Variable	nd
Phenolics	Extensin	<5	nd
	Lignin	nd	20
	Phenolic acid	0.3	nd

Table 3. Major polymers of the growing and mature plant cell walls in dicotelydons. Table compiled from [17].

2.3. Anti nutritional factors

The occurrence and amount of ANFs and their effect on protein and energy utilisation limits the inclusion of vegetable proteins in diets for pigs and poultry. In general, ANFs among legume species are similar, however the actual amounts of ANFs varies widely between different species and cultivars [21]. The main ANFs include protease inhibitors, lectins, tannins, phytic acid and indigestible carbohydrates. In addition, lupins contain considerable amounts of alkaloids and lima beans contain increased amounts of cyanogens as their dominant ANFs [22,23]. Oilseeds, and subsequently oilseed meals, have more specific ANFs depending on the actual species. Rapeseed meal contains glucosinolates and soybean is particularly high in trypsin and chymotrypsin inhibitors while cottonseed meal contains gossypol [24]. Based on what is commonly known in the feed industry and academia it may be assumed that the content of ANFs may differ depending on growing conditions and, especially, the heat processing of the feed ingredients.

The best characterized ANFs of soybean are protease inhibitors, lectins and phytate.

The Kunitz inhibitor (KSTI) together with the Bowman-Birk inhibitor (BBI) are the most abundant protease inhibitors in soybeans, and are commonly referred to as trypsin inhibitors even though they may also inhibit e.g. chymotrypsin and other proteases belonging to the serine family. The mechanism of trypsin inhibitors is to bind to the active site of the protease and thereby cause inactivation of the enzyme, which then cannot proceed to degrade protein. When the level of protease activity in the gut is depressed the pancreas responds in a compensatory fashion by producing more of the digestive enzymes. In some species this has been shown to be related to an enlargement of the pancreas [25,26]. When animals are fed SBM with a high level of protease inhibitors the digestive proteases trypsin and chymotrypsin are inactivated leading to impaired animal performance. This has been exemplified e.g. by Sklan *et al.* [27] who showed that chicks fed raw soy had significantly reduced bodyweight gain compared to a control group fed heated soy. Furthermore, the chicks fed the raw soy had increased pancreas weight and reduced trypsin activity in the small intestine.

Lectins are carbohydrate binding proteins generally considered to have an anti nutritional effect [28,29]. This has been exemplified e.g. by Schulze *et al.* who showed that inclusion of purified soybean lectin into a pig diet increased the amount of dry matter, nitrogen (N) and AA passing into the terminal ileum [30].

Phytic acid, also commonly referred to as phytate, is a well described anti nutritional factor that under physiological pH conditions binds minerals and protein thereby preventing utilization of these nutrients by the animal [31,32].

Furthermore, the high molecular weight soy proteins glycenin and β-conglycenin act as potential antigenic factors leading to the formation of serum antibodies in particular in young animals, e.g. early-weaned piglets [33].

2.4. Impact of processing on nutritive value

As opposed to the cereal grains the composition and nutritive value of oilseed meals for animal feed not only depends on the cultivar, the climatic condition and level of fertilisation, but is also influenced by the processing conditions during the oil extraction procedure. Oilseed meals are generally obtained after a pre-press solvent extraction process. The combined effects of seed preparation, de-hulling pre-conditioning, cooking and solvent extraction will determine the nutritive value of the meal.

The processing of soybeans (Figure 1) and in particular the final heating step (toasting) is critical to the quality of the resulting SBM. The initial processing of soybeans includes cleaning, drying and cracking of the beans to remove the hulls. The dehulled soybeans are the raw material for production of full fat soybean meal which may or may not be heat treated to inactivate enzymes. When processing regular SBM the dehulled beans are conditioned at 65-70°C followed by flaking, which prepares the beans for oil extraction. The oil is usually extracted using a solvent such as hexane. Finally the resulting cake is treated in a toaster in order to remove the solvent and to heat the meal sufficiently to optimize its nutritional value. In this process control of the processing conditions such as temperature, moisture, pressure and processing time is highly important to maintain a high solubility of the SBM product. [34] After toasting a fraction of the hulls will often be transferred back into the meal to produce SBMs with different protein contents.

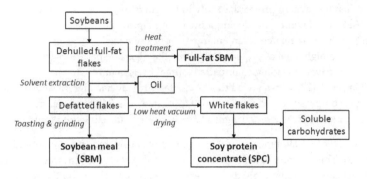

Figure 1. Overview of the processing of soybeans to obtain soybean meal (SBM), full fat SBM or Soy protein concentrate (SPC).

As mentioned earlier the nutritional value of SBM is limited by the presence of several ANFs interfering with feed intake and metabolism. ANFs are concentrated in large amounts in the hull fractions of oilseeds and de-hulling will consequently reduce the level of these substances in the meal. Furthermore, heat processing of the SBM acts to destroy the heat sensible ANFs such as protease inhibitors and lectins. However, in case SBM is heated excessively the occurrence of Maillard reaction will increase. Maillard reactions occur between the amino group of the amino acids and the reducing sugars eventually leading to a decrease in energy and amino acid digestibility [35]. Hence, the conditions applied during processing to ensure a high quality of the protein fraction are a compromise between sufficiently inactivating the ANFs and avoiding destruction of essential available nutrients.

An important problem faced by the feed industry is the lack of good techniques to correctly evaluate the quality of commercial SBMs. The available methods, of which determination of the protein dispersibility index (PDI) [36], KOH protein solubility [37,38] and urease activity [39] are the most commonly used, are based on changes in the physical and chemical properties of SBM occurring during heat treatment, and have shown not to be fully reliable [40,41]. Another means to estimate SBM quality is the determination of trypsin inhibitor content, but this method is tedious and also shows inconsistency [43].

The defatted soy flakes can also be processed to obtain a soy protein concentrate (SPC), which is a higher value protein product compared to SBM. This processing takes place by removing the solvent (e.g. hexane) by low heat vacuum drying and then removing the soluble carbohydrates yielding a final product with ~90 % protein [43]. SPC has a much lower level of ANFs than SBM and is therefore particularly well suited for young animals that do not tolerate normal SBM well, e.g. piglets. Lately the use of SPC in salmon and trout diets to replace fishmeal is also being evaluated [43].

3. Enzymes for improving the nutritive value of soybean meal

The use of exogenous microbial enzymes is today a mature concept in the animal feed industry and is used on a routine basis to improve the nutritive value of feed ingredients. The logical implication of improving the nutrient utilization is a reduced amount of nutrients in manure, which is highly beneficial for the environment especially in areas with intense animal production. The environmental benefits of using enzymes in animal diets has been exemplified in a series of published life cycle assessment studies investigating the effects of xylanase [44], phytase [45] and protease [46] when used in either pig or poultry diets. These studies together demonstrate the huge potential impacts on global warming, eutrofication and acidification that can be achieved by employing feed enzymes in animal diets to improve the utilization of nutrients.

3.1. Protein degrading enzymes

With protein being a quite expensive ingredient in animal diets, improving the nutritional value of the SBM protein fraction is an obvious target for enzyme application. The apparent ileal digestibility (AID) of CP in SBM is typically around 80-85 %, but lower values have also been reported. In an investigation of 6 different SBMs originating from South America, US and Spain the AID of CP and AA was shown to vary considerably between the batches with the US SBM having the highest digestibility value (82.3 %) followed closely by the Spanish SBM (81.8 %) and with the South American SBM's having considerably lower (75.2-76.8 %) digestibility values [10]. These results serve to demonstrate the impact of differences in SBM quality when formulating diets to achieve the necessary protein content and AA availability. Since it is both difficult and laborious to investigate SBM quality, in practice diets are often formulated to contain higher levels of nutrients than required, thereby providing a safety margin.

Early attempts to improve the nutritive value of SBM for pigs and poultry aimed at pretreating SBM in the presence of a protease to increase protein solubilisation and obtain a decrease in antigenicity. Using this approach it was demonstrated that treating SBM with either an acidic protease or an alkaline protease increased the amount of soluble α-amino N concentration and reduced the antigenic protein concentration, more so with the acidic protease compared to the alkaline protease treated SBM [47]. Furthermore, feeding studies in piglets [47] and broilers [48] showed that feeding SBM treated with the acidic protease instead of the non-treated SBM as part of a cereal-based diet led to performance improvements in both species as well as to improved N digestibility and reduced serum antisoya antibodies in broilers. In contrast feeding SBM treated with the alkaline protease reduced performance in piglets [47], and even though Ghazi et al. did observe reductions in chick serum antisoya antibodies upon feeding SBM treated with alkaline protease only the acid protease treated SBM resulted in a positive effect on performance and digestibility parameters [48]. Differences in digestibility may occur either directly or indirectly due to hydrolysis of ANFs interfering with the digestive process. Protease treatment in above studies did not influence the already low level of trypsin inhibitor and lectin. Hence, it was

concluded that the increase in performance and N digestibility by treatment with the acidic protease was a result of general improvement in digestion of SBM protein rather than
inactivation of ANFs [47,48].

More recent studies have shown that direct addition of a pure protease from *Nocardiopsis
prasina* can lead to significant increases in CP and AA digestibility in broilers fed SBM or full
fat SBM [49,50]. It was concluded that AA utilization was on average improved by about 5
% in SBM and 6 % in full fat SBM. Furthermore, the same protease has been demonstrated in
several studies to have a positive impact on growth performance and N digestibility of
broilers fed complete corn-SBM based diets [51,52,53]. The efficiency of the *Nocardiopsis prasina* protease to improve digestion of the SBM protein fraction is supported by internal laboratory studies at Novozymes A/S demonstrating the ability of the protease to improve
protein hydrolysis in different SBM batches as well as in full fat SBM (Figure 2).

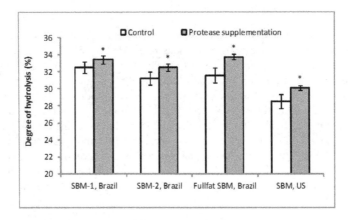

Figure 2. Increase in degree of protein hydrolysis (DH, %) by a pure protease from *Nocardiopsis prasina* dosed at 100
mg purified enzyme protein per kg of SBM (n = 5). The Various SBM batches were incubated in an *in vitro* digestion
system and protease effect on top of endogenous enzymes (pepsin and pancreatic enzymes) was analysed as increase
in DH (method published in [53]). Error bars indicate standard deviation and asterisks indicate a significant impact of
the protease (P<0.05; Tukey HSD test). Unpublished data, Novozymes A/S.

The way by which a protease increases hydrolysis and digestion of the SBM protein fraction
may be related to both general hydrolysis of the SBM proteins and to degradation of various
ANFs present in SBM. A general hydrolysis of SBM protein would presumably increase the
availability of the protein for further hydrolysis and absorption in the gastro intestinal tract.
On the other hand, degradation of ANFs will improve the natural digestion and utilization
of protein as the adverse effects of the ANFs are reduced.

In this context internal studies have shown that the *Nocardiopsis prasina* protease is capable
of degrading various anti nutritional proteins from soybean. As exemplified in Figure 5 the
protease efficiently degraded both the Kunitz inhibitor (KSTI) and lectin (Figure 3), leaving
only 10-20 % KSTI and around 15 % lectin intact, while purified porcine trypsin and chymo-

trypsin could not degrade these proteins to nearly the same extent. The ability of a feed protease to degrade ANFs presents an interesting possibility to alleviate the negative impacts of including raw soy or under processed SBM in e.g. poultry or swine diets.

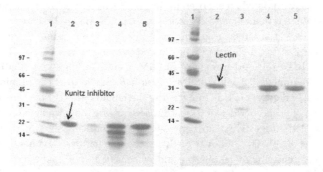

Figure 3. Degradation of purified Kunitz trypsin inhibitor (left) and lectin (right) both from *Glycine max* by purified proteases as analysed by SDS-page. Lane 1: low molecular weight marker (kDa), lane 2: no protease, lane 3: *Nocardiopsis prasina* protease, lane 4: porcine trypsin, lane 5: porcine chymotrypsin. Inhibitors (purchased from Sigma-Aldrich) and proteases were incubated in a 10:1 ratio on mg protein basis for 2 hours at 37°C, after incubation protein was precipitated with TCA, re-suspended in SDS-page sample buffer and analysed by gel electrophoresis. Unpublished data, Novozymes A/S.

3.2. Carbohydrate degrading enzymes

For degradation of the major dietary fibre constituents of importance in cereals there are many enzyme products available on the market and approved by the EU authorities based on proven efficacy in animal trials. The beneficial effects on animal performance, especially in broiler chickens, are assumed to be caused by a combination of depolymerisation of viscous arabinoxylans and a degradation of the indigestible cell wall. The resulting viscosity reduction improves nutrient absorption and the degradation of the cell walls improves the liberation of nutrients (e.g. starch and protein) enclosed by the indigestible cell walls [54,55]. This effect is commonly referred to as the cage effect. For oil seed meals the expected outcome of degradation of the cell walls is less obvious since they generally do not have an intact cell wall architecture due to extensive processing and thereby there is no cage effect. Still the water holding capacity of the material could be reduced and this could have a positive influence on nutrient absorption. In addition liberation of galactose could provide additional energy, at least from SBM. In rape seed, which has less galactan side chains associated with the pectin matrix, an anticipated energy benefit would be limited. Another possibility would be to degrade the oligosaccharides of the raffinose series and the most common attempts to improve the nutritive value of SBM have targeted these. A successful degradation of this fraction could release galactose, and also sucrose, which is the molecule from which the raffinose series is built. Raffinose, stachyose and verbascose contains (1-6)-α-galactopyranosyl units with variable chain lengths of which stachyose (two galactose units in the chain linked to sucrose) is the pre-dominant in SBM. The oligosaccharides of the raffinose series can be broken down using an α-galactosidase that liberates the galactose from the sucrose molecule. This is a straightforward enzymatic application that has been tested *in vitro* showing that α-galactosidase can degrade the oligosaccharides of the raffinose series in SBM [56,57]. Internal trials at Novozymes A/S (Figure 4) has indicated a large variability in the efficacy of α-galactosidase when exposed to the conditions prevailing in the upper gastro intestinal tract simulated in an *in vitro* digestion system. The average release of galactose from the raffinose series was about 6.7 mg/g diet and since maize only contains minor amounts of galactose the main part of the galactose originated from the SBM and about 16 g of galactose could theoretically be released from 1 kg of SBM, corresponding to a potential AME value increase of approximately 0.3 MJ/ kg SBM. The *in vitro* model does not provide information of any additional beneficial effects of degrading the raffinose series. Still the application is of interest since the oligosaccharides of the raffinose series are indigestible but readily fermented and this may cause digestive disturbances with gas production and rapid digesta passage rates in poultry [58,59]. The removal of these oligosaccharides by ethanol extraction has been shown to increase the metabolisable energy content as well as the transit time in adult roosters [60]. However, this procedure alters the general composition of the SBM by also extracting other ethanol soluble components, resulting in a meal with higher CP content and improved nutritive value. When using oligosaccharide or non-starch polysaccharide degrading enzymes the SBM composition is not altered and the effects observed can be attributed to the enzyme *per se*. Trials in pigs have generated high digestibilities in the small intestine even without α-galactosidase supplementation [61], while poultry data indicate a positive enzyme effect on the digestion of the raffinose series but not on performance [62]. Based on these and similar data it may be concluded that performance data are not strong enough to justify, from an economical point of view, the additional supplementation of this enzyme in broiler chicken diets that already may contain xylanases and phytases.

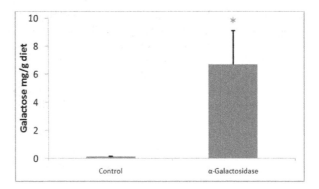

Figure 4. Release (mg/g) of galactose by α-galactosidase from 5 replicates of a model diet containing 600 g maize and 400 g soybean meal per kg and incubated at 40°C in an *in vitro* system mimicking the stomach (pH 3±0.2 and 3000 U of pepsin/g diet) and small intestine (pH 7±0.2 and 8 mg/g diet of pancreatin) for a total period of 6 h. Standard deviations are indicated by error bars and asterisks indicate a statistically significant difference (P<0.05; Tukey HSD test) compared to blank. Unpublished data, Novozymes A/S.

The use of galactanase has shown that also the galactose side chains of the rhamnogalacuronans may to some extent be degraded [63]. As opposed to many other feed enzyme products on the market a classical wild type fermentation product of *Aspergillus aculeatus* (RONO-ZYME® VP, DSM Nutritional Products, Basel, Switzerland), is able to significantly impact the NSP fraction of SBM. In an *in vitro* trial comparing this product with a selection of feed enzymes developed for targeting the cereal ingredients in animal feed the preparations were dosed at 100 times the commercial recommendation in a buffer system. After incubation the solubilised material was removed by centrifugation and the pellets containing the residual insoluble NSP were analysed according to Theander *et al.* [4]. The effect of the *A. aculeatus* product on NSP containing arabinose, although not statistically significant from the control, was the highest compared to all other enzyme treatments with a reduction of about 6.5 % (Figure 5). In addition the *A. aculeatus* product gave a significant reduction (P<0.05) of galactose by 9.6 % compared to the control treatment, whereas the other enzyme treatments did not provide a reduction. The residual glucose content was significantly lower for the *A. aculeatus* product and the blend product compared to the control, 22 and 13 %, respectively, and these effects were statistically significant compared to all other treatments (Figure 3). The results indicate that in order to degrade the complex cell wall matrix of SBM several enzymatic activities are required at a high activity level and these are not generally found in single commercial products developed for targeting type II cereal cell walls. It is also obvious that only one single enzymatic activity is not enough to provide a sufficient degradation of the different polysaccharide structures of importance (Table 3).

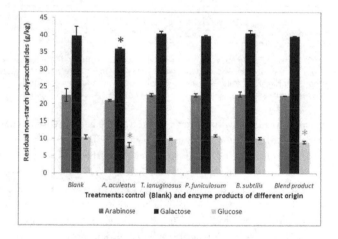

Figure 5. Residual insoluble content (g/kg) of arabinose, galactose and glucose non-starch polysaccharide residues in a soybean meal after incubation at 40°C in an acetate buffer (0.1 M pH 5.0) for 4 hours and with enzymes dosed at 100 times the commercial recommendation. The products used were the Aspergillus aculeatus wild type fermentation or different products containing only xylanase activity derived from Thermomyces lanuginosus or mainly xylanase and ß-glucanase activity in products derived from Penicillium funiculosum or Bacillus subtilis and a commercial blend product containing xylanase, ß-glucanase and α-amylase activities. Standard deviations are indicated by error bars and asterisks indicate a statistically significant difference (P<0.05; Tukey HSD test) compared to blank. Unpublished data, Novozymes A/S.

3.3. Phytic acid degrading enzymes

The content of phytic acid in SBM is typically around 1.3-1.4 % of dry matter and constituting around 50 % of the total phosphorus pool [64-66]. Phytic acid is not absorbed in the gastro
intestinal tract, but may be degraded by a phytase to render the phosphorus free and available for absorption. Phytases are enzymes that cleave of the phosphate groups from the inositol ring of phytic acid, thereby rendering free phosphorus to be utilized by the animal and also
lowering the anti nutritional effect of phytic acid on mineral and protein availability. Phytase activity is present in most seeds but the activity in oilseed meals including SBM is relatively
low [65], hence degradation of SBM phytic acid necessitates the presence of phytase either
from cereals in the diet or from a microbial phytase source. Since a large part of the feed for
poultry and swine is treated at high temperatures e.g. in a pelleting process in order to sanitize
the feed from *Salmonella* infections etc., cereal phytases are often inactivated in the final feed
(e.g. [67]). Hence, the use of microbial phytases in feed formulation is extensive.

Feedstuff	Residual phytic acid (%)
Wheat	21.5 ± 1.3 ab
Maize	24.5 ± 1.3 b
Barley	21.3 ± 1.3 ab
Soybean meal	47.9 ± 1.3 c
Rapeseed meal	21.3 ± 1.3 ab
Soybean meal-maize blend	17.2 ± 1.5 a

Table 4. abc: Different letters indicate significant differences (P<0.05), tested by Tukey HSD. Residual phytic acid (%) after incubation of feedstuffs with a commercial phytase (RONOZYME® HiPhos, DSM Nutritional Products, Basel, Switzerland) for 30 minutes (pH 4, 40°C). All feedstuffs were heat treated and additional calcium (6 g kg⁻¹ dry matter) was added. Values are representative of the sum of inositol hexaphosphate and inositol pentaphosphate. [68]

Degradation of phytic acid in SBM has been demonstrated e.g. by Brejnholt et al. [68] showing a 50 % degradation of phytic acid upon incubation of SBM with a bacterial phytase at
pH 4 (30 minutes, 40°C). Interestingly this study indicated that it was more difficult to degrade phytic acid in SBM compared to phytic acid in cereal meals and rape seed meal (Table
4). This difference might be related to the content of protein in the feedstuffs, as protein is
known to form insoluble complexes with phytic acid at low pH [69]. In support of this hypothesis it has been shown that phytic acid in SBM is much less soluble than phytic acid in
maize meal at low pH (Figure 6). Furthermore, internal data produced at Novozymes A/S
also show that treatment of a SBM-maize mixture with pepsin to degrade the protein fraction has a positive impact on phytic acid solubility (Figure 7). In the digestive tract endogenous digestive proteases will always be present to degrade protein and thereby improve the
availability of phytic acid for hydrolysis by phytase. Supporting this there are a huge
amount of *in vivo* studies demonstrating that phytases effectively releases phosphate in animals fed diets containing SBM e.g. Aureli *et al.* [70] showing phytase efficacy in broilers.

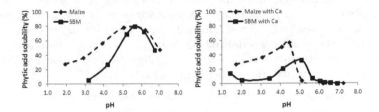

Figure 6. Phytic acid solubility (%) in maize and SBM (left) and in maize and SBM with additional 5 g calcium kg⁻¹ dry matter. Error bars represent standard deviations (2xSD) of 3 replicates. Modified from [71].

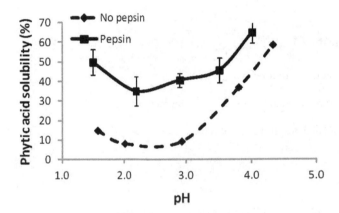

Figure 7. Phytic acid solubility (%) in a SBM-maize (30:70) mixture with additional 5 mg calcium g⁻¹ dry matter in the presence or absence of pepsin (3000 U g⁻¹ dry matter). Error bars represent standard deviations (2xSD) of 3 replicates. Unpublished data, Novozymes A/S.

4. Conclusions

SBM is the most important protein source in animal feed and it is estimated that approximately half of the SBM produced for animal feed is used in poultry diets, while another 25 % is used in pig diets. SBM is primarily added due to its high content of protein and favourable composition of AAs, while the low content of metabolisable energy and a high content of NSP may provide problems when incorporating this oil seed meal at high levels. The presence of several ANFs such as trypsin inhibitors and lectins may represent a severe problem that can restrict animal performance when feeding a meal that has not been properly processed, however there are indications that these short comings can be overcome by proper heat inactivation of the SBM as well as enzyme supplementation of the diet.

Research results indicate that the nutritive value of SBM may be enhanced by adding exogenous microbial enzymes such as carbohydrate degrading enzymes, proteases and phytases. However, there is a need for improved carbohydrate degrading enzymes to better target the oligosaccharides of the raffinose series and also to reduce the negative effects of high levels of complex NSP constituents, that evidently are not degraded by common xylanases or ß-glucanases used for improving the feeding value of cereals.

The use of protease to improve protein digestibility and reduce the presence of anti nutritional proteins represents a novel and promising application that has an environmental impact by improving protein digestibility and thereby reducing nitrogen excretion from farm animals. In a similar way phytase supplementation is already a well established environmentally friendly application that reduces phosphorus excretion when feeding SBM rich in phytic acid.

Author details

D. Pettersson* and K. Pontoppidan

*Address all correspondence to: kpon@novozymes.com

Department of Feed Applications, Novozymes A/S, Denmark

References

[1] Soystats. (2012). http://www.soystats.com/2011/accessed 6 May).

[2] Bach, Knudsen. K. E. (1997). Carbohydrate and lignin contents of plant materials used in animal feeding. *Animal Feed Science and Technology*, 67-319.

[3] Sauvant, D., Perez-M, J., & Tran, G. (2004). Tables of Composition and Nutitional Value of Feed Materials. *INRA editions, Wageningen Publishers.*

[4] Theander, O., Åman, P., Westerlund, E., Andersson, R., & Pettersson, D. (1995). Total dietary fibre determined as neutral sugar residue, uronic acid residue and Klason lignin (The Uppsala Method): Collaborative study. *Journal of AOAC International*, 74-1030.

[5] Weier, E. T., Stocking, R. C., & Barbour, M. G. (1970). Botany: An introduction to plant biology. *New York: John Wiley & Sons Inc.*

[6] Ravindran, V., Hew, L. I., & Bryden, W. L. (1998). Digestible amino acids in poultry feedstuffs. *Rural Industries Research & Development Corporation, RIRDC publication* [98].

[7] NRC. (1994). Nutrient requirements of poultry. *9th revised edition. Washinton: National Academic Press.*

[8] Clarke, E., & Wiseman, J. (2005). Effects of variability in trypsin inhibitor content of soja bean meals on true and apparent ileal digestibility of amino acids and pancreas size in broiler chicks. *Animal Feed Science and Technology*, 121-125.

[9] Hermida, M., Valencia, D. G., Serrano, M., & Mateos, G. G. (2008). Influence of soybean meal origin on its nutritive value and quality parameters. *Poultry Science*, 87-29.

[10] De Coca-Sinova, A., Valencia, D. G., Jiménez-Moreno, E., Lázaro, R., & Mateos, G. G. (2008). Apparent ileal digestibility of energy, nitrogen, and amino acids of soybean meals of different origin in broilers. *Poultry Science*, 87-2613.

[11] Minic, Z., & Jouanin, L. (2006). Plant glycoside hydrolases involved in cell wall polysaccharide degradation. *Plant Physiology and Biochemistry*, 44-435.

[12] Huismann, M. M. H., Schols, H. A., & Voragen, A. G. J. (1998). Cell wall polysaccharides from soybean (Glycine max) meal. Isolation and characterisation. *Carbohydrate Polymers*, 37-87.

[13] Siddiqui, R. I., & Wood, P. J. (1977). Carbohydrates of rapeseed: a review. *Journal of the Science of Food and Agriculture*, 28-530.

[14] Mc Ginnis, J. (1983). Carbohydrate utilization in feedstuffs. *In: Proceedings of the Minnesota nutrition conference, St. Paul, Minnesota*, 106-107.

[15] Longstaff, M., & Mc Nab, J. M. (1989). Digestion of fibre polysaccharides of pea (pisum sativum) hulls, carrot and cabbage by adult cockerels. *British Journal of Nutrition*, 62-563.

[16] Carré, B., Derouet, L., & Leclercq, B. (1990). The digestibility of cell wall polysaccharides from wheat (bran or whole grain), soya bean meal and white lupin meal in cockerels, Muscovy ducks and rats. *Poultry Science*, 69-623.

[17] Ishii, T. (1997). Structure and functions of feruloylated polysaccharides. *Plant Science*, 127-111.

[18] Mc Cann, M. C., & Roberts, K. (1991). The cytoskeletal basis of plant growth and form. *New York, USA: Academic Press.*

[19] Carpita, N. C., & Gibeaut, D. M. (1993). Structural models of primary cell walls in flowering plants: consistency of molecular structure with the physical properties of the walls during growth. *Plant Journal*, 3-1.

[20] Whitaker, J. R., Voragen, A. G. J., & Wong, D. W. S. (2003). Handbook of food enzymology. *New York, USA: Marcel Dekker.*

[21] Warenham, C. N., Wiseman, J., & Cole, D. J. A. (1994). Processing and antinutritive factors in feedstuff. *In: Cole DJA, Wiseman J, Varley MA (eds.) Principles of Pig Science. Nottingham: Nottingham Press*, 1, 141-167.

[22] Liener, I. E. (1989). Antinutritional factors. *In: Matthews RH (ed.) Legumes: Chemistry, technology and human nutrition. New York: Marcel Dekker*, 340-366.

[23] Liener, I. E. (1989). Antinutritional factors in legume seeds: State of the art. *In: Huisman J, van der Poel TFB, Liener IE (eds.) Recent Advances of Research in Antinutritional Factors in Grain Legume seeds: proceedings of the 1st International Workshop on Antinutritional Factors in Legume Seeds, 13-25 November 1988, Wageningen*, 6-13.

[24] Dongmo, T., Pone, D. K., & Ngoupayou, J. D. N. (1989). Cottonseed cake in breeder hens diets: effects of supplementation with lysine and methionine. *Archiv für Geflügelkunde*, 53-231.

[25] Di Pietro, C. M., & Liener, I. E. (1989). Heat inactivation of the Kunitz and Bowman-Birk soybean protease inhibitors. *Journal of Agricultural and Food Chemistry*, 37, 39-44.

[26] Liener, I. E. (1994). Implications of antinutritional components in soybean foods. *Critical Reviews in Food Science and Nutrition*, 34-31.

[27] Sklan, D., Hurwitz, S., Budowski, P., & Ascarelli, I. (1975). Fat digestion and absorption in chicks fed raw soy or heated soybean meal. *Journal of Nutrition*, 105-57.

[28] Grant, G., & van Driessche, E. (1993). Legume lectins: physicochemical and nutritional properties. *In : van Driessche E, van der Poel AFB, Huisman J, Saini HS (eds.) Recent advances of research in antinutritional factors in legume seeds. Wageningen, Netherlands: Wageningen Press*, 219-233.

[29] Kennedy, J. F., Palva, P. M. G., Corella, M. T. S., Cavalcanti, M. S. M., & Coelho, L. C. B. B. (1995). Lectins, versatile proteins of recognition: a review. *Carbohydrate Polymers*, 26-219.

[30] Schulze, H., Butts, C. A., Verstegen, M. W. A., Moughan, P. J., & Huisman, J. (1995). Purified soybean lectins affect ileal nitrogen and amino acid flow in pigs. *Proceedings of the 2nd European Conference on Grain Legumes, Copenhagen*, 300-301.

[31] Kirchgessner, M., & Windisch, W. (1995). Zum einfluss von mikrobieller phytase auf zootechnische leistungen und die verdauungsquotienten von phosphor, calcium, trockenmasse und stickstoff bei abgestufter Ca-versorgung in der ferkelaufzucht. *Agribiological Research*, 48-309.

[32] Selle, P. H., Ravindran, V., & Caldwell, Bryden. W. L. (2000). Phytate and phytase: consequences for protein utilization. *Nutrition Research Reviews*, 13-255.

[33] Li, D. F., Nelssen, J. L., Reddy, P. G., Blecha, F., Klemm, R., & Goodband, R. D. (1991). Interrelationship between hypersensitivity to soybean proteins and growth performance in early-weaned pigs. *Journal of Animal Science*, 69-4062.

[34] Snyder, H. E., & Kwon, T. W. (1987). Processing of soybeans. *In: Soybean utilization. New York, USA: Avi Book.*

[35] Qin, G. X., Verstegen, M. W. A., & Van der Poel, A. F. B. (1998). Effect of temperature and time during steam treatment on the protein quality of full-fat soybean meal from different origins. *Journal of the Science of Food and Agriculture*, 77-393.

[36] Batal, A. B., Douglas, M. W., Engram, A. E., & Parsons, C. M. (2000). Protein dispersibility index as an indicator of adequately processed soybean meal. *Poultry Science*, 79-1592.

[37] Araba, M., & Dale, N. M. (1990). Evaluation of protein solubility as an indicator of overprocessing of soybean meal. *Poultry Science*, 69-76.

[38] Araba, M., & Dale, N. M. (1990). Evaluation of protein solubility as an indicator of underprocessing of soybean meal. *Poultry Science*, 69-1749.

[39] Americal Oil Chemists Society. (2000). Official Methods and Recommended Practices. *5th edition. Urbana, IL: American Oil Chemists Society.*

[40] Valencia, D. G., Serrano, M. P., Lázaro, R., Latorre, M. A., & Mateos, G. G. (2008). Influence of micronization (fine grinding) of soya bean meal and full fat soya bean meal on productive performance and digestive traits in young pigs. *Animal Feed Science and Technology*, 147-340.

[41] Valencia, D. G., Serrano, M. P., Centeno, C., Lázaro, R., & Mateos, G. G. (2008). Pea protein as a substitute of soya bean protein in diets for young pigs: Effects on productivity and digestive traits. *Livestock Science*, 188-1.

[42] Rackis, J. J., Mc Ghee, J. E., Liener, I. E., Kakade, M. L., & Puski, G. (1974). Problems encountered in measuring trypsin inhibitor activity of soy flour. *Report of a collaborative analysis. Cereal Science Today*, 19-513.

[43] Peisker, M. (2001). Manufacturing of soy protein concentrate for animal nutrition. *In: Brufau J. (ed.) Feed Manufactoring in the Mediterranean Region. Improving safety: From Feed to Food. Zaragoza, CIHEAM-IAMZ, Reus, Spain*, 1036-1107.

[44] Nielsen, P. H., Dalgaard, R., Korsbak, A., & Pettersson, D. (2008). Environmental assessment of digestibility improvement factors applied in animal production: Use of Ronozyme® WX CT xylanase in Danish pig production. *The International Journal of Life Cycle Assessment*, 13-49.

[45] Nagaraju, R. K., & Nielsen, P. H. (2011). Environmental advantages of phytase over inorganic phosphate in poultry feed. *Poultry Punch*, 27-57.

[46] Oxenböll, K. M., Pontoppidan, K., & Nji-Fru, F. (2011). Use of a protease in poultry feed offers promising environmental benefits. *International Journal of Poultry Science*, 10-842.

[47] Rooke, J. A., Slessor, M., Fraser, H., & Thomson, J. R. (1998). Growth performance and gut function of piglets weaned at four weeks of age and fed protease-treated soya-bean meal. *Animal Feed Science and Technology*, 70-175.

[48] Ghazi, S., Rooke, J. A., Galbraith, H., & Bedford, M. R. (2002). The potential for the improvement of the nutritive value of soya-bean meal by different proteases in broiler chicks and broiler cockerels. *British Poultry Science*, 43-70.

[49] Bertechini, R. L., Carvalho, J. C. C., Mesquita, F. R., Castro, S. F., Meneghetti, C., & Sorbara, J. O. B. (2009). Use of a protease to enhance the utilization of soybean meal amino acids by broilers. *Poultry Science*, 88(1), 69.

[50] Bertechini, R. L., Carvalho, J. C. C., Mesquita, F. R., Castro, S. F., Remolina, D. F., & Sorbara, J. O. B. (2009). Use of a protease to enhance the utilization of full fat soybean amino acids by broilers. *Poultry Science*, 88(1), 70.

[51] Angel, C. R., Saylor, W., Vieira, S. L., & Ward, N. (2011). Effects of a monocomponent protease on performance and protein utilization in 7- to 22-day-old broiler chickens. *Poultry Science*, 290-281.

[52] Freitas, D. M., Vieira, S. L., Angel, C. R., Favero, A., & Maiorka, A. (2011). Performance and nutrient utilization of broilers fed diets supplemented with a novel monocomponent protease. *Journal of Applied Poultry Research*, 20-322.

[53] Fru-Nji, F., Kluenter, A. M., Fischer, M., & Pontoppidan, K. (2011). A feed serine protease improves broiler performance and increases protein and energy digestibility. *Journal of Poultry Science*, 48-239.

[54] Pettersson, D., & Åman, P. (1989). Enzyme supplementation of a poultry diet containing wheat and rye. *British Journal of Nutrition*, 62-139.

[55] Steenfeldt, S., & Pettersson, D. (2001). Improvements in nutrient digestibility and performance of broiler chickens fed a wheat-and-rye based diet supplemented with enzymes. *Journal of Animal Feed Science*, 10-143.

[56] Slominski, B. A. (1994). Hydrolysis of galactooligosaccharides by commercial preparations of α-galactosidase and ß-fructofuranosidase: potential for use as dietary additives. *Journal of the Science of Food and Agriculture*, 65-323.

[57] Ao, T., Cantor, A. H., Pescatore, A. J., Pierce, J. L., & Dawson, K. A. (2010). Effects of citric acid, alpha-galactosidase and protease inclusion on in vitro nutrient release from soybean meal and trypsin inhibitor content in raw whole soybeans. *Animal Feed Science and Technology*, 162-58.

[58] Leske, K. L., Jevne, C. J., & Coon, C. N. (1993). Effect of oligosaccharide additions on nitrogen corrected true metabolizable energy of soy protein concentrate. *Poultry Science*, 72-664.

[59] Leske, K. L., & Coon, C. N. (1999). Nutrient content and energy digestibilities of ethanol-extracted, low α-galactoside soybean meal as compared to intact soybean meal. *Poultry Science*, 78-1177.

[60] Coon, C. N., Leske, K. L., Akavanichan, O., & Cheng, T. K. (1990). Effect of oligosaccharide-free soybean meal on true metabolizable energy and fiber digestion in adult roosters. *Poultry Science*, 69-787.

[61] Gdala, J., Johansen, H. N., Bach-Knudsen, K. E., Knap, I. H., Wagner, P., & Joergensen, O. B. (1997). The digestibility of carbohydrates, protein, and fat in the small and

large intestine of piglets fed non-supplemented and enzyme supplemented diets. *Animal Feed Science and Technology*, 65, 15-33.

[62] Graham, K. K., Kerley, M. S., Firman, J. D., & Allee, G. L. (2002). The effect of enzyme treatment of soy-bean meal on oligosaccharide disappearance and chick growth performance. *Poultry Science*, 81, 1014-1019.

[63] Vahjen, W., Busch, T., & Simon, O. (2005). Study on the use of soya bean polysaccharide degrading enzymes in broiler nutrition. *Animal Feed Science and Technology*, 120-259.

[64] Nelson, T. S., Ferrara, L. W., & Storer, N. L. (1968). Phytate phosphorus content of feed ingredients derived from plants. *Poultry Science*, 47, 1372-1374.

[65] Eeckhout, W., & De Paepe, M. (1994). Total phosphorus, phytate phosphorus and phytase activity in plant feedstuffs. *Animal Feed Science and Technology*, 47-19.

[66] Ravindran, V., Ravindran, G., & Sivalogan, S. (1994). Total and phytate phosphorus contents of various foods and feedstuffs of plant origin. *Food Chemistry*, 50-133.

[67] Jongbloed, A. W., & Kemme, P. A. (1990). Effect of pelleting mixed feeds on phytase activity and the apparent absorbability of phosphorus and calcium in pigs. *Animal Feed Science and Technology*, 28, 233-242.

[68] Brejnholt, S. M., Dionisio, G., Glitsoe, V., Skov, L. K., & Brinch-Pedersen, H. (2011). The degradation of phytate by microbial and wheat phytases is dependent on the phytate matrix and the phytase origin. *Journal of the Science of Food and Agriculture*, 91-1398.

[69] Hill, R., & Tyler, C. (1954). The reaction between phytate and protein. *Journal of Agricultural Science*, 44-324.

[70] Aureli, R., Umar, Faruk. M., Cechova, I., Pedersen, P. B., Elvig-Joergensen, S. G., Fru, F., & Broz, J. (2011). The efficacy of a novel microbial 6-phytase expressed in Aspergillus oryzae on the performance and phosphorus utilization in broiler chickens. *International Journal of Poultry Science*, 10, 160-168.

[71] Pontoppidan, K. (2009). Factors influencing phytate degradation in piglets: feed phytate behaviour and degradation by microbial phytases. *PhD thesis. Department of Chemical and Biological Engineering Chalmers University of Technology Göteborg Sweden.*

Chapter 3

Variability for Phenotype, Anthocyanin Indexes, and Flavonoids in Accessions from a Close Relative of Soybean, *Neonotonia wightii* (Wright & Arn. J.A. Lackey) in the U.S. Germplasm Collection for Potential Use as a Health Forage

J.B. Morris, M.L. Wang and B. Tonnis

Additional information is available at the end of the chapter

1. Introduction

The closely related soybean species, *Neonotonia wightii* Wright & Arn. J.A. Lackey is in the *Fabaceae* family and originates from several tropical countries (NPGS, 2012; Cook *et al.*, 2005). The plants produce vines with slender stems (2-3 cm in diameter) consisting of glabrous to densely pubescent trichomes and a strong taproot. The leaves are pinnately trifoliolate with elliptic, ovate, or rhombic ovate, acute to obtuse (1.5-15 cm long, and 1.3-12.5 cm wide), glabrous to densely pubescent leaflets. The stipules are lanceolate (4-6 mm long) and the petiole is 2.5-13 cm long. The inflorescence is axillary with dense or lax racemes which are 2-35 cm long on peduncles (3-12.5 cm long) and comprises 20-150 flowers. Each flower is 4.5-11 mm long with white to mauve-blue standards, however small violet streaks are noticeable on the lower part, and will change to yellow or orange at senescence. The 1.5-4 cm long by 2.5-5 mm wide, glabrous to densely pubescence with grey to reddish brown trichomes on the pods are linear, oblong, straight or slightly curved at the apex and transversely grooved with a weak septa between the seeds. Each pod contains 3-8 oblong with rounded corners, laterally compressed, olive green to reddish brown (occasionally mottled, aril white), 2-4 mm long, 1.5-3 mm wide, and 1-1.5 mm thick seeds (Cook *et al.*, 2005). *Neonotonia wightii* consists of diploid (2n = 22) and tetraploid (2n = 44) genotypes which are self-pollinated (cleistogamous) and low outcrossing (Cook *et al.*, 2005).

Neonotonia wightii has several common names including glycine (Australia, Kenya); soja perene (Brazil); soya perenne forrajera, soya forrajera, soya perenne (Colombia, Mexico); soja perenne (French); ausdauernde soja (German); soja-perene (Portuguese); Rhodesian kudzu (Taiwan); fundo-fundo (Tanzania); and thua peelenian soybean (Thai) [Cook *et al.*, 2005]. This species also has many synonyms including *Glycine javanica* auct., *G. javanica* L. var. *paniculata* Hauman, *G. albidiflora* De Wild., *G. claessensii* De Wild., *G. javanica* sensu auct., *G. javanica* L. var. *claessensii* (De Wild.) Hauman, *G. javanica* L. var. *longicauda* (Schweinf.) Baker, *G. javanica* L. subsp. *micrantha* (A. Rich.) F.J. Herm., *G. javanica* L. var. *mearnsii* (De Wild.) Hauman, *G. longicauda* Schweinf., *G. mearnsii* De Wild., *G. micrantha* A. Rich., *G. moniliformis* A. Rich., *G. petitiana* Hermann pro parte, *G. pseudojavanica* Taub., *G. wightii* (Wight & Arn.) Verdc. var. *longicauda* (Schweinf.) Verdc., *G. wightii* (Wight & Arn.) Verdc. subsp. *petitiana* (A. Rich.) Verdc. var. *mearnsii* (De Wild.) Verdc., *G. wightii* (Wight & Arn.) Verdc. subsp. *petitiana* (A. Rich.) Verdc., *G. wightii* (Wight & Arn.) Verdc. subsp. *pseudojavanica* (Taub.) Verdc., *G. wightii* (Wight & Arn.) Verdc. subsp. *wightii*, *Johnia wightii* (Wight & Arn.) Wight & Arn., and *Notonia wightii* Wight & Arn. (Cook *et al.*, 2005).

The plant is used as pasture for grazing, hay, and silage (Cook *et al.*, 2005). However, Viswanathan *et al.*, 2001 indicated that *N. wightii* seeds are used as food by Malayali tribes in Kollihills of the Namakkal District, Tamil Nadu, India. They found several essential amino acids, fatty acids, potassium, magnesium, manganese, and copper in *N. wightii* seeds. In Kenya, *N. wightii* produced abundant organic matter contributing to soil fertility and tolerated defoliation (Macharia *et al.*, 2010). A Brazilian study showed that *N. wightii* was one of several legumes with high crude protein, low NDF, and low phenolic concentrations for use as a ruminant feed (Valarini and Possenti, 2006). Mtui *et al.* (2006) found that *N. wightii* should be a component in dairy cow diets because of its high mineral concentration. *Neonotonia wightii* plants contain low levels of tannins and alkaloids as well (Mbugua *et al.*, 2008). Tauro *et al.* (2009) found that *N. wightii* could restore productivity to soils in Zimbabwe that have been cultivated continuously and low in nutrient levels. *Neonotonia wightii* has also been found to contribute the greatest green manure effect in the absence of fertilization in Zambia (Steinmaier and Ngoliya, 2001).

Anthocyanins are chemicals responsible for natural plant colors found in leaves, stems, and flowers. An anthocyanin meter with a 520 nm LED has been used to measure the absorbance near the wavelength at which free anthocyanin aglycones, cyanidin and pelargonidin monogluscosides absorb (Macz-Pop et al., 2004). Several studies have shown potential health benefits of anthocyanins in humans. Chokeberry (*Aronia meloncarpa* E.) anthocyanins (cyanidin derivatives) have been shown to be very potent inhibitors of colon cancer cells (Zhao *et al.*, 2004). When several anthocyanins including cyanidin-3,5-diglucoside and cyanidin-3-glucoside are ingested, apoptosis effects were observed and may have potential for human hepatitis B-associated hepatoma (Shin *et al.*, 2009). Lacombe *et al.* (2010) found that cyanidin-3-glucoside caused disintegration of E. coli outer membranes. Both cyanidin-3-glucoside and pelargonidin-3-glucoside showed potential for prevention of atherosclerosis (Paixao *et al.*, 2011). Corn silage containing anthocyanins may have nutritional value as a ruminant feed (Hosoda *et al.*, 2011). *Neonotonia wightii* has also been found to be rich in crude protein content and amino acids, but low amounts of the sulfur containing amino acid, methionine (Tokita *et al.*, 2006).

Isofalvones have been associated with reducing sheep fertility (Waghorn and McNabb, 2003). The isoflavone, genistein is a secondary metabolite found in many legumes including *N. wightii* (Ingham *et al.*, 1977; Keen *et al.*, 1989). Genistein can cause reduced fertility in sheep, however after 7-10 days of adaptation, sheep rumen microbes degrade genistein and other oestrogenic compounds to non-oestrogenic metabolites. Therefore, the effects of genistein on sheep fertility is short lived (Waghorn and McNabb, 2003). However, genistein has been shown to protect against mammary and prostate cancer by regulating receptors and growth signaling pathways (Lamartiniere *et al.*, 2002).

2. Materials and methods

2.1. Phenotyping

Neonotonia wightii is a photoperiod and frost-sensitive species requiring seed regeneration in a greenhouse. Twenty-two *N. wightii* accessions from countries throughout the world were used in this study (Table 1). There was not enough room remaining in the greenhouse to accommodate the regeneration of 7 additional *N. wightii* accessions, therefore fourteen accessions were planted in 27.5 cm x 27.5 cm plastic pots containing potting soil grown in a greenhouse from August 1, 2010 – April 1, 2011 each year. Three to four seedlings per pot were maintained for plant production. Due to the vigorous growth habit, all *N. wightii* accessions were grown with trellises. The experimental design was a randomized complete block with 4 replications assigned to *N. wightii* accessions. Phenotypic descriptors including branching, foliage, plant height, plant width, and relative maturity were recorded when each accession reached 50% maturity based on visual observation, while seed numbers were counted at the end of the growing season. Branching and foliage were based on a scale of 1-5 where, 1 = > 90%, 2 = 80-89%, 3 = 70-79%, 4 = 60-69%, and 5 = 50-59%, 6 = 40-49%, 7 = 30-39%, 8 = 20-29%, and 9 = 10-19% of each plant producing branches and/or foliage based on visual observations. Relative maturity dates were based on a scale of 5 to 9 where 5 = mid-season and 9 = very late.

2.2. Anthocyanin indexes

An Opti-Sciences CCM-200 chlorophyll content meter was converted to a hand-held anthocyanin meter. The manufacturer replaced the 655 nm light emitting diode (LED) of the CCM with a 520 nm LED to measure the absorbance near the wavelength at which free anthocyanin aglycones, cyanidin and pelargonidin monogluscosides absorb (Macz-Pop et al., 2004). Anthocyanin indexes were determined by inserting each leaflet between the meter and the LED diode, followed by gently pressing the LED directly on to the leaflet and recording from each of three leaflets of 15 *Neonotonia wightii* accessions growing in the greenhouse on 14 February 2008 and 18 March 2009.

Accession (PI)	Origin
156055	Zimbabwe
189613	South Africa
213256	India
213257	India
224976	South Africa
224977	South Africa
224978	South Africa
224979	South Africa
224980 (cultivar-Tropic Verde)	Zimbabwe
224981	Zimbabwe
230324	South Africa
233148	Rhodesia
234874	Congo
235287	Zimbabwe
247677	Congo
258381	Australia
259541	Unknown
259544	South Africa
259545	Brazil
277889	Zimbabwe
314847	South Africa
612241	Taiwan

Table 1. Origin for *N. wightii* accessions used in this study.

2.3. Genistein

As plants matured in the greenhouse, most leaflets were pre-disposed to senescence and only 7 *N. wightii* accessions could be evaluated for genistein variation during 2008. Therefore, preliminary research investigating variability for leaflet weight and genistein content was conducted among these 7 *N. wightii* accessions. Leaflet tissue from each *N. wightii* accession was ground to a fine powder with liquid nitrogen, dried at room temperature, and stored at -20°C until extraction. Approximately 0.15- 0.3 g of dried tissue from each accession was placed into 5 ml test tubes, and their weights were recorded to the nearest 0.001 g. Three ml of extraction solvent consisting of 80% HPLC-grade methanol with 1.2 M HCl was added to each test tube. Each tube with extraction solvent were vortexed and incubated at 80°C for 2 hr with occasional mixing by inversion. An additional 2.0 ml of 5% methanol was added to each test tube, resulting in a final concentration of 50% methanol. A portion of the extract was filtered through a 0.45 μm membrane prior to injection. Analytes were separated and identified by high performance liquid chromatography (HPLC). The stationary phase consisted of a Zorbax Eclipse XDB-C18 column (4.6 x 150 mm, 5μM particle size) (Agilent Technologies) at 40°C with a C18 guard column. The mobile phase consisted of 20% HPLC-grade acetonitrile (B) and 80% filtered water with 0.1% formic acid, pH 2.5 (A) at 2.0 ml/min. A gradient flow was used with the following profile: 15% B for 3 min, then 15% to 40% B from 3 min to 20 min. The column was washed with 95% B for 5 min, and then equilibrated for 7 min at 15% B between injections. The injection volume was 10 μl. Flavonoid peaks were monitored at 260 and 370 nm with a diode array detector. The standards for peak identification and quantification consisted of kaempferol, quercetin, myricetin, genistein, and daidzein (Sigma-Aldrich Chemical Co.). Each standard was dissolved in extraction solvent and diluted to the following concentrations: 1, 5, 10, 25, and 50 ng/μl.

Phenotype, anthocyanin index, and isoflavonoid data were subjected to an analysis of variance using SAS (SAS Institute, 2008). Mean separations were conducted using Duncan's multiple range test (P < 0.05, P < 0.01) and correlations were accomplished using Pearson's correlation in SAS (SAS Institute, 2008). Principal component analysis using PROC PRINCOMP (SAS Institute, 2008) were then used for multivariate analysis of the data. Eigenvalues, the percentage of variances explained by each principal component, and eigenvectors were also determined. Clustering was then performed on the data by entering the similarity matrix into PROC CLUSTER for cluster analysis with the unweighted paired group method using mathematical averages (UPGMA) by specifying the AVERAGE option (SAS Institue, 2008).

3. Results and discussion

3.1. Phenotype

Significant variability for morphological, plant maturity, and seed number characteristics observed among 14 *N. wightii* accessions are reported in Table 2. Only PI 224978 from South Africa produced significantly less branching and foliage production than most of the other accessions. Many plants extended beyond the top of the trellis which allowed us to measure actual plant heights. Plant height ranged from 99.3 to 116.3 cm with both PI 224979 (South Africa) and PI 224981 (Zimbabwe) producing significantly shorter plants (averaging 99.7 cm) than the other accessions. The other twelve accessions averaged 108.4 cm tall. The accessions PI 224976 (S. Africa), PI 224977 (S. Africa), PI 230324 (S. Africa), PI 213257 (India), and PI 224981 produced the significantly narrowest plants averaging 47.6 cm while all other accessions averaged 70.0 cm wide. The accession PI 213256 from India matured the earliest while PI 224976, PI 224977, and PI 213257 averaged maturity at mid season while all others matured late or very late (Fig. 1). The significantly highest seed producing accession was PI 213256 (2214 seeds) followed by the Congonese accession, PI 234874 (725 seeds) and PI 224977 (663 seeds). The significantly lowest seed producers were PI 224979, PI 224980 (cultivar, Zimbabwe), and PI 230324 averaging 70 seeds while all other accessions averaged 307 seeds. Branching significantly correlated with foliage ($r^2 = 0.84^{***}$) and foliage had a significant negative correlation with plant width ($r^2 = -0.28^*$). Plant width was significantly correlated with maturity ($r^2 = 0.57^{***}$) and maturity had a significant negative correlation with seed number ($r^2 = -0.39^{**}$). Phenotypic variation for the *N. wightii* accessions can be explained by plant selection leading to the potential development of cultivated varieties or breeding material.

			Plant			Seed
Accession	Branching	Foliage	ht. (cm)	wd. (cm)	Maturity	no.
224978	2.5a	2.3a	106.3abcd	52.5bc	7.5b	201b
224979	1.5b	2.0ab	99.3d	51.8bc	7.5b	89b
156055	1.0b	1.0b	104.0cd	70.3a	9.0a	405b
224976	1.0b	1.5ab	108.8abcd	50.0c	6.0c	304b
224977	1.0b	1.0b	110.0abcd	47.5c	6.0c	663b
213256	1.0b	1.0b	103.3cd	51.7bc	5.0c	2214a
213257	1.0b	1.0b	107.5abcd	47.5c	6.0c	461b
224980	1.0b	1.0b	116.3a	62.3ab	7.5b	78b
224981	1.0b	1.0b	100.0cd	50.5c	7.8ab	131b
230324	1.0b	1.3ab	107.5abcd	42.5c	8.3ab	42b
234874	1.0b	1.0b	105.5bcd	71.0a	9.0a	725b
235287	1.0b	1.0b	106.3abcd	69.5a	9.0a	293b
258381	1.0b	1.0b	115.0ab	71.3a	9.0a	327b
259544	1.0b	1.0b	110.3abc	66.3a	9.0a	336b

Means followed by the same letter are not significantly different.

Table 2. Phenotype and seed reproduction variability.

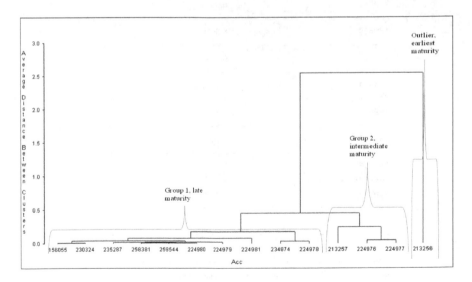

Figure 1. Dendrogram of distance between clusters based on morphological, plant maturity, and seed number differences. Accession numbers are given (Acc). Values on the baseline indicate average phenotypic distances between accessions. Two distinct clusters and one outlier (earliest maturity) for relative plant maturity can be distinguished.

3.2. Anthocyanin index, leaf weight, and genistein

Significant variation for leaf anthocyanin indexes among 15 diverse *N. wightii* accessions were observed also (Table 3). The accession, PI 213257 produced the significantly highest anthocyanin index (7.5) while all other accessions produced an average anthocyanin index of 6.1. This accession also produced significantly higher leaflet weight (0.228 g) than the other 6 accessions (averaging 0.142 g) (Table 4). However, PI 612241 from Taiwan produced significantly more genistein (90.03 μg g^{-1} of leaflet tissue) than all other accessions which averaged 51.3 μg g^{-1} (Table 4). There were no significant correlations among these traits. The flavonoids kaempferol, quercetin, myricetin, and daidzein were minutely or not detected.

Accession (PI)	N	Mean anthocyanin index
213257	4	7.50a
277889	4	7.13ab
314847	4	7.08abc
224976	4	7.00abc
247677	4	6.75abcd
259541	1	6.60abcd
189613	2	6.60abcd
224981	5	6.18abcde
234874	5	6.10bcde
259545	5	5.76cde
235287	5	5.64de
213256	5	5.46de
156055	5	5.44de
233148	5	5.14e
612241	4	5.10e

Means followed by the same letter are not significantly different.

Table 3. Preliminary leaf anthocyanin index variability among 15 diverse *N. wightii* accessions combined over 2 years (2008 and 2009).

Accession (PI)	N	Leaflet wt. (g)	Genistein (µg g⁻¹)
213257	4	0.228a	66.28ab
224976	4	0.175b	80.10ab
314847	4	0.162bc	59.10ab
277889	4	0.160bc	28.90b
247677	4	0.136bcd	37.95ab
189613	2	0.118cd	35.65b
612241	4	0.101d	90.03a

Table 4. Preliminary leaflet weight (g) and genistein variability among 7 *N. wightii* accessions during 2008.

3.3. Principal component analysis

Phenotypic, maturity, and seed number principal component analysis accounted for 44% of the total variation at the first principal component (Table 5). The amount of variation accounted for, cumulatively, by adding principal components 2 through 4 was 75, 88, and 96%, respectively. The first principal component was most correlated with plant width and maturity (Table 6). The second principal component accounted for 31% of the variation and was mostly due to branching and foliage while the third principal component explained 13% of the variation and was composed of primarily plant height. The fourth principal component accounted for 8% of the variation and was most correlated with seed number. Therefore, potential exists to develop cultivars with improved architecture, early or late maturity, and high or low seed yield. Anthocyanin index, leaflet weight, and genistein accounted for 63% of the total variation at the first principal component (Table 7). The cumulative amount of variation for components 2 through 3 was 98 and 100%, respectively. The first and second principal components were mostly correlated with anthocyanin index and genistein, respectively, while the third principal component correlated with both anthocyanin index and leaflet weight (Table 8). Potential exists to develop *N. wightii* cultivars with improved anthocyanin indexes, genistein content, and leaflet weight. Since all traits tested are quantitative, the variability among *N. wightii* accessions is attributed to genetic differences primarily since they were regenerated in a greenhouse.

	Principal		
component	Eigenvalue	% Variability	% Cummulative
1	2.6400	44.00	44.00
2	1.8319	30.53	74.53
3	0.8088	13.48	88.01
4	0.4535	7.56	95.57

Table 5. Eigenvalues and the proportion of total phenotypic, maturity, and seed reproduction variability among 14 *N. wightii* accessions (2010, 2011) as explained by the principal components.

Principal components	1	2	3	4	5	6
Branching	-0.26	0.58	0.06	0.49	-0.53	-0.22
Foliage	-0.35	0.52	0.23	0.01	0.72	0.15
Plant ht. (cm)	0.33	-0.16	0.88	0.27	-0.03	0.11
Plant width (cm)	0.54	0.06	-0.26	0.46	0.40	-0.49
Maturity	0.50	0.37	-0.24	0.07	-0.10	0.72
Seed no.	-0.37	-0.46	-0.19	0.67	0.11	0.37

Table 6. Eigenvectors, principal components for 6 phenotypic, maturity, and seed traits in 14 *N. wightii* accessions (2010-2011).

component	Principal Eigenvalue	% Variability	% Cummulative
1	1.8958	63.20	63.20
2	1.0540	35.13	98.33
3	0.0501	1.67	100.00

[1]Anthocyanin indexes were based on 15 *N. wightii* accessions.

[2]Leaflet weight and genistein were based on 7 *N. wightii* accessions.

Table 7. Eigenvalues and the proportion of total leaf anthocyanin index[1], leaflet weight (g)[2] (2009) and genistein[2] (2008) variability among *N. wightii* accessions as explained by the principal components.

Principal components	1	2	3
Anthocyanin index[1]	0.71	-0.03	0.69
Leaf wt. (g)[2]	0.62	0.47	-0.61
Genistein[2]	-0.30	0.87	0.36

[1]Anthocyanin index based on 15 *N. wightii* accessions.

[2]Leaflet weight and genistein based on 7 *N. wightii* accessions.

Table 8. Eigenvectors, principal components for two phytochemical traits and leaf weight in *N. wightii* accessions (2008, 2009)

4. Cluster analysis

Average distance cluster analysis grouped the original 14 *N. wightii* accessions into well defined phenotypes with two distinct relative plant maturity groups and one outlier (Fig. 1). Cluster or group 1 represents 10 late maturing *N. wightii* accessions and group 2 consists of three intermediate or mid-season maturing accessions. The outlier, PI 213256 represents the earliest maturing accession. The *N. wightii* accessions clustered in group 2 showed relatively closer genetic relationships than those in group 1. Using the distance values indicated in Fig. 1, the groupings at any similarity level can be identified. For example, PI 224976 and PI 224977 originate from South Africa with a phenotypic distance index of 0.0473, which indicates their close morphological similarities.

Average distance cluster analysis grouped 7 *N. wightii* accessions into well defined phenotypes with three distinct genistein producing accessions and one outlier (Fig. 2). Group 1 represents 2 intermediate genistein producing accessions while group 2 consists of two high genistein producing accessions. Group 3 is representative of two very high genistein producing accessions and one outlier (PI 277889) accession producing low amounts of genistein. Overall, *N. wightii* accessions showed similar genetic relationships in all groups and one outlier. However, PI 612241 from Taiwan and PI 224976 from South Africa had a phenotypic distance index of 0.3030, which indicates their close morphological similarities.

These results show substantial variability for various phenotypic traits, maturity, seed reproduction, and genistein in these *N. wightii* accessions regenerated in a greenhouse. However, additional studies are warranted for investigating if similar results will occur when *N. wightii* accessions are grown in field conditions over multiple years.

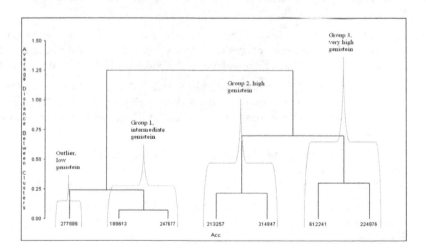

Figure 2. Dendrogram of distance between clusters based on anthocyanin indexes and genistein differences. Accession numbers are given (Acc). Values on the baseline indicate average phytochemical distances between accessions. Three distinct clusters and one outlier (low) for genistein can be distinguished.

Author details

J.B. Morris, M.L. Wang and B. Tonnis

USDA, ARS, Plant Genetic Resources Conservation Unit, Griffin, GA, USA

References

[1] Cook, Bruce, Pengelly, Bruce, Brown, Stuart, Donnelly, John, Eagles, David, Franco, Arturo, Hanson, Jean, Mullen, Brendan, Partridge, Ian, Peters, Michael and Schultze-Kraft, Rainer. (2005). Tropical Forages: Neonotonia wightii. CSIRO Sustainable Eco-systems (CSIRO), Department of Primary Industries and Fisheries (DPI&F Queensland), Centro Internacional de Agricultura Tropical (CIAT) and International Livestock Research Institute (ILRI). Online version: http://www.tropicalforages.info.

[2] Hosoda, Kenji, Eruden, Bayaru, Matsuyama, Hiroki and Shioya, Shigeru. Effect of anthocyanin-rich corn silage on digestibility, milk production and plasma enzyme activities in lactating dairy cows. Animal Science Journal, 2011 doi: 10.1111/j. 1740-0929.2011.00981.x.

[3] Ingham, John L., Keen, Noel T. and Hymowitz, Theodore. A new isoflavone phytoalexin from fungus-inoculated stems of Glycine wightii. Phytochemistry, 1977 16:1943-1946.

[4] Keen, N.T., Ingham, J.L., Hymowitz, T., Sims, J.J. and Midland, S. The occurrence of glyceollins in plants related to Glycine max (L.) Merr. Biochemical Systematics and Ecology, 1989 17:395-398.

[5] Lacombe, A., Wu, V.C., Tyler, S. and Edwards, K. Antimicrobial action of the American cranberry constituents; phenolics, anthocyanins, and organic acids, against Escherichia coli O157:H7. International Journal of Food Microbiology, 2010 139:102-107.

[6] Lamartiniere, Coral A., Cotroneo, Michelle S., Fritz, Wayne A., Wang, Jun, Mentor-Marcel, Roycelynn and Elgavish, Ada. Genistein chemoprevention: Timing and mechanisms of action in murine mammary and prostate. The Journal of Nutrition, 2002 132:5525-5585.

[7] Macharia, P.N., Kinyamario, J.I., Ekaya, W.N., Gachene, C.K.K., Mureithi, J.G., and Thuranira, E.G. Evaluation of forage legumes for introduction into natural pastures of semi-arid rangelands of Kenya. Grass and Forage Science, 2010 65:456-462.

[8] Macz-Pop, G.A., Rivas-Gonzalo, J.C., Perez-Alonso, J.J. and Gonzalez-Paramas, A.M. Natural occurrence of free anthocyanin aglycones in beans (Phaseolus vulgaris L.). Food Chemistry, 2004 94:448-456.

[9] Mbugua, David M., Kiruiro, Erastus M. and Pell, Alice N. In vitro fermentation of intact and fractionated tropical herbaceous and tree legumes containing tannins and alkaloids. Animal Feed Science and Technology, 2008 146:1-20.

[10] Mtui, D.J., Lekule, F.P., Shem, M.N., Rubanza, C.D.K., Ichinohe, T., Hayashida, M. and Fujihara, T. Seasonal influence on mineral content of forages used by smallholder dairy farmers in lowlands of Mvomero District, Morogoro, Tanzania. Journal of Food, Agriculture and Environment, 2006 4:216-221.

[11] National Plant Germplasm System. (2012). Germplasm Resources Information Network (GRIN). Database Management Unit (DBMU), National Plant Germplasm System. US Department of Agriculture, Beltsville, MD.

[12] Paixao, J., Dinis, T.C. and Almeida, L.M. Dietary anthocyanins protect endothelial cells against peroxynitrite-induced mitochondrial apoptosis pathway and Bax nuclear translocation: an in vitro approach. Apoptosis, 2011 16:976-989.

[13] SAS. (2008). Cary, NC, SAS Institute.

[14] Shin, D.Y., Ryu, C.H., Lee, W.S., Kim, D.C., Kim, S.H., Hah, Y.S., Lee, S.J., Shin, S.C. Kang, H.S. and Choi Y.H. Induction of apoptosis and inhibition of invasion in human hepatoma cells by anthocyanins from meoru. Annals of the New York Academy of Science, 2009 1171:137-148.

[15] Steinmaier, N. and Ngoliya, A. Potential of pasture legumes in low-external-input and sustainable agriculture (Leisa). 1. Results from green manure research in Luapula Province, Zambia. Experimental Agriculture, 2001 37:297-307.

[16] Tauro, T.P., Nezomba, H., Mtambanengwe, F. and Mapfurro, P. Germination, field establishment patterns and nitrogen fixation of indigenous legumes on nutrient-depleted soils. Symbiosis, 2009 48:92-101.

[17] Tokita, N., Shimojo, M. and Masuda, Y. Amino acid profiles of tropical legumes, cooper (Glycine wightii), Tinaroo (Neonotonia wightii) and Siratro (Macroptilium atropurpureum), at pre-blooming and blooming stages. Asian-Australasian Journal of Animal Science, 2006 19:651-654.

[18] Valarini, M.J. and Possenti, R.A. Research note: Nutritive value of a range of tropical forage legumes. Tropical Grasslands, 2006 40:183-187.

[19] Viswanathan, M.B., Thangadurai, D. and Ramesh, N. Biochemical and nutritional evaluation of Neonotonia wightii (Wight & Arn.) Lackey (Fabaceae). Food Chemistry, 2001 75:275-279.

[20] Waghorn, Garry C. and McNabb, Warren C. Consequences of plant phenolic compounds for productivity and health of ruminants. Proceedings of the Nutrition Society, 2003 62:383-392.

[21] Zhao, Cuiwei, Giusti, M. Monica, Malik, Minnie, Moyer, Mary P. and Magnuson, Bernadene A. Effects of commercial anthocyanin-rich extracts on colonic cancer and nontumorigenic colonic cell growth. Journal of Agricultural and Food Chemistry, 2004 52:6122-6128.

Chapter 4

The Effects of Hydrogenation on Soybean Oil

Fred A. Kummerow

Additional information is available at the end of the chapter

1. Introduction

Soybeans are very versatile, both as a food product and an ingredient in many industrial products. The oil produced by soybeans is contained within many foods we eat every day. Natural soybean oil contains several essential fatty acids that our body needs to work properly, including linoleic and linolenic acids. However, much of the soybean oil consumed in many parts of the world has been partially hydrogenated; that is, it's chemical composition has been changed. This hydrogenation removes the necessary essential fatty acids contained within the original oil. Some of the partially hydrogenated soybean oil has been converted to trans fatty acids.

Trans fatty acids have been shown to increase the risk of atherosclerosis and coronary heart disease due to their in vivo effects in two ways. They effect the levels of prostacyclin and thromboxane, which increases the risk of thrombosis, and they increase sphingomyelin production by the body, which then causes calcium influx into the arterial cells to increase, leading to atherosclerosis. Consumption of partially hydrogenated soybean oil can be harmful to the body.

2. Soybeans

Soybeans have many uses. When processed, a 60-pound bushel will yield around 11 pounds of crude soybean oil and 47 pounds of meal. Soybeans are about 18% oil and 38% protein. Because soybeans are high in protein, they are a major ingredient in livestock feed. Most soybeans are processed for their oil and protein for the animal feed industry. A smaller percentage is processed for human consumption and made into products including soy milk, soy protein, tofu and many retail food products. Soybeans are also used in many non-food (industrial) products [1].

Biodiesel fuel for diesel engines can be produced from soybean oil by a process called trans-esterification. Soy biodiesel is cleaner burning than petroleum-based diesel oil. Its use reduces particle emissions, and it is non-toxic, renewable and environmentally friendly. Soy crayons are made by replacing the petroleum used in regular crayons with soy oil, making them non-toxic and safer for children. Candles made with soybean oil burn longer but with less smoke and soot. Soy ink is superior to petroleum-based inks because soy ink is not toxic, renewable and environmentally friendly, and it cleans up easily. Soy-based lubricants are as good as petroleum-based lubricants, but can withstand higher heat. More importantly, they are non-toxic, renewable and environmentally friendly [1]. Soy can also be used in paint and plasticizers, and used in bread, candy, doughnut mix, frozen desserts, instant milk drinks, gruel, pancake flour, pan grease extender, pie crust, and sweet goods. Non-food items made with soybeans include anti-corrosives, anti-static agents, caulking compounds, core oils, diesel fuel, disinfectants, electrical insulation, epoxies, fungicides, herbicides, printing inks, insecticides, oiled fabrics, and waterproof cement [2].

Soybean oil is normally produced by extraction with hexane. The production consists of the following steps. The soybeans are first cleaned, dried and de-hulled prior to extraction. The soybean hulls need to be removed because they absorb oil and give a lower yield. This de-hulling is done by cracking the soybeans and a mechanical separation of the hulls and cracked soybeans. Magnets are used to separate any iron from the soybeans. The soybeans are also heated to about 75° C to coagulate the soy proteins to make the oil extraction easier. To extract the oil, first the soybeans are cut into flakes, which are put in percolation extractors and emerged in hexane. Counter flow is used as extraction system because it gives the highest yield. After removing the hexane, the extracted flakes only contain about 1% of soybean oil and are used as livestock feed, or to produce food products such as soy protein. The hexane is recovered and returned to the extraction process. The hexane free crude soybean oil is then further purified [3].

World production of soybean oil in 2010-2011 rose 8.0% to a new record high of 41.874 million metric tons. The U.S. accounts for 20.6% of world soybean oil production, while Brazil produces 15.8% and the European Union accounts for 5.8%. The consumption of soybean oil rose 9.2% worldwide in 2010-2011, with the U.S. accounting for 18.6%, Brazil accounting for 12.4%, India accounting for 6.9%, and the European Union accounting for 6.4% of demand [4].

3. Uses for soybean oil

Of the total of 18 million pounds of soybean oil consumed in 2011, approximately 9 million pounds was used for cooking and salad oil. 3.75 million pounds was used for baking, and 3.6 million pounds on industrial products. The remaining 900,000 pounds is used in various other edible products. The high smoke point of soybean oil makes it often used as a frying oil. If overused, however, it causes the formation of free radicals.

Soybean oil contains 52.5% linoleic (18:2 $\Delta^{9,12}$) acid, which is also known as 18:2n^6 or omega–6. It also contains 7.5% linolenic (18:3 $\Delta^{9,12,15}$) acid also known as 18:3n^3 or omega-3. The designation 18:2 $\Delta^{9,12}$, and 18:3 $\Delta^{9,12,15}$ means that these two fatty acids have double bonds (points of unsaturation) at position 9 and 12 or 9,12 and 15 at which hydrogen can be added. In the late 1800s, a French chemist discovered that an unsaturated fatty acid can be converted to a saturated fatty acid by bubbling hydrogen through a heated vegetable oil in a closed vessel. If completely hydrogenated, they become stearic acid. The commercial use of partially hydrogenation of soybean oil began in the early 1900s. The exact fatty acid composition of the partially hydrogenated soybean oil was essentially unknown until the development of gas chromatography (GC) by James and Martin in 1952. The Food and Drug Administration, using the American Oil Chemists Society method, labeled the isomers in partially hydrogenated fat as only one peak (elaidic acid). It is only with a GC equipped with a 200 meter column that it is possible to further separate the fatty acid isomers of partially hydrogenated fat into at least 14 separate isomeric fatty acids [5].

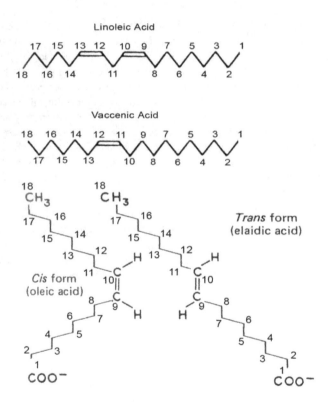

During hydrogenation, the double bond at any of these 9,12 or 9, 12, 15 positions can be shifted to form new cis and trans unsaturated fatty acid isomers not present in soybean oil.

The double bond of the cis-natural linoleic and linolenic fatty acids can also change the configuration from cis to trans, creating a geometric isomer like trans Δ^{11}-18:1 vaccenic acid in butter fat. Oleic acid, the largest percentage of the natural fatty acid in the human body, is cis Δ^9-18:1 (the number after delta indicates the position of the double bond at the 18 carbon atom chain counting from the carboxyl group).

Oleic acid goes through geometrical isomerisation during hydrogenation to trans Δ^9-18:1 acid known as elaidic acid; thus the "natural" oleic acid is turned into elaidic acid during the hydrogenation process, and becomes an "unnatural" fatty acid. It twists into a new form and can be both a cis and/or a trans fatty acid. In addition to geometrical isomerisation, the double bond of either cis or trans fatty acids can theoretically migrate along the 18 carbon chain of either oleic, linolenic, and linoleic acid, changing their position from Δ^9, $\Delta^{9,12,15}$, or $\Delta^{9,12}$, creating five monoene cis positional isomers, 6 trans monoene isomers and 3 trans diene positional isomers. Thus hydrogenated soybean oil contains 24.1% trans monoenes, 6.2% trans dienes and 9.4% cis monoene isomers or a total of 39.7% isomeric fatty acids. They were identified as cis and trans octadecenoic and octadecadienoic isomers on a GC equipped with a 200 meter column and by their mixed melting points with authentic octadecenoic and octadecadienoic acids. None of these fatty acids are present in natural soybean oil. The 14 isomers in hydrogenated fat can be used as a source of energy but they cannot substitute for EFA because they do not have the required double bond structure [5].

4. Nutrition

It was unknown until 1930 that linoleic (18:2 n^6) and linolenic (18:3 n^3) acids were essential fatty acids (EFA), and like the nine essential amino acids and the vitamins, cannot be synthesized in the human body; they must come from a diet that includes natural fats and oils. In one study, pregnant rats were fed linoleic, linolenic, and arachidonic acids by dropper. This was a sufficient amount for the mother rats to wean their young, but those pups from mothers fed only linolenic acid died before weaning. Although linolenic acid is considered an essential fatty acid, these data indicate that it may not be an essential fatty acid [6].

An increase in the sales of soy food is largely credited to the Food and Drug Administration's approval of soy as a cholesterol-lowering food [7]. A 2001 literature review argued that these health benefits were poorly supported by available evidence, and noted that data on soy's effect on cognitive function of the elderly existed [8].

The FDA issued the following claim for soy: "25 grams of soy protein a day, as part of a diet low in saturated fat and cholesterol, may reduce the risk of heart disease." [9]. Solae also submitted a petition on the grounds that soy can help prevent cancer. On February 18, 2008, Weston A. Price Foundation submitted a petition for the removal of this health claim. 25 g/day soy protein was established as the threshold intake because most trials used at least this much protein and not because less than this amount is inefficacious [10]. An American Heart Association review of a study of the benefits of soy protein casts doubt on the FDA claim for soy protein. However, AHA concludes "many soy products should be beneficial to cardiovascular and overall health because of their high content of polyunsaturated fats, fiber, vitamins, and minerals and low content of saturated fat" [11].

EFA are required to synthesize the eicosanoids that are needed to regulate blood flow in the arteries and veins. Linoleic acid (n-6) is synthesized into arachidonic acid, and linolenic acid (n-3) is synthesized into eicosapentaenoic acid. Both in turn are made into prostacyclin or thromboxane. Prostacyclins are synthesized in the endothelial cells that line the blood vessel wall. Thromboxanes are synthesized in the platelets in the blood. The balance between prostacyclin for flow and thromboxane for clotting is a very delicate one and can be changed by different diets and different drug prescriptions. Fish have already converted the linolenic acid they get from seaweed into eicosapentaenoic acid. Hence, fish oil is often recommended as a dietary supplement. Prostacyclin and thromboxane can be made from linoleic acid as well. The least expensive source of omega-3 and omega-6 is soybean oil, which is sold as vegetable oil in a supermarket [12].

However, this vegetable oil is stripped of Vitamin E, which is then sold in capsules. The removal of Vitamin E leaves the oil more susceptible to oxidation, which harms the natural fatty acids that are needed for good health.

How soybean oil is used in modern humans was developed in prehistoric humans to assure their survival. There must have been long periods of time between meals, that is fasting periods, and there were times in which they had food available, the "fed" period. During this fed period, carbohydrates were used within two hours as a quick source of energy. Extra carbohydrates were stored first as glycogen in the muscles and liver and then any excess converted to fat and stored in the adipose tissues (the fat around your middle and elsewhere). This stored fat was then available for energy during the long fasting periods. Modern humans have inherited this way of handling these fed and fasting periods. This process assured the survival of prehistoric humans but has now become one way that obesity is developing in humans today. Too much food is available all hours of the day and night, and eating it is a pleasure.

To avoid adding fat to your body, any carbohydrates you eat should be used up as a calorie source before the next meal. Any carbohydrates that have already turned into fat and any fat in your diet itself should be used for energy within the cell during the fasting period. Eating a snack between meals means adding additional carbohydrates into the system before any of the fat from the previous meal has been used for energy. It ends up adding to your adipose tissue. If you weighed yourself before a hearty meal and again the next day, you may find you have gained a pound or two, the amount depending on how much food you ate and the fat you stored. As such a meal may also contain excess salt, some of the weight gain can be due to excess water you stored. Millions of dollars are spent to try to get rid of this stored fat, and the government is planning to spend millions more dollars to solve the obesity problem. Prehistoric humans had no choice in controlling the time between fasting and fed periods because they had no refrigerators, fast food outlets, or supermarkets to run to. Modern humans do have this choice. More time between the fed periods, that is between meals, may help with the obesity problem [12].

The fat in the intestinal tract is first converted into tiny droplets of fat (chylomicrons) by the intestinal cells. The intestinal tract is not just a through highway, but is actively involved in the process of metabolizing fat so that the body can use it. The chylomicrons diffuse from the intestinal tract into the lymph system and into the veins through the thoracic duct and end up in the blood. The blood, during the fed period, carries these chylomicrons for deposit where they are resynthesized into adipose tissue and stored fat around the stomach, hips, and other locations. The fat (triglycerides) in adipose tissue is "mobilized" when the glycogen in the muscle and liver has been reduced.

The glycerin portion goes to the liver. The free fatty acids take a different route and are combined with a protein named albumin. Therefore, there must be enough albumin in the blood to carry the free fatty acids in the blood. This fatty acid albumin complex is water-soluble enough in the blood to be carried to cells of all kinds that use the fatty acid portion as an energy source. Any excess fatty acid goes to the liver and is remade into triglycerides. The cellular organelle (the endoplasmic reticulum) in the liver cells participates in coating the very small triglyceride droplets with protein and adds phospholipid and cholesterol to produce very low density lipoprotein (VLDL), which furnishes the fatty acid for the approximately 50 thousand trillion cells in the body [12].

Correction of the inhibition of lipoprotein lipase by protein binding of free fatty acids permits normal protein transport of FFA into the cellular mitochondrial oxidative phosphorylative cycle with the resultant production of high-energy phosphate which is the cellular fuel. Without this fuel, in addition to oxygen, the life process comes to a halt. Bacteria have used this method of providing energy for at least two billion years (Ratz).

5. Fried foods

Another issue with fats is the preparation of foods by frying them in fat. There are problems with deep fat fried food that affect our nutrition. These problems occur because of chemical alterations in the fat that happen as a consequence of deep fat frying food. This frying process is as follows:

1. Food picks up oxygen from the air during frying that negatively alters the fat composition.

2. The foods fried in these fats pick up those altered fats.

3. These altered foods have a direct, negative influence on the nutritional value of the fat.

The changes in the fat are dependent on at least four factors:

1. The length of time it was exposed to heat—in commercial operations, the length of time a food is fried leads to how much fat is absorbed on the cooked food item;

2. The temperature of the fat;

3. The exact composition of the fat used, such as corn oil, cottonseed oil, soybean oil, beef tallow, or hydrogenated fat, and

4.   What is being fried, e.g., chicken or fish.

Feeding the fats fried at varying lengths of time led to very different outcomes in the nutrition of animals. Those fed the fats fried the shortest period of time were healthier than those fed the fats fried for the longest times. Those fed fats heated at higher temperatures were not as healthy as those fed on fat heated to lower temperatures. It was interesting also that animals fed on heated margarine did not grow as well as those on fresh margarine and that their plasma cholesterol level increased. Those fed on heated butter oil grew as well as those on fresh butter oil.

Oil from commercial fat fryers was used in a set of experiments that clearly showed that poor nutrition resulted. This is important because used fat from commercial operations is typically collected and fed to animals, such as pigs, to provide energy for rapid growth. When we conducted experiments feeding the commercially used fat for frying to rats, they did not do well. When we added protein to their diets, the effect of the "bad" heated fat was countered because the added protein provided more adequate nutrition. We tried to fortify the diets with adequate vitamins, but that could not counter the growth-depressing effect of the heated oil. A few vitamins, such as riboflavin, helped a bit.

Fish contain high amounts of polyunsaturated fat that are not present in the fat of chicken or beef. Thus, when fish are fried, the polyunsaturated fat in them can leak into the frying fat, causing the fat to be changed more radically into a less healthy version. Chicken and hamburger have less of this polyunsaturated fat and thus are healthier choices to fry.

Eating excessive amounts of fried food also slows down digestion. People may get stomachaches as a result. As early as 1946, a link that heated fats may lead to cancer was shown. What we don't know yet is whether heated fats by themselves lead to cancer or whether the heated fat combined with specific foods cause cancer. Animals fed heated fat combined with a known carcinogen developed cancer, whereas those fed fresh fat combined with a known carcinogen did not. Thus the heated fat was a co-carcinogen.

Commercial frying of food has increased worldwide since our studies on heated fats. In Germany, fat fryers are required by law to test their frying fat for its freshness by a method approved by the German government. In the U.S. a test is also available, but its use is not mandatory [12].

6. Free radicals

Free radicals are produced from oxidized linoleic (n-6) and linolenic acid (n-3); they are fragments of unsaturated fatty acids. This is especially likely to happen when the essential fatty acids are heated, especially the n-3 variety. All oils change structures when they are heated, but hose high in n-3 fatty acids have more problems than those high in n-6. Free radicals provide another reason to avoid fried food. The first sign of fats becoming free radicals is that they are rancid, and they begin to smell "off" and their taste becomes bitter. Roasted peanuts, for example, can become rancid and then shouldn't be eaten.

Free radicals are "bad" since they destroy vitamins A, D, C, and E, thus preventing these vitamins from doing positive things in the body. Free radicals also destroy both the essential fatty acids and the essential amino acids. They oxidize the LDL into something called oxidized low density lipoproteins (oxLDL). These oxLDL are very powerful components in the blood that have been considered since about 1990 as involved in the development of heart disease [12].

Essential fatty acids do more than regulate the blood; they are also a key to reproduction. Since the 1930's, we've known that reproduction always fails on fat-free diets. In studies on rats, reproduction continues under low fat conditions because the rats have enough linoleic acid stored in their bodies. They manufacture arachidonic acid from the linoleic acid in their own fat, so they can reproduce healthy young even after a fat-free diet. If the rats did not have enough linoleic acid stored in their bodies (such as rats born to mothers on fat-free diets), we found they could not make enough of the arachidonic acid needed for healthy reproduction, and their young die. Women need the essential fatty acids for reproduction. The easiest way to supply them is from plant oils [5].

Data from ADM shows the composition of three different hydrogenated fats, based on a serving size of 14 grams. The first two were made of enzymatically interesterified soybean oil, and contained 0 grams of trans fat per serving. The third was made of partially hydrogenated soybean and/or cottonseed oil, and contained 4.5 grams of trans fat per serving. The take away message is that due to effective food industry lobbying, food labeling rules allow foods with up to half a gram of trans fat per serving to be labeled "0 trans fat". So look for "partially hydrogenated vegetable oil" on the label.

Several researchers have documented the effects of foods without trans fat and their positive effects on lowering CHD. Mozaffarian et al. showed that n-3 PUFAs from both seafood and plant sources may reduce CHD risk, with little apparent influence from background n-6 PUFA intake. They found lower death rates among those with high seafood and plant-based diets. Plant-based n-3 PUFAs may particularly reduce CHD risk when seafood-based n-3 PUFA intake was low, which has implications for populations with low consumption or availability of fatty fish. Kris-Etherton et al. found that nuts and peanuts routinely incorporated in a healthy diet with a composite of numerous cardioprotective nutrients reduced the risk of CHD. They also suggested that higher intake of trans fat could adversely affect endothelial function, which might partially explain why the positive relationship between trans fat and cardiovascular risk is greater than one would predict based solely on its adverse effects of plasma lipids [12].

7. Two mechanisms involved in coronary heart disease

Two mechanisms may be involved in CHD: One, the oxidation of the fatty acids and cholesterol in LDL leading to a change in sphingomyelin concentration in the arteries, which is a process that occurs over a life time; two, the deposition of trans fat in the cardiovascular system. Trans fat calcifies both the arteries and veins and causes blood clots. Trans fat leads to the reduction of prostacyclin that is needed to prevent blood clots in the coronary arteries. A blood clot in any of the coronary arteries can result in sudden death.

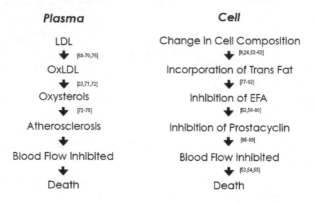

Plasma	Cell
LDL	Change In Cell Composition
⬇ [68-70,76]	⬇ [9,24,52-62]
OxLDL	Incorporation of Trans Fat
⬇ [23,71,72]	⬇ [77-92]
Oxysterols	Inhibition of EFA
⬇ [72-76]	⬇ [53,56-80]
Atherosclerosis	Inhibition of Prostacyclin
⬇	⬇ [85-89]
Blood Flow Inhibited	Blood Flow Inhibited
⬇	⬇ [53,54,55]
Death	Death

8. Mechanism one

When sufficient biological antioxidants are not present in the plasma, the LDL is oxidized to oxLDL and cholesterol is oxidized to oxysterol. Oxysterols incorporated into the endothelial layer of the arteries and veins can change the phospholipid cell membrane composition so that more sphingomyelin incorporates into the membrane which becomes "leaky" to calcium infiltration. Oxysterols were present at higher concentrations in the plasma of patients who had coronary artery bypass grafting (CABG) surgery. These patients had 40 times more calcium in their bypassed veins than normal veins in the same patient. When purchased oxysterols were added to plasma from patients who did not need CABG surgery, endothelial cells cultured in their blood and tested with radioactive calcium the incorporation of radioactive calcium did not differ from that of plasma from CABG patients. This indicates that oxysterols stimulated calcification. When endothelial cells were cultured with oxysterols in a standard culture media, the cells became calcified in a similar way to those of the CABG patient. The oxidation of cholesterol and deposition of calcium is the primary cause for the development of atherosclerosis in the arteries and veins.

In a review article entitled "The pathogenesis of atherosclerosis: Perspectives for the 1990s" Ross stated "Atherosclerosis of the extremities is most apparent at branching points of the arterial tree where blood flow is irregular with current and back currents. The cellular events that occur during the progression of lesions in hypercholesterolemic animals are almost exactly mirrored by those observed in human atherosclerotic coronary arteries in hearts removed in transplant operations" [13]. De Bakey et al. have noted similar atherosclerosis (thickening) at branching and bifurcation during coronary artery bypass grafting (CABG) surgery [14].

Keaney stated that the gene expression pattern in the arterial wall is subject to influence by modified forms of LDL [15], which altered both scavenger reception (CD36) expression and the expression of pro-inflammatory genes [16]. The disturbed laminar flow pattern of fluids occurs near branch points [17], bifurcations, at major curves and at arterial geometries [18] that are typically associated with the earliest appearance (and subsequent progression) of atherosclerotic lesions [19]. An endothelial receptor for oxLDL, a designated lectin-like oxLDL receptor (LOX-1) [20], was identified [21]. The transient application of shear stress showed that the initial stimulation of shear stress was sufficient for induced expression of LOX-1 and that sustained application of shear stress was not required [22]. The over-expression of LOX-1 receptors at the bifurcation and the higher level of modified LDL and oxysterols in the plasma of persons needing CABG surgery could lead to a higher uptake of modified LDL, resulting in a greater delivery of oxysterols to the endothelial cells at the bifurcations. The levels of sphingomyelin in plasma have been shown to be higher in patients with coronary heart disease and those with left ventricular dysfunction [23]. Furthermore, it was found that sphingomyelin levels in the blood correlate with and can be used to accurately predict coronary artery disease [24]. Sphingomyelin has long been known to accumulate in atheromas of both humans and animals, and contributes to the formation of atherosclerosis [25].

Thickening [26] was noted in the branching arteries in aging porcine on a non-cholesterol diet. It did not differ significantly in sphingomyelin composition from that of the non branching adjacent tissue of porcine at 6 months of age. By 18 and 48 months of age, however, the sphingomyelin content was significantly higher at the thickened branching areas than at the non thickened segment of the arteries. This indicated that during aging of the arteries, there was a striking increase in the amount of sphingomyelin in the membrane of the cells at the branching points of arteries [26]. Lipid extracted from both porcine and human arteries indicated that aging is a factor that increased sphingomyelin. There was more sphingomyelin in the aging arteries of both porcine and human arteries.

The non branching segment of the aorta obtained, on autopsy, from six men 21-27 years of age contained four times more sphingomyelin than in arteries isolated from human umbilical cords, indicating that the sphingomyelin content of arteries increases with age. Aging is not the only factor that increased the sphingomyelin composition of arterial cells. Women and men under 40 years of age who had been subjected to CABG surgery contained the same high percentage of sphingomyelin in their non atheromatous arterial cells as those over 40 years of age. Therefore, heart disease itself seemed to have caused an increase in non atheromatous arterial cells in sphingomyelin composition prematurely in CABG patients, pointing to a fundamental disturbance in phospholipid metabolism in their arterial cells.

The phospholipid composition of a normal arterial cell has less sphingomyelin, and this amount increases until half the artery is sphingomyelin. That is, the more sphingomyelin was in the arterial cells, the more Ca^{2+} was identified. This is because the hydroxyl group and amide group of sphingomyelin act as both donors and acceptors of hydrogen bonds [27]. Furthermore, Lehninger found that sphingomyelin's long, 18-to-26 carbon atoms chain fatty acids altered the positioning of other phospholipids. Dipalmitoylphosphatidylcholine has no amide bond [28]. As both sphingomyelin and dipalmitoylphosphatidylcholine are largely on the extracellular side of the membrane [29,30], such bilayer asymmetry would enhance binding. These in vitro results showed that sphingomyelin-Ca^{2+} binding goes beyond an isolated individual membrane binding Ca^{2+}, to lattice type matrix binding among adjacent membranes [31]. These results in vitro were simulated in vivo Ca^{2+} deposition (calcification) in arteries and veins.

9. The in vivo effect of sphingomyelin on the composition of the vascular membrane

Patients who had CABG surgery sometimes needed a second CABG surgery because the vein used in the first surgery had been occluded. During this second surgery, an unoccluded vein from the same patient was used to replace the occluded vein. The occluded veins contained, on average, significantly more sphingomyelin and Ca^{2+} than the unoccluded veins [32]. The unoccluded veins contained 24% sphingomyelin and 182 ppm of Ca^{2+} as compared to 48% of sphingomyelin and 6,345 ppm of Ca^{2+} in the occluded veins that had been used in the first CABG surgery. The increased sphingomyelin and Ca^{2+} concentrations in the occluded veins were responsible for the initial formation of atherosclerosis in these patients.

10. Oxysterols increased sphingomyelin and Ca^{2+} deposition in patients with CABG surgery

Ridgway found that 25-hydroxycholesterol stimulated sphingomyelin synthesis in Chinese hamster ovary cells [33]. Similarly in humans, an oxysterol increased sphingomyelin synthesis during the development of atherosclerosis. A significant increase in the concentrations of oxysterols, phospholipids, and Ca^{2+} were noted in patients who had CABG surgery [26, 32]. Patients who had cardiovascular disease had increased oxysterol levels in their plasma compared with the controls; that is, by comparison to cardiac catheterized patients with no stenosis [32]. The plasma from CABG patients had a higher concentration of oxysterols than was present in the controls. Human endothelial cells were cultured for 72 hours in a medium containing plasma obtained from CABG patients, or from controls patients with addition of 5 types of oxysterols (7-keto-cholesterol, cholestane-3β, 5α, 6β-triol, 7β-hydroxycholesterol, β-epoxy cholesterol, and 7α-hydroxycholesterol). These added oxysterols increased the total oxysterol level in the controls equivalent to that in the CABG plasma.

Phospholipid	Human		Porcine	
(%)	younger	older	3 weeks	2 years
Phosphatidylcholine	34.1	19.2	44.74	33.91
Phosphatidylethanolamine	8.8	2.4	25.18	24.76
Sphingomyelin	44.8	68.8	16.06	23.72
Phosphatidylinositol	+			
Phosphatidylserine	5.0	1.6	11.35	14.55
Phosphatidic acid	1.0	0.6		
Lysolecithin	3.9	8.0	trace	1.28

Table 1. Data from Kummerow F.A.. 1987. Factors which may alter the assembly of biomembranes so as to influence their structure or function In Membrane Biogenesis. Op den Kamp J. A. F., editor. Springer-Verlag. 95.Phospholipid composition of human and porcine arterial tissues

Oxysterols stimulated sphingomyelin synthesis and inhibited sphingomyelin metabolism [34, 23, 24]. When radioactive Ca^{2+} ($^{45}Ca^{2+}$) influx was measured, significantly higher influx of $^{45}Ca^{2+}$ was noted in the endothelial cells cultured with added oxysterols indicating that oxysterols increased Ca^{2+} influx into endothelial cells [34]. By using a radiolabeled choline, the time- and dose-dependent effects of 27-hydroxycholesterol on sphingomyelin synthesis could be observed. The increased radioactivity in sphingomyelin, which was accompanied by decreased radioactivity in phosphatidylcholine in 27-hydroxycholesterol-treated cells, was higher than that in control cells. This result indicated that 27-hydroxycholesterolincreasedthetransferofcholinefromphosphatidylcholinetosphingomyelin. An interesting finding was that the increased radioactivity in sphingomyelin by 27-hydroxycholesterol was detected first, followed by enhanced Ca^{2+} uptake and the accumulation of cytosolic free Ca^{2+}. Moreover, decreased activities of neutral and acid sphingomyelinase, which hydrolyze sphingomyelin, were also detected in 27-hydroxycholesterol treated cells [35]. Therefore, the cause for calcification was related to the structure and location of sphingomyelin in the cell membrane.

11. The concentration of cholesterol and lipid oxidation products in the plasma of cardiac catheterized patients was also determined [36]

The concentration of cholesterol, lipid oxidation products and total antioxidant capacity in the plasma of 2000 cardiac catheterized patients with 0, 10–69 and 70–100% stenosis of their arteries were analyzed. The results showed that lipid oxidation products increased with the severity of stenosis, they were 2.92 mmol/L at 0% stenosis, 3.19 mmol/L at 10–69% stenosis and 3.48 mmol/l at 70–100% stenosis. The total antioxidant capacity decreased with the severity of stenosis. The plasma cholesterol concentration, however, was not significantly different between these groups of patients. It was 201.9 mg/dL at 0% stenosis, 203.2 mg/dL at

10–69% stenosis and 207.5 mg/dL at 70–100% stenosis. Therefore, the concentration of oxidation products, rather than the concentration of cholesterol in the plasma, increased with the severity of atherosclerosis [36]. In all age groups, all of the women and men with cardiovascular atherosclerosis also had increased individual and total oxysterol levels in their plasma as compared with the controls.

The *in vivo* oxidation was enhanced by sphingomyelin. The oxidation could come from the consumption of too many polyunsaturated fatty acids in soybean oil [32, 36]. Polyunsaturated fats in vegetable oil could provide more oxidized LDL and more oxidized sterols into the plasma, which would increase the possibility of atherosclerosis. Sphingomyelin accumulates in the arterial system of humans and animals, and these increased levels mean an increased likelihood of atherosclerosis formation.

12. Mechanism two

Trans fatty acids are available on every continent. There are at least six hydrogenation plants in the United States alone; there is one in Texas, four in Illinois, and one in New Jersey. The FDA has estimated that daily intake of trans fatty acids in northern Europe to be at around 4.5g-17g/capita, and 1.34-4.9 in southern Europe. In India, 2.7-4.8g/capita/day was estimated, and only 2.7-4.8g/day in Australia and New Zealand. The least amount of trans fatty acids is consumed in Hong Kong, Japan, Korea, and China at 1.5-3g/capita/day. A large hydrogenation plant is located in a suburb of Tokyo that uses both fish and vegetable oils, as well as one in Beijing. These trans fatty acid-filled oils are liquid at room temperature, and similar to olive oil that has been used for centuries in southern Europe as an important source of fat in the diet. Butter, lard and beef tallow are saturated fats that have been used for centuries as a fat source in the diet in northern Europe [37].

The second mechanism that may be involved in CHD is trans fat. Trans fat calcifies both the arteries and veins and causes blood clots. Trans fat inhibits COX-2, an enzyme that converts arachidonic acid to prostacyclin that is needed to prevent blood clots in the coronary arteries. A blood clot in any of the coronary arteries can result in sudden death. The American Heart Association has stated that 42% of victims of a sudden heart attack do not reach a hospital still alive.

A study in 2004, with piglets from mothers fed hydrogenated soybean oil showed that their arteries contained less linoleic acid converted to arachidonic acid than the arteries of piglets from mothers fed butterfat or corn oil. This indicated that the trans fat in hydrogenated soybean oil inhibited the metabolic conversion of linoleic to arachidonic acid. Furthermore, an analysis of the fat embedded in the arteries of the piglets from mothers fed partially hydrogenated soybean oil showed that they contained 3% trans fat incorporated into their phospholipids by 48 days of age [38].

If a mother is breast-feeding her child and also eating foods containing trans fat, she would have a substantial amount of trans fat in her milk supply and pass those to her infant. Pregnant porcine fed hydrogenated fat contained 11.3% trans fat in their milk at the birth of their piglets, which decreased during lactation to 4% in 21 days. The plasma of the piglets increased from 5% trans fat three days after birth to 15.3% at six weeks of age. Transferring this result to humans, a human mother would also transfer the trans fat in her milk supply to her infant. The infant would incorporate the trans fat into his/her arterial cells inhibiting arachidonic acid synthesis and prostacyclin secretion [4].

Furthermore, calcium deposition into the endothelial cells could be enhanced. To date, the FDA has not considered the daily intake of trans fat relevant to the health of small children since they do not exhibit overt heart disease. In cases where children have died of unknown causes and had been autopsied, 99% of them showed the beginning stages of hardening (calcifications) of the arteries, which ultimately can lead to heart disease [39].

13. The effects of trans fatty acids on calcium influx into human arterial epithelial cells

The influence of trans fatty acids and magnesium on cell membrane composition and on calcium influx into arterial cells. The percentage of fatty acids incorporated into the endothelial cells was proportional to the amount added to the culture medium. Adequate magnesium was crucial in preventing calcium influx into endothelial cells. Without an adequate amount of magnesium in the culture medium, linoelaidic and elaidic acids, even at low concentrations, increased the incorporation of $^{45}Ca^{2+}$ into the cells, whereas stearic acid and oleic acid did not. A diet inadequate in magnesium combines with trans fat may increase the risk of calcification of endothelial cells [40].

Vaccenic acid in butter did not inhibit the metabolic conversion of linoleic to arachidonic acid. Epidemiological studies of intake of ruminant trans fat and risk of coronary heart disease (CHD) indicated that the intake of ruminant trans fatty acid was innocuous or even protective against CHD. Thus a study with an animal model has shown that trans-fat decreased synthesis of arachidonic acid from linoleic acid. This study was carried a step further with endothelial cells in the first layer of the artery. They were cultured in a medium that contained the fatty acids of soybean oil or in a medium that contained the fatty acids of hydrogenated soybean oil. The latter cells contained trans-fat in their membrane phospholipid and significantly less arachidonic acid and secreted less prostacyclin than endothelial cells that had been cultured with the fatty acids from unhydrogenated soybean oil [5].

We found that in the cells cultured with trans fat, the free arachidonic acid released by phospholipase action was shunted to metabolism by another pathway leaving less free arachidonic acid available as substrate for prostacyclin synthesis. Cyclooxygenase (COX) is the enzyme that is necessary to make prostacyclin to keep the blood flowing, thus lowering the potential for a heart attack. Vane et al. have shown that COX is the enzyme that converts arachidonic acid to prostaglandin H_2, is further metabolized to prostanoids. Vane et. al. stated two isoforms of COX existed, a constitutive (COX-1) and an inducible (COX-2) enzyme. COX-2 may be the enzyme that recognizes the isomers produced during hydrogenation as a foreign substrate and reacts to them by causing inflammation and reduction of prostacyclin. COX-2 is the inducible isoform of COX. COX-1 is present constitutively while COX-2 is expressed primarily after the inflammatory insult [41].

The ability to form prostacyclin from arachidonic acid was assayed using a radioimmunoassay kit. Trans-fat depressed the synthesis of prostacyclin. The addition of an excess amount of linoleic acid to this hydrogenated soybean oil fatty acids did not increase the secretion of prostacyclin in endothelial cells. The concentration of trans fatty acid rather than the concentration of linoleic acid was therefore responsible for regulating the synthesis and secretion of prostacyclin in endothelial cells. The trans fat in hydrogenated fat not only depressed the synthesis of prostacyclin that regulated the clotting of blood but also, could not serve as precursors for prostacyclin synthesis. The trans fat "incorporated" into the membrane lipids of blood vessels and muscle tissues and displaced the essential linoleic, linolenic and arachidonic acids.

In another study, rats were fed either corn oil, butter, hydrogenated vegetable oil, or coating fat for 10 weeks at 10g/100g diet. In the group fed coating fat, arachidonic acid was found to be significantly lower in the phospholipid fatty acid content of the platelets, aorta, and heart. The ratio of 20:3(n-9)/20:4(n-6) was greater than in the groups fed corn oil, butter, or hydrogenated vegetable oil, indicating that the group fed coating fat was essential fatty acid deficient. The composition of coating fat was 33% trans fat and only 0.3% linoleic acid, whereas hydrogenated oil was made up of 18% trans fat and 32.8% linoleic acid. It was then concluded that the consumption of hydrogenated fats high in trans 18:1 acids with adequate amount of linoleic acid had no effect on the amount of thromboxane or prostacyclin by platelet or aorta in vitro. The coating fat is dangerous because of its lack of linoleic acid [42].

To demonstrate the process of calcification, endothelial cells cultured with/without trans fat showed that trans fatty acid calcify arterial cells. One with a trans fatty acid added as the "unnatural" elaidic acid (t18:1 n^9) and the other with a cis fatty acid added as the "natural" oleic acid (cis 18:1 n^9) and testing with radioactive calcium. More radioactive calcium infiltration occurred into the endothelial cells cultured with elaidic acid than with oleic acid. An autopsy of 24 human specimens showed that human subjects that had died of heart disease contained up to 12.2% trans fat in their adipose tissue, 14.4% in liver, 9.3% in heart tissue, and 8.8% in aortic tissue and in atheroma.

14. The trans fatty acids in partially hydrogenated fat can cause blood clots

Partially hydrogenated soybean oil contained 14 cis and trans isomers that were formed during hydrogenation [4, 5]. They inhibited cyclooxygenase, an enzyme required for the conversion of arachidonic acid to prostacyclin, a molecule which prevents blood clots [43]. Moreover, oxidized fat enhanced thromboxane synthesis [44, 45], which caused the formation of a blood clot. Trans fatty acids in partially hydrogenated vegetable oil decreased prostacyclin synthesis by inhibiting cyclooxygenase. Oxysterols enhanced thromboxane synthesis [44, 45]. Both prostacyclin and thromboxane are involved in sudden cardiac death.

According to WebMD, "sudden cardiac death (SCD) is a sudden, unexpected death caused by loss of heart function (sudden cardiac death). It is the largest cause of natural death in the U.S., causing about 325,000 adult deaths in the United States each year. SCD is responsible for half of all heart disease deaths. SCD occurs most frequently in adults in their mid-30s to mid-40s, and affects men twice as often as it does women." [46]

Under the current Food and Drug Administration mandate [47], food items with any amount of trans fatty acids are allowed, as long as they are labeled. Products containing less than 0.5g/serving can be labeled as "trans free" or 0%. This is misleading, because it is easy to circumvent this rule by making the serving size listed on a label small enough to meet the 0.5g threshold. The food industry has taken advantage of this rule by making the serving sizes small enough to contain less than 0.5g/serving of trans fat. Fifteen foods labeled "trans fat free" were analyzed for fat content. Two contained 0% trans fatty acid, two contained higher than 0.5g/serving and the rest contained between 0.014 to 0.25g/serving. If the serving size is increased, foods would contain more than 0.5g of trans fatty acids. In 2003, the daily intake of trans fatty acids for men was estimated by the Food and Drug Administration to be nearly 7 grams per day, and almost 5 grams per day for women [47]. It is possible for people to eat the same amount of trans fatty acids today as in earlier periods, even though they have supposedly been removed from the food supply. A recent article in JAMA, "Levels of Plasma trans-fatty acids in Non-Hispanic White Adults in the United States in 2000 and 2009" listed levels in the year 2000 at 38.0, and in 2009 as 14.0µ/ml, which was considered significant [48].

15. Environmental impact of soybean use

Epidemiological data collected by the Center for Disease Control (CDC) further illustrate the potential harmful effects of trans fat. These data showed that, death from CHD in the USA increased from 265.4/100,000 in 1900 to 581/100,000 population by 1950. During this time period, both margarine and shortening had a high percentage of trans fat (ranging from 39-50%) and a low percentage of linoleic acid (ranging from 6-11%) according to the technical director of the Institute of Shortening and Edible Oils. In 1968 Dr. Campbell Moses, medical director of the AHA, appointed a five member subcommittee on fats of the AHA nutrition committee to revise the 1961 version of "Diet and Heart Disease." At the time it was known that an increase in EFA composition of a dietary fat would lower plasma cholesterol levels and there was strong evidence that trans fatty acids increased plasma cholesterol levels. The first revised version by the AHA committee stated:

"Partial hydrogenation of polyunsaturated fats results in the formation of trans forms which are less effective than cis, cis forms in lowering cholesterol concentrations. It should be noted that many currently available shortenings and margarines are partially hydrogenated and many contain little polyunsaturated fat of the natural cis, cis form." The members of the

Institute of Shortening and Edible Oils Inc objected to this version. The second revised and distributed version, omitted references to hydrogenated fat and cis fatty acids stated: "Margarines that are high in polyunsaturates usually can be identified by the listings of a liquid oil first among the ingredients. Margarines and shortenings that are heavily hydrogenated or contain coconut oil, which is quite saturated, are ineffective in lowering the serum cholesterol." The industry agreed to lower the trans fatty acids and increase the level of EFA in shortenings and margarine. Dr. R.I. Levy, director of the National Heart, Lung, and Blood Institute at the time, believed 1968 a watershed, as the incidence of CHD has steadily declined in the US since 1968. Why it decreased remained unknown in 1968.

On October 24th, 1978, ten years after the reformulation of hydrogenated fat, the National Institute of Health (NIH) held a conference in Bethesda, Maryland, on the Decline in CHD Mortality. A recent editorial in Circulation cited this symposium. Three major conclusions reached were;

1. The decrease in CHD mortality was real and not a result of artifacts or changes in death certificate coding,

2. Both primary prevention through changes in risk factor fundamentals and clinical research leading to better medical care probably have contributed to but did not fully explain the decline, and

3. A precise quantification of the causes requires further studies.

" In hindsight, the reformulation of hydrogenated fat with its lowering of the trans fatty acids and raising of linoleic acid could have also been responsible for the decline. The per capita consumption of hydrogenated fat continued to increase after 1950. However, the increase in the linoleic acid content in the reformatted 1968 fat and the increasing use of soybean oil in salad dressing and other food items could have helped to keep a decreasing death rate from CHD. The death rate from heart disease dropped substantially during the next decades even though the consumption of hydrogenated fat kept increasing and animal fat was decreasing. Lower trans fat and increased linoleic acid are possible explanations for this change.

The death rate from CHD declined after 1968 from 588.8/100,000 to 217/100,000 in 2004 in the USA. According to AHA data, 451,300 Americans died of CHD in 2004. Heart disease is still the number one cause of death. However, in a population of approximately 300 million, today the deaths would have been 1,480,000 at the 1950 rate according to the National Institute of Health (NIH). A recent study based on the autopsy of young men showed the CHD rate has been increasing since 2004. The recent reformulation of hydrogenated fat raises the trans fatty acid levels from 20% to almost 40%.

In 2003, the metabolism of the trans fat in hydrogenated oil was assumed to follow the same pathway as the natural ruminant trans fat in butterfat. The Food and Drug Administration has stated that the main reason for the trans fat in partially hydrogenated oil to remain in the diet in the USA rested on the generally held belief that trans fat is metabolized the same way as the natural trans (vaccenic acid) in butterfat. The FDA allowed the isomeric fatty acids in hydrogenated vegetable oils to remain in food products because they assumed that some of that trans fat may be from the natural vaccenic acid that has no harmful effects. Approximately 2.6% of the total daily fat intake is from trans fat and that 50% of the trans may be from vaccenic acid ($18:1n^{11}$).

16. Conclusion

The oil produced by soybeans is widely used by manufacturers of both food products and industrial manufactured goods. Crude soybean oil contains essential fatty acids that our body needs to work properly. However, much of the soybean oil consumed today has been partially hydrogenated. This hydrogenation removes the necessary essential fatty acids contained within the original oil. Additionally, some of the partially hydrogenated soybean oil has been converted to trans fatty acids.

There are two mechanisms that have been shown to lead to heart disease involving the consumption of trans fatty acids. They effect the levels of prostacyclin and thromboxane, which increases the risk of thrombosis, and they increase sphingomyelin production by the body, which then causes calcium influx into the arterial cells to increase, leading to atherosclerosis. Soybeans can be an excellent source of protein, but partially hydrogenated soybean oil can be detrimental to health.

NC Soybean Producers Assn. How soybeans are used. Retrieved from http://www.ncsoy.org/ABOUT-SOYBEANS/Uses-of-Soybeans.aspx

Author details

Fred A. Kummerow*

Address all correspondence to: fkummero@uiuc.edu

University of Illinois, USA

References

[1] Pedersen, P. (2007). Soy Products. http://extension.agron.iastate.edu/soybean/uses_soyproducts.htmlaccessed 1 August 2012).

[2] Soya. (2012). Information about soy and soya products. http://www.soya.be/soybean-oil-production.phpaccessed 31 July).

[3] Kummerow, F. A. (2005). Improving hydrogenated fat for the world population. *Prevention and Control*, 1, 157-164.

[4]   Kummerow, F. A. (2009). The negative effects of hydrogenated trans fats and what to do about them. *Atherosclerosis*, 205, 458-465.

[5]   Quackenbush, F. W., Steenbock, H., Kummerow, F. A., & Platz, B. R. (1942). Linoleic acid, pyridoxine and pantothenic acid in rat dermatitis. *Journal of Nutrition*, 24, 225-234.

[6]   Wansink, B. (2003). How do front and back package labels influence beliefs about health claims? *Journal of Consumer Affairs*, 37, 305-316.

[7]   Sirtori, C. R. (2001). Risks and benefits of soy phytoestrogens in cardiovascular diseases, cancer, climacteric symptoms and osteoporosis. *Drug Safety*, 24, 665-682.

[8]   Henkel, J. (2000). Health claims for soy protein, questions about other components. *FDA Consumer*, 34, 13.

[9]   Messina, M. (2003). Potential public health implications of the hypocholesterolemic effects of soy protein. *Nutrition*, 19, 280-281.

[10] Sacks, F. M., Lichtenstein, A., Van Horn, L., et al. (2006). Soy Protein, Isoflavones, and Cardiovascular Health: An American Heart Association Science Advisory for Professionals from the Nutrition Committee. *Circulation*, 113, 1034-1044.

[11] Kummerow, F., & Kummerow, J. (2008). Cholesterol Won't Kill You, But Trans Fat Could. *Bloomington: Trafford Publishing*.

[12] Ross, R. (1993). The pathogenesis of atherosclerosis. *New England J Med*, 297, 369-377.

[13] De Bakey, M. E., Dietrich, E. B., Garrett, H. E., & Mc Cutchen, J. J. (1967). Surgical treatment of cerebrovascular disease. *Postgrad Med J*, 42, 218-226.

[14] Keaney, J. (2000). Atherosclerosis from lesion formation to plaque activation and endothelial disfunction. *Molecular Aspects of Medicine*, 21, 118.

[15] Nicholson, A., Febbraio, M., Han, J., Silverstein, R., & Hajjar, D. (2000). CD36 in atherosclerosis, the role of class B macrophage scavenger receptor. *Ann. NY Acad. Sci*, 902, 128-131.

[16] Leschziner, M., & Dimitriadis, K. (1989). Computation of three-dimensional turbulent flow in non-orthogonal junctions by a branch-coupling method. *Computers & Fluids*, 17, 371-396.

[17] Koenig, W., & Ernst, E. (1992). The possible role of hemorheology in atherothrombogenesis. *Atherosclerosis*, 94, 93-107.

[18] Gimbrone, M., Topper, J., Nagel, T., Anderson, K., & Garcia-Cardena, G. (2000). Endothelial dysfunction, hemodynamic forces, and atherogenesis. *Ann. NY Acad. Sci*, 902, 230-240.

[19] Kataoka, H., Kume, N., Miyamoto, S., Minami, M., Moriwaki, H., Murase, T., Sawanmura, T., Masaki, T., Hashimoto, N., & Kita, T. (1999). Expression of lectin-like oxidized LDL receptor-I human atherosclerosis lesions. *Circulation*, 99, 3110-3117.

[20] Li, D., & Mehta, J. (2000). Antisense to LOX-1 inhibits oxidized LDL-mediated upregulation of monocyte chemoattractant protein-1 and monocyte adhesion to human coronary artery endothelial cells. *Circulation*, 101, 2889-2895.

[21] Murase, T., Kume, N., Korenaga, R., Ando, J., Sawamura, T., Masaki, T., & Kita, T. (1998). Fluid shear stress transcriptionally induces lectin-like oxidized low density lipoprotein receptor-1 in vascular endothelial cells. *Circ. Res*, 83, 328-333.

[22] Chen, X., Sun, A., Yunzeng, Z., et al. (2011). Impact of sphingomyelin levels on coronary heart disease and left ventricular systolic function in humans. *Nutrition & Metabolism*, 8, 25.

[23] Jiang, X., Paultre, F., Pearson, T., et al. (2000). Plasma sphingomyelin level as a risk factor for coronary artery disease. *Arteriosclerosis Thromb. & Vasc. Biol*, 20, 2614-2618.

[24] Nelson, J. C., Jiang, X. C., Tabas, I., et al. (2006). Plasma Sphingomyelin and Subclinical Atherosclerosis: Findings from the Multi-Ethnic Study of Atherosclerosis. *Am. J. Epidemiol*, 163, 903-912.

[25] Kummerow, F. A., Przybylski, R., & Wasowicz, E. (1994). Changes in arterial membrane lipid composition may precede growth factor influence in the pathogenesis of atherosclerosis. *Artery*, 21, 63-75.

[26] Bittman, R. (1988). Sterol exchange between mycoplasma membranes and vesicles. *Yeagle PL, editor. Boca Raton, FL: Biology of Cholesterol CRC Press*.

[27] Lehninger, A. L. (1975). Biochemistry. *New York: Worth Publishers Inc*.

[28] Bergelson, L. D., & Barsukov, L. I. (1977). Topological asymmetry of phospholipids in membranes. *Science*, 197, 224-230.

[29] Devaux, P. F. (1991). Static and dynamic lipid asymmetry in cell membranes. *Biochemistry*, 30, 1163-1173.

[30] Holmes, R. P., & Yoss, N. L. (1984). Hydroxysterols increase the permeability of liposomes to Ca2+ and other cations. *Biochim Biophys Acta*, 770, 15-21.

[31] Kummerow, F. A., Cook, L. S., Wasowicz, E., & Jelen, H. (2001). Changes in the phospholipid composition of the arterial cell can result in severe atherosclerotic lesions. *J Nutr Biochem*, 12, 602-607.

[32] Ridgway, N. D. (1995). Hydroxycholesterol stimulates sphingomyelin synthesis in Chinese hamster ovary cells. *J Lipid Res*, 36, 1345-1358.

[33] Zhou, Q., Wasowicz, E., Handler, B., et al. (2000). An excess concentration of oxysterols in the plasma is cytotoxic to cultured endothelial cells. *Atherosclerosis*, 149, 191-197.

[34] Zhou, Q., Band, M. R., Hernandez, A., & Kummerow, F. A. (2004). 27-Hydroxycholesterol inhibits neutral sphingomyelinase in cultured human endothelial cells. *Life Sci*, 75, 1567-1577.

[35] Kummerow, F. A., Olinescu, R., Fleischer, L., Handler, B., & Shinkareva, S. (2000). The relationship of oxidized lipids to coronary artery stenosis. *Atherosclerosis*, 149, 181-190.

[36] Kummerow, F. A. (2005). Improving hydrogenated fat for the world population. *Prevention & Control*, 1, 157-164.

[37] Kummerow, F. A., Zhou, Q., & Mahfouz, MM. (2004). Trans fatty acids in hydrogenated fat inhibited the synthesis of the polyunsaturated fatty acids in the phospholipid of arterial cells. *Life Sciences*, 74, 2707-2723.

[38] Stary, H. (1999). Atlas of atherosclerosis progression and regression. *New York: Parthenon Publishing Group.*

[39] Kummerow, F. A., Zhou, Q., & Mahfouz, MM. (1999). Effect of trans fatty acids on calcium influx into human arterial endothelial cells. *American J Clin Nutr*, 70, 832-838.

[40] Vane, J. R., & Moncada, S. (1977). The discovery of prostacyclin: a fresh insight into arachidonic acid metabolism, biochemical aspects of prostaglandins and thromboxanes. *New York: Academic Press.*

[41] Mahfouz, M. M., & Kummerow, F. A. (1999). Hydrogenated fat high in trans monoenes with an adequate level of linoleic acid has no effect on prostaglandin synthesis in rats. *Journal of Nutrition*, 129, 15-24.

[42] Kummerow, F. A., Mahfouz, MM, & Zhou, Q. (2007). Trans fatty acids in partially hydrogenated soybean oil inhibit prostacyclin release by endothelial cells in presence of high level of linoleic acid. Prostaglandins Other Lipid Mediat; , 84, 138-153.

[43] Mahfouz, MM, & Kummerow, F. A. (1998). Oxysterols and TBARS are among the LDL oxidation products which enhance thromboxane A_2 synthesis by platelets. *Prostaglandins Other Lipid Mediat*, 56, 197-217.

[44] Mahfouz, MM, & Kummerow, F. A. (2000). Oxidized low-density lipoprotein (LDL) enhances thromboxane A_2 synthesis by platelets, but lysolecithin as a product of LDL oxidation has an inhibitory effect. *Prostaglandins Other Lipid Mediat*, 62, 183-200.

[45] Maddox, T. (2012). Heart disease and sudden cardiac death. http://www.webmd.com/heart-disease/guide/sudden-cardiac-deathaccessed 31 July).

[46] FDA. (2003). Food labeling: trans fatty acids in nutrition labeling, nutrient content claims, and health claims. *Final rule. Fed Regist*, 68, 41433-41506.

[47] Vesper, H. W., Kuiper, H. C., Mirel, L. B., Johnson, C. L., & Pirkle, J. L. (2012). Levels of plasma trans-fatty acids in non-Hispanic white adults in the United States in 2000 and 2009. *Journal of American Medical Association*, 307, 562-563.

Chapter 5

Soybean: Non-Nutritional Factors and Their Biological Functionality

A. Cabrera-Orozco, C. Jiménez-Martínez and
G. Dávila-Ortiz

Additional information is available at the end of the chapter

1. Introduction

Legumes are important for the diet of a significant part of the world´s population; they are a good source of protein, carbohydrates, minerals and B-complex vitamins. In this sense, the soybean is an important legume because it has a high protein (35-48%) with a nutritionally balanced amino acid profileso their products are commonly used as a source of vegetable protein worldwide and a great proportion of high-quality oil (15-22%)[1].

The accessible price and stable supply are favourable factors for legumes to emerge as an important source of protein for human food [2]. However, the nutritional value of soybeans is lower than expected due to the presence of various non-nutritive compounds that hinder or inhibit the uptake of nutrients and produce adverse physiological and biochemical effects in humans and animals; since these could be toxic in some cases, they are referred as anti-nutritional factors [3, 4].

Recently it has been found that legumes, in the appropriate proportion, may have a beneficial role for health. It seems clear that, in many cases, the same interaction that causes legumes to be considered as anti-nutrients is responsible for its beneficial effects. Thereby,these compounds are called non-nutritional compounds or nutritionally bioactive factors, because while they lack nutritive value, are not always harmful [5]. Available data indicate that the balance between harmful and beneficial effects of these compounds is a function of concentration, exposure time and interaction with other dietary components. However, the threshold concentration at which the beneficial and harmful effects occur has not been evaluated in most cases [6].Moreover, they are compounds that do not appear equally in all legumes, and their physiological effects in humans and other animals are different as well [7, 8].

These compounds have an important role in secondary metabolism of legumes, as reserve compounds for the biosynthesis of endogenous compounds, which accumulate in seeds and are used during the germination process, and as mechanisms of defense against bacteria, viruses, fungi, insects and animals [9].

From a biochemical point of view, these compounds have diverse nature. They may be proteins (protease inhibitors, α-amylase inhibitors and lectins), carbohydrates (α-galactosides, vicine, convicine, saponins), non-protein amino acids (L-DOPA, β-ODAP), polyphenols (condensed tannins, isoflavones), alkaloids, inositol phosphates, etc., so their extraction and quantification methodology is very specific. In soybeans, the non-nutritional factors are mainly; inositol phosphates, saponins, protease inhibitors, isoflavones, lectins, oligosaccharides and taninis[10, 11].

1.1. Phytic acid and inositol phosphates

Phytic acid (myo-Inositol hexakisphosphate or 1, 2, 3, 4, 5, 6-hexakis dihydrogen phosphate myo-Inositol), also abbreviated as $InsP_6$ or IP_6, is the main form of storage of phosphorus and inositol in seeds of cereals, legumes and oilseeds. However for humans and monogastric animals phosphorus is not available in that form, because these are not provided with sufficient activity of endogenous phosphatases (phytases) that are capable of releasing the phosphate group from phytic acid or inositol phosphates lighter phosphorylated [12].

This molecule is formed from the esterification of phosphate groups to each of the six hydroxyl groups in a molecule known as myo-Inositol (Figure 1). Usually, it represents 65 to 85% of total phosphorus in seeds while forming insoluble salts with mono and divalent cations. By releasing H+ ions from the phosphate groups, allows the molecule to interact with the ions Mn^{2+}, Fe^{2+}, Zn^{2+} and K^+ to produce the corresponding salts, which are known as phytates. The name phytin has been used to designate a mixture of salts with Ca^{2+} and Mg^{2+}. Phytates and phytins usually bind to proteins in the protein bodies, the latter are membrane-limited structures where storage proteins are deposited. Salts of phytic acid are accumulated in seeds during the maturation period and are distributed uniformly in the cotyledons and embryonic axis in legumes [13, 14].

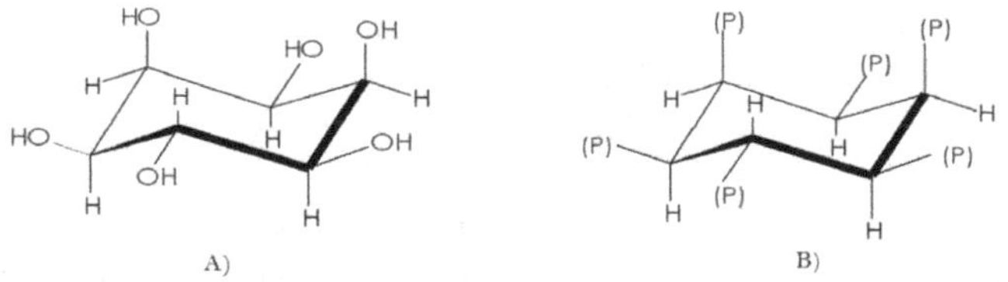

Figure 1. A) Chemical structure of myo-Inositol, B) Phytic acid structure (P) = H_2PO_4 [15]

In the soybean (*Glycine max*) phytic acid is uniformly distributed in the cotyledons, in the same way as in most legumes, probably as a soluble potassium phytate, which constitutes approximately 1.5% of the total weight of the cotyledon. One gram of soybeans contains about 9.2-16.7 mg of phytate, which represents 57% of organic phosphorus and 70% of total phosphorus [16, 17].

1.1.1. Synthesis and Function

Phytic acid is sythesized from 1D-myo-Inositol 3-phosphate (Ins$_3$P$_1$); in turn, the latter is formed from D-glucose-6-phosphate by action of synthase Ins$_3$P1, and from myo-Inositol by action of myo-Inositol kinase; this reaction represents the first step in the metabolism of inositol and in the phytic acid biosynthesis. Subsequently, the phosphatidylinositol kinases catalyze the gradual phosphorylation of Ins$_3$P$_1$ to produce myo-Inositol di-, tri-, tetra-, penta- and hexaphosphate (Figure 2) [14, 18, 19].

During germination, phosphorus and cations are released from phytates by the increased activity of an enzyme called phytase, then, they become available for use during the seedling growth. The enzyme phytase (myo-Inositol-hexaquisfosfatophosphohydrolase) is capable of sequentially hydrolyzing phytic acid to myo-Inositol, which produces intermediate products with a lower number of phosphate ester groups (IP$_5$, IP$_4$, IP$_3$, and possibly di- and mono- phosphateinositols) and inorganic phosphate [20, 21].

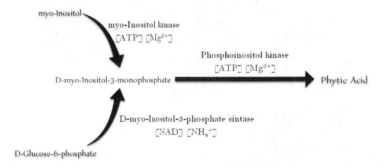

Figure 2. Biosynthesis of phytic acid [14].

A clear role of phytic acid in the seed tissues metabolism is the storage and recovery of phosphorus, minerals, and myo-Inositol during germination and growth [22]. Another physiological functions of phytic acid in plants, is the inhibition of the metabolism, since, by binding to multivalent cations required for cellular processes, the metabolism is slower, so it could be a latency-inducing molecule. Furthermore, the antioxidant capacity of phytic acid increases the time of seed latency, as it prevents lipid peroxidation [18, 23].

As well, phytates and also less-phosphorylated forms of phytic acid regulate diverse cellular functions such as DNA repair, chromatin remodelling, endocytosis, nuclear export of mRNA, and is an important hormonal marker for the development of seedlings and seeds [24-28].

Less-phosphorylated molecules of myo-Inositol are presentin free form in nature, in small amounts, as transient intermediates in biochemical reactions. The mono-, bi-, tri- myo-Inositol phosphates are important components of a group of phospholipids, known as phosphoinositides, which are present in many plants and animal tissues [18]. Raboy (2009),reported a very detailed description of the synthesis and metabolism of phytic acid and myo-Inositol phosphates in plants (Figure 3).

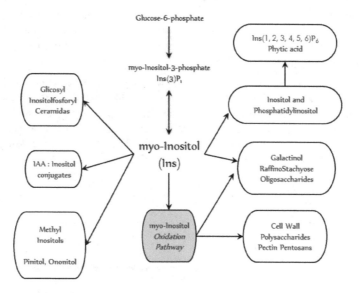

Figure 3. Pathways in plant biology that utilize myo-Inositol [29].

1.1.2. Bioavailability of minerals

Most studies on the interaction between phytic acid/inositol phosphates and minerals reveal the existence of an inverse relationship between the absorption of these micronutrients and inositol phosphates, although there are substantial differences in individual behaviour of each mineral element [16].

The interaction of phytic acid with minerals and other nutrients is pH-dependent [30], since the degree of protonation of the phosphate groups is a function of pH [31]. The molecule works in a wide region of pH as a highly negatively-charged ion, so its presence in the diet has an adverse impact on the bioavailability of mineral monovalent, divalent and trivalent ions, such as Zn^{2+}, Fe^{2+} / Fe^{3+}, Ca^{2+}, Mg^{2+}, Mn^{2+} and Cu^{2+}[32-34]; these complexes are more soluble at low or acid pH and insoluble at high or basic pH [35].

Another important aspect to be considered is that the interaction of phytic acid with minerals is due to its several phosphate groups, thereby, minerals may bind to one, two, or more phosphate groups of one, two or more phytic acid molecules [36]. Other studies have shown that the inhibitory effect of the absorption of InsP depends on the degree of phosphorylation of inositol, when it is high (5 or 6 phosphates) the absorption of Ca and Zn is significantly inhibited, however at lower levels of phosphorylation this effect is not observed [34].

The solubility of the complexes formed depends also on the InsP-mineral ratio; for instance, the solubility of the InsP-Ca complex, is extremely low in 1/8 ratios, but other ratios show higher solubility [37]. The complexes hexa-, penta-, tetra- and tri- Ca are insoluble, while complexes mono- and di- Ca are soluble [38]. Nevertheless, the absorption of Cafrom soluble complexes InsP-Ca is very low, because these complexes do not undergo passive transport in gut due to the high electric charge they have [39].

The inositol phosphates directly or indirectly interact with various minerals in the diet to reduce their bioavailability, in this context, the synergistic effect of the secondary cations (Ca^{2+}) has been widely demonstrated. Two cations may, when present at the same time, act together to increase the amount of insoluble precipitate in salt form, *i. e.* a mineral has a higher affinity for certain complex (InsP-mineral) which generates more insoluble salts[40]. For instance, the phosphate inositol bound to Ca shows higher affinity to Zn, which decreases its reabsorption [41].

1.1.3. Bioavailability of proteins

The degree of interaction between phytic acid (and its phosphate inositols) and depends on the protein, net charge, conformation and interactions with minerals at a given pH.At low pH, below the isoelectric point of proteins, phytic acid phosphate esters bind to the cationic group of basic amino acids, for example, arginine, histidine and lysine, may form InsP-protein complexes.

At a pH above the isoelectric point of proteins, since the charge of proteins as well as that of the phytic acid is negative, the interaction would be impossible, however, interaction occurs through the formation of complexes with divalent such as Ca^{2+} or Mg^{2+}. This binding takes place via the formation of ionized carboxyl groups and the deprotonated imidazole group of histidine, which requires a minimum concentration of these cations to maintain these complexes. At this pH, some binary complexes may exist because lisyl and arginyl residues of the proteins are still positively charged. At high pH the interaction between proteins and phytic acid decreases, arginyl and lysyl groups lose their charge, and therefore its ability to form binary complexes [36]; as well, they may form complexes such as protein-InsPand protein-mineral-InsP(Figure 4), which reduces their bioavailability.Such complexes may affect the protein structure, which may decrease the enzyme activity, function, solubility, absorption and protein digestibility [16, 36]. Particularly, the ability to inhibit proteolytic(pepsin, trypsin, chymotrypsin), amylolytic (amylase) and lipolytic (lipase) enzymes, is responsible of their anti-nutritional properties. This inhibition may be due to the nonspecific nature of InsP-protein interactions and the chelation of calcium ions, which are essential for the activity of trypsin and α-amylase [16]. Furthermore, phytic acid can also bind to starch, either directly via hydrogen bonds or indirectly through the proteins to which it is associated [42].

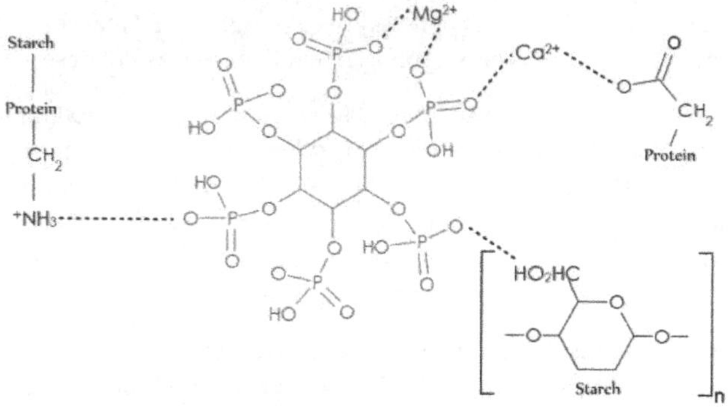

Figure 4. InsPinteractions with minerals, proteins and starch [36].

1.1.4. Pharmacological properties

In vivo and *in vitro* studies have shown that phytic acid (InsP$_6$)has prevention and therapeutic properties against cancer. Several mechanisms have been suggested to explain its anticarcinogenic effect:

a. Experiments have shown that this compound induces apoptosis in cancer cells, causes differentiation of malignant cells and its reversion to normal phenotype, and increases the activity of natural killer cells of the immune system. In addition, IP$_3$ and IP$_4$ compounds have an important role in cellular signal transduction, regulating functions, cell growth and differentiation [43].

b. A second way in which phytic acid reduces the risk of cancer, is by chelation of Fe^{3+} and the suppression of the formation of radicals (•OH), which also originates antioxidant properties. Fe^{3+} is an effective catalyst for many biological functions, in which this ion is reduced to Fe^{2+}. The oxidation of Fe^{2+}to Fe^{3+} leads to the formation of O$^{2-•}$ that spontaneously generates O$_2$ and H$_2$O$_2$. The Fenton's reagent (Fe^{2+} + H$_2$O$_2$) quickly generates •OH, a highly-reactive oxyrradical which indiscriminately attacks most of the biomolecules. By blocking the redox cycle of Fe, which is necessary in many oxidation reactions, the lipid peroxidation and DNA damage are inhibited [37 , 44].

c. Zn is involved in DNA synthesis and cell proliferation as a cofactor for many enzymes like thymidine kinase. So, by binding to Zn^{2+}, the phytic acid indirectly reduces cell proliferation [45].

d. Phytic acid can reduce the starch digestibility and cause low absorption, so that starch remains available in the colon to be fermented by bacteria, producing short chain fatty acids, which have protective activity against cancer [46].

Phytic acid can retard the digestion and absorption of starch in several ways: by direct binding to the polysaccharide, by its binding to the α-amylase, or by chelation of Ca^{2+} needed for activation of α-amylase. Through these mechanisms a delay occurs in the glycemic response, therefore, due to lower blood glucose, insulin is required in less amount and this reduces the risk of diabetes [47].

Respect to prevention of kidney stones and treatment of hypercalciuria, experimental evidences demonstrate that di- and tri- inositol phosphates ($InsP_2$ and $InsP_3$), are effective to prevent the formation of hydroxyapatite crystals *in vitro*, which act as the core for the formation of some kidney stones [24].

At levels from 0.2 to 9% of phytic acid in diet, the plasmatic levels of cholesterol and triglycerides are significantly reduced [48]. This seems to be related with the capability of phytic acid to be bound to Zn, which reduces the serum levels of Zn and the Zn/Cu ratio, since high values of this ratio tend to increase the risk of cardiovascular diseases, for instance, hypercholesterolemia [42].

1.2. Saponins

Saponins (Figure 5) are a big group of glycosides which are known by their surfactant properties and are widely distributed in green plants [49] .The name 'saponin' derives from the Latin word *sapo* which means soap, due to their property of generating foam in agitated aqueous solutions [50]. These substances are amphiphilic glycosides, wherein the polar constituents are sugars (pentoses, hexoses or uronic acids) that are covalently linked to a nonpolar group, which consists of an aglycone, called sapogenin, which can be either steroidal or triterpenoid. This combination of polar and nonpolar components in their molecular structure explains their surfactant property in aqueous solutions [51].

As mentioned above, the saponins are secondary metabolites that can be classified into two groups based on the nature of the aglycone skeleton. The first group consists of steroidal saponins, which are present almost exclusively in monocotyledons angiosperms. The second group is composed of triterpenoidsaponins, which occur mainly in dicototyledonous flowering plants [52]. Steroidal saponins comprise a steroidal aglycone, a spirostane skeleton of 27 carbons (C_{27}), which generally comprises a six-ring structure. In some cases, the hydroxyl group at position 26 is used to form a glycosidic bond, so that the structure of the aglycone becomes a pentacyclic structure; this structure is known as furostano skeleton. The triterpenoidsaponins have an aglycone with a backbone of 30 carbons (C_{30}), which form a pentacyclic structure (Figure 5).

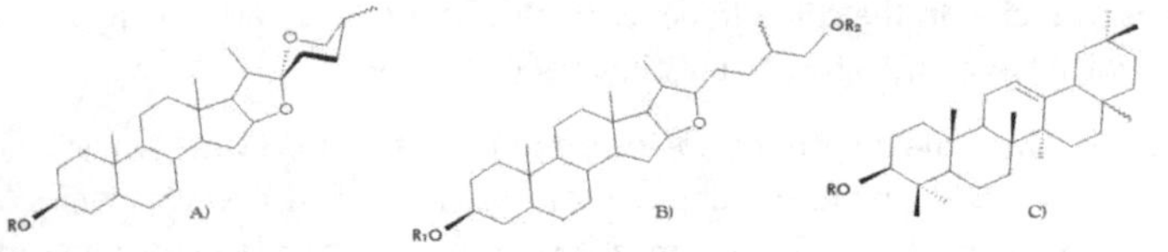

Figure 5. Skeletons of aglycone: (A) steroidal spirostane, (B) steroidal furostane (C) triterpenoid. R = sugar residue.

It has been identified that soy contains saponins with triterpenoid-type aglycones, this kind of aglycones are subdivided into five major groups; soysapogenol A, B, C, D and E (Figure 6), and their glycosides are correspondingly called as saponins of group A, group B, and so on[53, 54].From this classification, four aglycones (soysapogenol A, B, C and E) [55]were isolated after hydrolisis of soy saponins, specifically five saponins were identified 5 with two distinct types of aglycones: soysapogenin I (the main component), soysapogenins II and III, which contain soysapogenol B, and soysaponins A1, A2 and A3, which contain soysapogenol A[55].The saponins containing soyasapogenol C and E have not been found in soybeans, so these aglycones could be formed as a product during the hydrolysis of saponins[56].Another study reported the isolation and characterization of soysaponin IV. The type of sugars attached to the aglycones found in soybeans have been identified as rhamnose, galactose, glucose, arabinose, xylose and glucuronic acid [55].

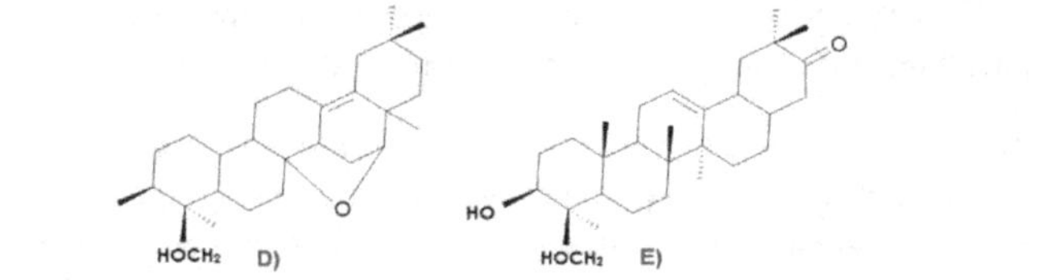

Figure 6. Groups of soyasapogenols *(Oakenfull, 1981)*.

The total content of saponins in the hypocotyl fraction of soybeans, where acetyl-soyasaponins A1 and A4, mainly, are synthesized, is approximately 0.62 to 6.16%[57]. Other works have reported 5-6% saponin content in soybeans [58]. Lower values, approximately 0.6%, have been reported as well [59].

1.2.1. Synthesis and Function

The capability to synthesize saponins is widespread among plants belonging to the *Magno-liophyta* division, which includes both dicotyledons and monocotyledons. However, most of saponin-producing species are within dicotyledonous plants. The biological function of sap-onins is not completely understood. They are usually considered as a part of the defense system of the plant, due to their antimicrobial, fungicide, allelopathic, insecticide and mol-luscicide activities [60]. The synthesized saponins are accumulated during the regular growth of plants. Nonetheless, their accumulation is influenced by several environmental factors such as bioavailability of nutrients and water, solar radiation or a combination of them[61]. Some studies on soy have shown a variation in the content of saponins in soy-beans with different degrees of maturity, however, the nature of this variation is not suffi-cient to influence on the saponindistribution in different varieties [57]. Little is known about the enzymes and biochemical pathways involved in the biosynthesis of saponins in plants [54]. However, two key aspects have been suggested for biosynthesis: the first one is the cyc-lization of the 2,3-oxidoscualene thought the isoprenoid pathway, which is a starting point for the biosynthesis of the sapogenin, and the second one is the glycosylation of sapogenins.

1.2.2. Membranolytic activity

Saponins have the ability to cleavege the erythrocytes. This hemolytic property is generally attributed to the interaction between saponins and sterols in the erythrocyte membrane. As a result, the membrane is broken, which causes an increase in its permeability and the con-sequent loss of hemoglobin. It has been investigated the effect of saponins in the membrane structure through human erythrocyte hemolysis[5, 62]. The results indicated that the frac-ture in the erythrocyte membrane was not closed again, so that the damage in the lipid bi-layer is irreversible. However, this toxic property is difficult to occur *in vivo*, since there is evidence that no complications are detected when saponins are ingested orally, which re-duces his hemolytic capability to *in vitro* studies [63]. The saponins have little anti-nutrition-al activity, given no damage is produced in humans when they are consumed in the amounts regularly found after food processing. However, high concentrations of saponins are also capable of breaking the membrane of other cells such as those of the intestinal mu-cosa, which modifies the cell membrane permeability, and then affects the active transport and the absorption of nutrients[45].

This ability to affect the cell membrane, depends on the structural characteristics of the sapo-nins, *i. e.* the structure of the aglycone, the number of sugars in the side chains and the side chains length [64]. In Figure 7, the interaction of saponins with cell membranes is schemati-cally shown.

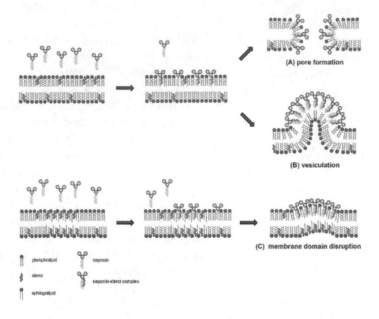

Figure 7. Schematic models of the molecular mechanisms of saponin activities towards membranes [65]Saponins integrate with their hydrophobic part (sapogenin) into the membrane. Within the membrane they form complexes with sterols, which subsequently, driven by interaction of their extra-membranous orientated saccharide residues, accumulate into plaques. Sterical interference of these saccharide moieties causes membrane curvature subsequently leading to (A) pore formation in the membrane [66] or (B) hemitubular protuberances resulting in sterol extraction via vesiculation[67]. Alternatively, after membrane integration saponins may migrate towards sphingolipid/sterol enriched membrane domains (C) prior to complex formation with the incorporated sterols, thereby interfering with specific domain functionalities [68]. Similarly to (B), accumulation of saponins in confined membrane domains has further been suggested to cause deconstructive membrane curvature in a dose-dependent manner.

1.2.3. Pharmacological properties

Many studies highlight the pharmaceutical properties of the soybean saponins, among which the anti-carcinogenic activity is mentioned, given by the membranolyticactivity these molecules have shown in human cells of colon carcinoma[69, 70]. Other studies have demonstrated hypocholesterolemic activity due to depletion of body cholesterol by preventing its reabsorption, thus increasing its excretion. Soluble fibers in legumes are known to increase the viscosity of gastric and intestinal contents, and may be one of the factors responsible for the lowering of cholesterol levels [71, 72]. Studies on health benefits of saponins suggest their hepatoprotective activity, but these studies are limited to cell culture and few animal studies. Studies on rats have shown soybean saponin to have an anabolic effect on bone components, suggesting its role as a nutritional factor in the prevention of osteoporosis [73]. Another activity that has been reported is anti-mutagenicity in breast cells [74].

1.3. Inhibitors of trypsin

Protease inhibitors are proteins widely distributed in the plant kingdom, have the ability to inhibit the proteolytic activity of digestive enzymes such as serine-proteases (trypsin and chymotrypsin) which are characteristic of the gastrointestinal tract of animals, though also may inhibit endogenous proteases and enzymes of bacteria, fungi and insects. These serine-protease inhibitors are proteins that form very stable complexes with digestive enzymes, which preventtheir catalytic activity [75].

Protease inhibitors have been classified into several families based on homology in the sequence of amino acids in the inhibitory sites. The molecular structure of the inhibitor affects both the force and the specificity of the inhibitor. The two main families of protease inhibitors found in legumes are the Kunitz inhibitor and the Bowman-Birk inhibitor, so named after its isolation [2, 51]. In the latter case, the characterization was carried out by *Birk*, [*76*], so this name was added.

Both types of proteases are found in soybeans (*Glycine max*); in other legume seeds, such as beans (*Phaseolus vulgaris*) and lentil (*Lens culinaris*), protease inhibitors have been characterized as members of the Bowman-Birk family. Both inhibitors are water soluble proteins (albumin) and constitute from 0.2 to 2% of total soluble protein of legumes [75, 77], particularly soybeans havereported 50 trypsin inhibitor units / mg of dry sample [78].

1.3.1. Kunitztype inhibitor

The first protease inhibitor to be isolated and characterized was the Kunitz inhibitor. It has a MW between 18 and 24 kDa and contains between 170 and 200 amino acid residues. These inhibitors have one head, i. e., one molecule of inhibitor inactivates one molecule of trypsin. It is a competitive inhibitor, binds to the active sites of trypsin in the same way the substrate of the enzyme does, resulting in the hydrolysis of peptide bonds between amino acids of the reactive site of the inhibitor or the substrate (Figure 8).

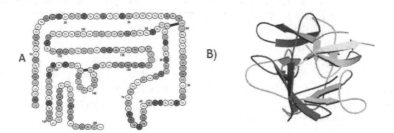

Figure 8. A) Primary structure of the Kunitz inhibitor from soybean [79]. Disulphide bonds are shown in black, B)Tridimentional structure Kunitz inhibitor from soybean [79].

Inhibitors differ from the substrate protein in the reactive site residues, which are linked via disulphide bonds. After hydrolysis, the modified inhibitor maintains the same conformation, due to the disulphide bonds. This generates a stable enzyme-inhibitor complex. This type of inhibitors are generally absent in seeds of *Phaseolus, Pisum, Vigna unguiculata* and *Glycine max* [80].

1.3.2. Bowman-Birk type inhibitor

These inhibitors are low molecular weight polypeptides (7 to 9 kDa) containing 60 to 85 amino acid residues (Figure 9). They have several disulphide bonds which make them stable to heat, acids and bases. These inhibitors have two heads (two separate sites of inhibition) and are competitive inhibitors. They can simultaneously and independently inhibit two enzymes, thus, there are trypsin/trypsin are trypsin/chymotrypsin inhibitors [74,77].

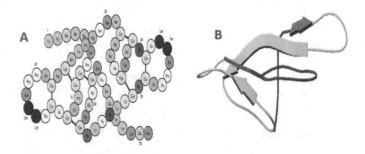

Figure 9. A) Primary structure of Bowman-Birk type inhibitor from soybean*(Odani y Ikenaka, 1973). Disulphide bond-sand active sites for trypsin (Lys16-Ser17) and chymotrypsin (Leu44-Ser45) are shown in black B)* Tridimentional structure Bowman-Birk inhibitor from soybean [81]

An example of this type of inhibitor is the Bowman-Birk inhibitor from soybeans, which is constituted by a polypeptide chain of 71 amino acids, containing seven disulphide bonds. It has a MW of 8 kDa and is called dual head inhibitor because it has independent binding sites for trypsin and chymotrypsin, so that the active site for trypsin is Lys16-Ser17, whereas for chymotrypsin is Leu44-Ser45 [82, 83].

1.3.3. Synthesis and Function

Protease inhibitors have a regulatory function; they are involved the in proteolytic self-regulation process of the protein deposited in the protein bodies before and during the seed germination by inhibiting endogenous proteases. They also participate as protective agents against insects and microorganisms[76].

1.3.4. Anti-nutritional properties

Protease inhibitors ingested within legumes have adverse effects in animals. First, these compounds form inactive complexes with trypsin/chymotrypsin, so that the levels of these free digestive enzymes are reduced, thus making difficult proteolysis and amino acid absorption. In addition, these enzyme-inhibitor complexes, which are rich in sulfur amino acids, are excreted.

Finally, these inhibitors cause hypertrophy/hyperplasia of the pancreas due to chronic hypersecretion of pancreatic enzymes [trypsin and chymotrypsin), which leads to deviation of the sulfur amino acids that were used to synthesize tissue proteins to the synthesis of these enzymes [84]. All this derives in reduction of the amount of essential amino acids, which inhibits animal growth and exacerbates an already critical situation with respect to the protein of legumes which is deficient in sulfur amino acids [3, 84-87].

The mechanism by which inhibitors of proteases stimulate pancreatic secretion is not entirely clear. There is a theory about this secretion would be regulated by a negative feedback mechanism, so that, when the content of trypsin/chymotrypsin in duodenum is reduced below a certain level, the endocrine cells of the duodenal mucosa release the hormone cholecystokinin, prompting the pancreas to synthesize more serine-proteases (Figure 20). The reduced levels of trypsin and chymotrypsin are produced when the protease inhibitors ingested reach the duodenum and bind to these digestive enzymes by forming complexes. Although this does not seem to be the only mechanism by which the pancreatic secretion is activated.

Recent studies have demonstrated that both states of protease inhibitors, free and enzyme-inhibitor complexes, bind to the duodenal mucosa and stimulate the release of cholecystokinin, thus increasing the pancreatic secretion of serine proteases [84, 88]. The action of the trypsin inhibitors on the human organism is not totally understood, since human trypsin has two forms: cationic, which is the main component of pancreatic juice and is weakly inhibited; and anionic, comprising about 10 to 20% of the total trypsin, which is completely inhibited [82, 84].

1.3.5. Pharmacological properties

Since the Bowman-Birk type inhibitors are proteins with a high amount of cysteine , these inhibitors make an important contribution to the content of sulfur amino acids, thus increasing the nutritional value of legumes[85, 89]. The Bowman-Birk inhibitor from soybeans as well as their counterparts present in other legumes, are involved in the prevention and treatment of cancer (colon, breast, liver, lung, prostate, etc.) by inhibiting chymotrypsin. One mechanism through which these compounds can prevent carcinogenesis is by reducing the protein digestibility and the bioavailability of amino acids such as leucine, phenylalanine or tyrosine, which are necessary for the development of cancer cells [90-92].

1.4. Isoflavones

Isoflavones are widely distributed in the plant kingdom, mainly in plants of the legume family, being soybeans the source with the highest content of these components [93]. Isoflavones are oxygen heterocycles containing a 3-phenylchroman skeleton that is hydroxylated at 40 and 7 positions *(Figura 10)* [94]. Based on the substitution pattern on carbons and 6, three aglycon forms of isoflavones commonly found in soybeans are daidzein, genistein, and glycitein. These

three isoflavones can also exist in conjugated forms with glucose (daidzin, geinstin, glycitin), malonylglucose (malonyldaidzin, malonylgeinstin, malonylglycitin), and acetylglucose
(acetyldaidzin, acetylgeinstin, acetylglycitin) units. Thus 12 free and conjugated forms of
isoflavones have been isolated from different soybean samples (Table 1) [95].

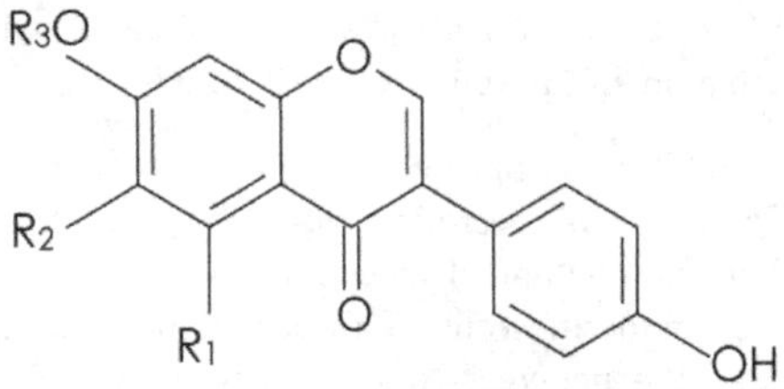

Figure 10. Chemical structureof an isoflavone [95].

Name	R_1	R_2	R_3
Daidzein	H	H	H
Glycitein	H	OCH_3	H
Genistein	OH	H	H
Daidzin	H	H	Glu
Glycitin	H	OCH_3	Glu
Genistin	OH	H	Glu
Acetyldaidzin	H	H	Glu-$COCH_3$
Acetylglycitin	H	OCH_3	Glu-$COCH_3$
Acetylgenistin	OH	H	Glu-$COCH_3$
Malonyldaidzin	H	H	Glu-$COCH_2COOH$
Malonylglycitin	H	OCH_3	Glu-$COCH_2COOH$

Table 1. Chemical structures of 12 isoflavones isolated from soybeans [95].

1.4.1. Synthesis and Function

The variation of the concentration of isoflavones in soybeans is mainly due to the soybean
variety, environment, location and growing conditions, such as year, area and temperature,
post-harvest storage and the methodology used to determine this concentration [95]. The
content of isoflavones in soybeans ranks 1.2 to 2.4 mg of total isoflavones per gram of sample [96], distributed in different concentrations in the tissues of the seed, being higher in the
embryo than in the endosperm [97].

In plants, these compounds play several roles, such as protection against UV light and phytopathogens, signal transduction during nodulation, attraction of pollinator animals and defense against insects and herbivores[98].

1.4.2. Pharmacological properties

The enriched extracts of isoflavones have been evaluated for prevention of a wide range of health problems associated with cardiovascular diseases, osteoporosis and breast cancer, prostate and colon [99]. Furthermore, soy isoflavones have a structure very similar to a phenolic estrogen known as phytoestrogen, so that these compounds have been used as a natural alternative for postmenopausal therapy [100].

1.5. Lectins

In soybeans, a class of proteins called lectins or phytohemagglutinins is present. These compounds can be defined as proteins or glycoproteins of non-immune origin, which can reversibly bind to specific sugar segments through hydrogen bonds and Van Der Waals interactions, with one or more binding sites per subunit [101]. Lectins are tetrameric proteins composed of two different types of subunits: E-type subunit (MW = 34 kDa) and L-type subunit (32 kDa). The first one has the characteristic of binding to erythrocytes, while the second one to lymphocytes[102].Therefore, it is possible to find 5 combinations of these four subunits, i. e. 5 isoforms, as follows: E_4, E_3L, E_2L_2, EL_3, and L_4. Soybean seeds show hemagglutinating activity at 2400 mg per mg of dry sample [78].

The name lectin [from Latin legere, which means to choose or to select), was adopted by Boyd for many years to emphasize the caability of some lectins to bind specifically to cells of the ABO blood groups [103]. Currently the name lectinis preferred over the haemagglutinin one and is widely used to denote all vegetable proteins that possess at least one non-catalytic domain, which binds reversibly to a specific mono- or oligosaccharide [104].

According to the overall structure of the plant lectins, these are subdivided into four main classes: Merolectins which are proteins having a single carbohydrate-binding domain; Hololectins, comprising all lectins having di- or multivalent carbohydrate-binding sites; Chimerolectins, proteins consisting of one or more carbohydrate-binding domain(s) plus an additional catalytic or another biological activity dependent on a distinct domain other than the carbohydrate- binding site; and Superlectins which also possess at least two carbohydrate-binding domains but differ from the hololectins because their sites are able to recognise structurally unrelated sugars [105].

Lectins can be divided according to the monosaccharide for which they show the highest affinity: D-mannose/D-glucose, D-galactose/N-acetyl-D-galactosamine, L-fucose and N-acetyl-glucosamineacid [106]. Thus depending on the specificity toward a given monosaccharide the lectin will selectively bind to one of these above sugars which are typical constituents of eukaryotic cell surfaces [101].

1.5.1. Synthesis and Function

The wide distribution of lectins in all tissues of plants and their ubiquitous presence in the plant kingdom suggest important roles for these proteins. One possible physiological function that has emerged is the defensive role of these carbohydrate-binding proteins against phytopathogenic microorganisms, phytophagous insects and plant-eating animals [102, 107]. Indeed it has been shown that plant lectins possess cytotoxic, fungitoxic, anti-insect and anti-nematode properties either in vitro or in vivo and are toxic to higher animals[63, 81, 104]. One of the most important features of plant lectins, compatible with the proposed defensive function, is the remarkably high resistance to proteolysis and stability over a large range of pH, even when they are out of their natural environment [103].

1.5.2. Anti-nutritional properties

Some of these were found to be toxic or antinutritional for man and animals. In general, nausea, bloating, vomiting and diarrhoea characterize the oral acute toxicity of lectins on humans exposed to them.

In experimental animals fed on diets containing plant lectins the evident symptoms are loss of appetite, decreased body weight and eventually death [84, 108].

As most lectins are not degraded during their passage through the digestive tract they are able to bind the epithelial cells which express carbohydrate moieties recognised by them. This event is undoubtedly the second one in importance for determining the toxicity of orally fed lectins. Indeed, lectins which are not bound by the mucosa usually induce little or no harmful antinutritive effect for the consumers [88]. Once bound to the digestive tract, the lectin can cause dramatic changes in the cellular morphology and metabolism of the stomach and/ or small intestine and activate a cascade of signals which alters the intermediary metabolism. Thus, lectins may induce changes in some, or all, of the digestive, absorptive, protective or secretory functions of the whole digestive system and affect cellular proliferation and turnover. In 1960, Jaffe' suggested that the toxic effects of ingested lectins were due to their ability to combine with specific receptor sites of the cells lining the small intestine and to cause a non-specific interference with absorption and nutrient utilisation [50].

Author details

A. Cabrera-Orozco, C. Jiménez-Martínez and G. Dávila-Ortiz*

*Address all correspondence to: crisjm_99@yahoo.com

Graduates in Food, National School of Biological Sciences.National Polytechnic Institute, Mexico

References

[1] Rex, Newkirk. P. (2010). SoyBean, Feed Industry Guide, CIGI. *Canadian International Grains Institute*, 1.

[2] Bowman, Y. (1944). Fractions derived from soybean and navy beans which retard tryptic digestion of casein. Proc Soc Exp Biol Med; , 57, 139.

[3] Grant, G. (1989). Anti-nutritional effects of soyabean: a review.

[4] E, L. I. (1989). Antinutritional factors. *In: Legumes: Chemistry, Technology, and Human Nutrition. Matthews RH, editor: M. Dekker.*

[5] Baumann, E., Stoya, G., Völkner, A., Richter, W., Lemke, C., & Linss, W. (2000). Hemolysis of human erythrocytes with saponin affects the membrane structure. *Acta Histochemica*, 102(1), 21-35.

[6] Silveira, Rodríguez. M. B., Monereo, Megías. S., & Molina, Baena. B. (2003). Alimentos funcionales y nutrición óptima: "Cerca o lejos" . *Revista Española de Salud Pública*, 77, 317-331.

[7] Khokhar, S., Frias, J., Price, K. R., Fenwick, G. R., & Hedley, C. L. (1996). Physico-Chemical Characteristics of Khesari Dhal (Lathyrus sativus): Changes in α-Galactosides, Monosaccharides and Disaccharides during Food Processing. *Journal of the Science of Food and Agriculture*, 70(4), 487-492.

[8] Grela, E. R., T. , S., & J., M. (2001). Antinutritional factors in seeds of Lathyrus sativus cultivated in Poland. *Lathyrus Lathyrism Newsletter*, 2, 101.

[9] Chung, T. K., Wong, T. Y., Wei, I. C., Huang , W. Y., & Lin, Y. (1998). Tannins and Human Health: A Review. *Critical Reviews in Food Science and Nutrition*, 38(6), 421-64.

[10] Liener, I. E. (1994). Implications of antinutritional components in soybean foods. *Critical Reviews in Food Science and Nutrition*, 34(1), 31-67.

[11] Osman, M. A., Reid, P. M., & Weber, C. W. (2002). Thermal inactivation of tepary bean (Phaseolus acutifolius), soybean and lima bean protease inhibitors: effect of acidic and basic pH. *Food Chemistry*, 78(4), 419-423.

[12] Holm, P. B., Kristiansen, K. N., & Pedersen, H. B. (2002). Transgenic Approaches in Commonly Consumed Cereals to Improve Iron and Zinc Content and Bioavailability. *The Journal of Nutrition*, 132(3), 514S-516S.

[13] Reddy, N. R., & Salunkhe, D. K. (1981). Interactions Between Phytate, Protein, and Minerals in Whey Fractions of Black Gram. *Journal of Food Science*, 46(2), 564-567.

[14] Loewus, F. A., & Murthy, P. P. N. (2000). myo-Inositol metabolism in plants. *Plant Science*, 150(1), 1-19.

[15] Raboy, V. (1997). Accumulation and storage of phosphate and minerals. *In: Cellular and Molecular Biology of Plant Seed Development: Kluwer Academic Publishers.*

[16] Cheryan, M., & Rackis, J. J. (1980). Phytic acid interactions in food systems. *C R C,Critical Reviews in Food Science and Nutrition,* 13(4), 297-335.

[17] Beleia, A., Thu, Thao. L. T., & Ida, E. I. (1993). Lowering Phytic Phosphorus by Hydration of Soybeans. *Journal of Food Science,* 58(2), 375-377.

[18] Cosgrove, D. J., & Irving, G. C. J. (1980). Inositol phosphates: their chemistry, biochemistry, and physiology. Elsevier Scientific Pub. Co.

[19] Honke, J., Kozłowska, H., Vidal-Valverde, C., Frias, J., & Górecki, R. (1998). Changes in quantities of inositol phosphates during maturation and germination of legume seeds. *Zeitschrift für Lebensmitteluntersuchung und-Forschung A,* 206(4), 279-83.

[20] Irving, G. C. J. (1980). Phytase. *In: Inositol phosphates: their chemistry, biochemistry, and physiology: Elsevier Scientific Pub. Co.*

[21] Nayini, N. R., & Markakis, P. (1986). Phytases. *In: Phytic acid: chemistry & applications: Pilatus Press.*

[22] Raboy, V. (2003). myo-Inositol-1,2,3,4,5,6-hexakisphosphate. *Phytochemistry,* 64(6), 1033-43.

[23] Graf, E., Empson, K. L., & Eaton, J. W. (1987). Phytic acid. A natural antioxidant. *Journal of Biological Chemistry,* 262(24), 11647-50.

[24] Zhou, J. R., & Erdman, J. W. (1995). Phytic acid in health and disease. *Critical Reviews in Food Science and Nutrition,* 35(6), 495-508.

[25] York, J. D., Odom, A. R., Murphy, R., Ives, E. B., & Wente, S. R. (1999). A Phospholipase C-Dependent Inositol Polyphosphate Kinase Pathway Required for Efficient Messenger RNA Export. *Science,* 285(5424), 96-100.

[26] Lemtiri-Chlieh, F., Mac, Robbie. E. A. C., & Brearley, CA. (2000). Inositol hexakisphosphate is a physiological signal regulating the K+-inward rectifying conductance in guard cells. *Proceedings of the National Academy of Sciences,* 97(15), 8687-92.

[27] Laussmann, T., Pikzack, C., Thiel, U., Mayr, G. W., & Vogel, G. (2000). Diphospho-myo-inositol phosphates during the life cycle of Dictyostelium and Polysphondylium. *European Journal of Biochemistry,* 267(8), 447-51.

[28] Shears, S. B. (2001). Assessing the omnipotence of inositol hexakisphosphate. *Cell Signal,* 13(3), 151-8.

[29] Raboy, V. (2009). Approaches and challenges to engineering seed phytate and total phosphorus. *Plant Science,* 177(4), 281-96.

[30] Reddy, N. R., & Sathe, S. K. (2002). Food Phytates. *CRC Press.*

[31] Nolan, K. B., Duffin, P. A., & Mc Weeny, D. J. (1987). Effects of phytate on mineral bioavailability. in vitro studies on Mg2+, Ca2+, Fe3+, Cu2+ and Zn2+ (also Cd2+) solubilities in the presence of phytate. *Journal of the Science of Food and Agriculture,* 40(1), 79-85.

[32] Fredlund, K., Isaksson, M., Rossander-Hulthén, L., Almgren, A., & Sandberg-S, A. (2006). Absorption of zinc and retention of calcium: Dose-dependent inhibition by phytate. *Journal of Trace Elements in Medicine and Biology*, 20(1), 49-57.

[33] Lopez, H. W., Leenhardt, F., Coudray, C., & Remesy, C. (2002). Minerals and phytic acid interactions: is it a real problem for human nutrition? *International Journal of Food Science & Technology*, 37(7), 727-739.

[34] Lönnerdal, B., Sandberg, A. S., Sandström, B., & Kunz, C. (1989). Inhibitory effects of phytic acid and other inositol phosphates on zinc and calcium absorption in suckling rats. *The Journal of Nutrition*, 119(2), 211-214.

[35] Torre, M., Rodriguez, A. R., & Saura-, Calixto. F. (1991). Effects of dietary fiber and phytic acid on mineral availability. *Critical Reviews in Food Science and Nutrition*, 30(1), 1-22.

[36] Thompson, L. (1987). Reduction of phytic acid concentration in protein isolates by acylation techniques. *Journal of the American Oil Chemists' Society*, 64(12), 1712-7.

[37] Graf, E., & Eaton, J. W. (1993). Suppression of colonic cancer by dietary phytic acid. *Nutrition and Cancer*, 19(1), 11-9.

[38] Gifford-Steffen, S. R. (1993). Effect of varying concentrations of phytate, calcium, and zinc on the solubility of protein, calcium, zinc, and phytate in soy protein concentrate. *Journal of food protection*, 56(1), 42.

[39] Schlemmer, U., & Müller, H. D. J. K. (1995). The degradation of phytic acid in legumes prepared by different methods. *Eur J Clin Nutr*, 49(3), S207-10.

[40] Sandberg, A. S., Larsen, T., & Sandström, B. (1993). High dietary calcium level decreases colonic phytate degradation in pigs fed a rapeseed diet. *The Journal of Nutrition*, 123(3), 559-66.

[41] Hardy, R. (1998). Phytate Aquaculture Magazine. 24(6), 77.

[42] Thompson, L. U. (1993). Potential health benefits and problems associated with antinutrients in foods. *Food Research International*, 26(2), 131-49.

[43] Shamsuddin, A. M. (2002). Anti-cancer function of phytic acid. *International Journal of Food Science & Technology*, 37(7), 769-82.

[44] Graf, E., & Eaton, J. W. (1985). Dietary suppression of colonic cancer fiber or phytate? *Cancer*, 56(4), 717-8.

[45] Gee, J. M., Price, K. R., Ridout, C. L., Wortley, G. M., Hurrell, R. F., & Johnson, I. T. (1993). Saponins of quinoa (Chenopodium quinoa): Effects of processing on their abundance in quinoa products and their biological effects on intestinal mucosal tissue. *Journal of the Science of Food and Agriculture*, 63(2), 201-9.

[46] Thompson, L. U., & Zhang, L. (1991). Phytic acid and minerals: effect on early markers of risk for mammary and colon carcinogenesis. *Carcinogenesis*, 12(11), 2041-5.

[47] Rickard, Sharon. E., & Thompson, Lilian. U. (1997). Interactions and Biological Effects of Phytic Acid. *Antinutrients and Phytochemicals in Food: American Chemical Society*, 294-312.

[48] Jariwallar, R. J. (1999). Inositol hexaphosphate (IP6) as an anti-neoplastic and lipid-lowering agent. *Anticancer research*, 19(5), 3699.

[49] Tyler, V. E., Brady, L. R., & Robbers, J. E. (1981). Pharmacognosy. *Lea & Febiger*.

[50] Jaffe', W. G. (1960). Studies on phytotoxins in beans. *Arzneimittel Forschung*, 10, 1012.

[51] Kunitz, M. (1945). Crystallization of a trypsin inhibitor from soybean. *Science*, 101(2635), 668-9.

[52] Bruneton, J. (1999). Pharmacognosy, Phytochemistry, Medicinal Plants. *Technique & Documentation*.

[53] Bondi, A., Birk, Y., & Gestetner, B. (1973). Forage saponins. *In Chemistry and biochemistry of herbage: Academic Press*, 17.

[54] Haralampidis, K., Trojanowska, M., & Osbourn, A. (2002). Biosynthesis of Triterpenoid Saponins in Plants. History and Trends in Bioprocessing and Biotransformation. *In: Dutta N, Hammar F, Haralampidis K, Karanth N, König A, Krishna S, et al., editors.: Springer Berlin / Heidelberg*, 31-49.

[55] Kitagawa, I., Yoshikawa, M., Wang, H., Saito, M., Tosirisuk, V., Fujiwara, T., et al. (1982). Revised structures of soyasapogenols A, B, and E, oleanene-sapogenols from soybean. *Structures of soyasaponins I, II, and III. Chemical & pharmaceutical bulletin*, 30(6), 2294-2297.

[56] Ireland, P. A., & Dziedzic, S. Z. (1986). Effect of hydrolysis on sapogenin release in soy. *Journal of Agricultural and Food Chemistry*, 34(6), 1037-1041.

[57] Shimoyamada, M., Kudo, S., Okubo, K., Yamauchi, F., & Harada, K. (1990). Distributions of saponin constituents in some varieties of soybean [Glycine max] plant. *Agricultural and Biological Chemistry*, 54(1), 77.

[58] Potter, B. V. L. (1990). Recent advances in the chemistry and biochemistry of inositol phosphates of biological interest. *Natural Product Reports*, 7(1), 1-24.

[59] Sodipo, O. A., & Arinze, H. U. (1985). Saponin content of some Nigerian foods. *Journal of the Science of Food and Agriculture*, 36(5), 407-408.

[60] Kerem, G. F., Makkar, Z. S. H. P., & Klaus, B. (2002). The biological action of saponins in animal systems: a review. *British Journal of Nutrition*, 88, 587-605.

[61] Szakiel, A., Pączkowski, C., & Henry, M. (2011). Influence of environmental abiotic factors on the content of saponins in plants. *Phytochemistry Reviews*, 10(4), 471-491.

[62] De Geyter, E., Swevers, L., Soin, T., Geelen, D., & Smagghe, G. (2012). Saponins do not affect the ecdysteroid receptor complex but cause membrane permeation in insect culture cell lines. *Journal of Insect Physiology*, 58(1), 18-23.

[63] Committee NRCFP. (1973). Toxicants occurring naturally in foods. *National Academy of Sciences*.

[64] Kenji, O., Matsuda, H., Murakami, T., Katayama, S., Ohgitani, T., & Yoshikawa, M. (2000). Adjuvant and haemolytic activities of 47 saponins derived from medicinal and food plants. *Biological Chemistry*, 381(1), 67-74.

[65] Augustin, J. M., Kuzina, V., Andersen, S. B., & Bak, S. (2011). Molecular activities, biosynthesis and evolution of triterpenoid saponins. *Phytochemistry*, 72(6), 435-457.

[66] Armah, C. N., Mackie, A. R., Roy, C., Price, K., Osbourn, A. E., Bowyer, P., et al. (1999). The membrane-permeabilizing effect of avenacin A-1 involves the reorganization of bilayer cholesterol. *Biophys J*, 76(1), 281-290.

[67] Keukens, E. A. J., de Vrije, T., van den Boom, C., de Waard, P., Plasman, H. H., Thiel, F., et al. (1995). Molecular basis of glycoalkaloid induced membrane disruption. *Biochimica et Biophysica Acta (BBA)- Biomembranes*, 1240(2), 216-228.

[68] Lin, F., & Wang, R. (2010). Hemolytic mechanism of dioscin proposed by molecular dynamics simulations. *Journal of Molecular Modeling*, 16(1), 107-118.

[69] Rao, A. V., & Sung, M. K. (1995). Saponins as anticarcinogens. *Journal Nutrition*, 125(3), 717s.

[70] Sung, M. K., Kendall, C. W. C., & Rao, A. V. (1995). Effect of soybean saponins and gypsophila saponin on morphology of colon carcinoma cells in culture. *Food and Chemical Toxicology*, 33(5), 357-366.

[71] Amarowicz, R., & Shimoyamada, M. K. O. (1994). Hypocholesterolemic effects of saponins. *Roczniki Państwowego Zakładu Higieny*, 45(1-2), 125.

[72] Rao, A. V., & Gurfinkel, D. M. (2000). Dietary saponins and human health. *In:Saponins in Food, Feedstuffs and Medicinal Plants: Kluwer Academic Publishers*, 291.

[73] Antinutrients and Phytochemicals in Food:. (1997). *American Chemical Society*, 348.

[74] Berhow, M. A., Wagner, E. D., Vaughn, S. F., & Plewa, M. J. (2000). Characterization and antimutagenic activity of soybean saponins. *Mutation Research/Fundamental and Molecular Mechanisms of Mutagenesis*, 448(1), 11-22.

[75] Shewry, P. R., & Casey, R. (1999). Seed Proteins:. *Kluwer Academic*.

[76] Birk, Y. (1968). Chemistry and nutritional significance of proteinase inhibitors from plant sources. *Annals of the New York Academy of Sciences*, 146(2), 388-399.

[77] Savage, G. P., Morrison, S. C., & Benjamin, C. (2003). TRYPSIN INHIBITORS. *Encyclopedia of Food Sciences and Nutrition. Oxford: Academic Press*, 5878-5884.

[78] Valdebouze, P., Bergeron, E., Gaborit, T., & Delort-Laval, J. (1980). Content and distribution of trypsin inhibitors and hemagglutinins in some legume seeds. *Canadian Journal of Plant Science*, 60(2), 695-701.

[79] De Meester, P., Brick, P., Lloyd, L. F., Blow, D. M., & Onesti, S. (1998). Structure of the Kunitz type soybean trypsin inhibitor (STI): implication for the interactions between members of the STI family and tissue-plasminogen activator. *Acta Crystallogr, Sect,* 54, 589-97.

[80] Sathe, S. K. (2002). Dry Bean Protein Functionality. *Critical Reviews in Biotechnology,* 22(2), 175-223.

[81] Voss, R. H., Ermler, U., Essen, L. O., Wenzl, G., Kim, Y. M., & Flecker, P. (1996). Crystal structure of the bifunctional soybean Bowman-Birk inhibitor at 0.28-nm resolution. Structural peculiarities in a folded protein conformation. *EurJBiochem,* 242, 122.

[82] Liener, I. E., & Kakade, M. L. (1980). Protease inhibitors. *In:Toxic constituents of plant foodstuffs: Academic Press.*

[83] N., G. (1991). Proteinase inhibitors. *In:Toxic Substances in Crop Plants: Royal Society of Chemistry.*

[84] L., I. E. (1989). Antinutritional factors. *In: Legumes: Chemistry, Technology, and Human Nutrition: M. Dekker.*

[85] Hedemann, M. S., Welham, T., Boisen, S., Canibe, N., Bilham, L., & Domoney, C. (1999). Studies on the biological responses of rats to seed trypsin inhibitors using near-isogenic lines of Pisum sativum L (pea). *Journal of the Science of Food and Agriculture,* 79(12), 1647-53.

[86] Carbonaro, M., Grant, G., Cappelloni, M., & Pusztai, A. (2000). Perspectives into Factors Limiting in Vivo Digestion of Legume Proteins: Antinutritional Compounds or Storage Proteins? *Journal of Agricultural and Food Chemistry,* 48(3), 742-749.

[87] Friedman, M., & Brandon, D. L. (2001). Nutritional and Health Benefits of Soy Proteins†. *Journal of Agricultural and Food Chemistry,* 49(3), 1069-1086.

[88] P., A., G., G., B., S., & A., M.-CM. (2004). The mode of action of ANFs on the gastointestinal tract and its microflora. *In: Recent Advances of Research in Antinutritional Factors in Legume Seeds and Oilseeds: Proceedings of the Fourth International Workshop on Antinutritional Factors in Legume Seeds and Oilseeds, Toledo, Spain, 8-10March: Wageningen Academic Publishers.*

[89] Sastry, M. C. S., & Murray, D. R. (1987). The contribution of trypsin inhibitors to the nutritional value of chick pea seed protein. *Journal of the Science of Food and Agriculture,* 40(3), 253-261.

[90] Troll, W., & Frenkel, K. R. W. (1987). Protease inhibitors: possible preventive agents of various types of cancer and their mechanisms of action. *Prog Clin Biol Res,* 239, 297-315.

[91] Tamir, S., Bell, J., Finlay, T. H., Sakal, E., Smirnoff, P., Gaur, S., et al. (1996). Isolation, characterization, and properties of a trypsin-chymotrypsin inhibitor from amaranth seeds. *J Protein Chem,* 15(2), 219.

[92] Kennedy, A. R. (1998). The Bowman-Birk inhibitor from soybeans as an anticarcino-genic agent. *The American Journal of Clinical Nutrition*, 68(6), 1406S-1412S.

[93] Klejdus, B., Mikelová, R., Adam, V., Zehnálek, J., Vacek, J., Kizek, R., et al. (2004). Liquid chromatographic-mass spectrometric determination of genistin and daidzin in soybean food samples after accelerated solvent extraction with modified content of extraction cell. *Analytica Chimica Acta*, 517(1-2), 1-11.

[94] Peñalvo, J. L., Nurmi, T., & Adlercreutz, H. (2004). A simplified HPLC method for total isoflavones in soy products. *Food Chemistry*, 87(2), 297-305.

[95] Luthria, D. L., Biswas, R., & Natarajan, S. (2007). Comparison of extraction solvents and techniques used for the assay of isoflavones from soybean. *Food Chemistry*, 105(1), 325-33.

[96] Rostagno, M. A., Palma, M., & Barroso, C. G. (2004). Pressurized liquid extraction of isoflavones from soybeans. *Analytica Chimica Acta*, 522(2), 169-177.

[97] Kim, J. S., Kim, J. G., & Kim, W. J. (2004). Changes in Isoflavone and Oligosaccharides of Soybeans during Germination. *Korean Journal of Food Science and Technology*, 36(2), 294.

[98] Koes, R., Verweij, W., & Quattrocchio, F. (2005). Flavonoids: a colorful model for the regulation and evolution of biochemical pathways. *Trends in Plant Science*, 10(5), 236-242.

[99] Anderson, J. J. B. G. C. S. (1997). Phytoestrogens and Human Function. *Nutrition Today*, 32(6), 232-239.

[100] Messina, M. (1999). Soy, soy phytoestrogens (isoflavones), and breast cancer. *The American Journal of Clinical Nutrition*, 70(4), 574-5.

[101] Lis, H. , & N., S. (1998). Lectins:Carbohydrate-Specific Proteins That Mediate Cellular Recognition. *Chem Rev*, 98(2), 637.

[102] Chrispeels, M. J., & Raikhel, N. V. (1991). Lectins, Lectin Genes, and Their Role in Plant Defense. *The Plant Cell*, 3, 1.

[103] Vasconcelos, I. M., & Oliveira, J. T. A. (2004). Antinutritional properties of plant lectins. *Toxicon*, 44(4), 385-403.

[104] J., P. W., & J., V. D. E. (1995). Lectins as plant defense proteins. *Plant Physiol*, 109(2), 347.

[105] Damme, E. J. M. V., Peumans, W. J., Barre, A., & Rougé, P. (1998). Plant Lectins: A Composite of Several Distinct Families of Structurally and Evolutionary Related Proteins with Diverse Biological Roles. *Critical Reviews in Plant Sciences*, 17(6), 575-692.

[106] Goldstein, I. J., & Poretz, R. D. (2007). Isolation, physicochemical characterization, and carbohydrate-binding specificity of lectins. *In: Lectins: Springer*.

[107] Harper, S. M., Crenshaw, R. W., Mullins, M. A., & Privalle, L. S. (1995). Lectin Bind-
       ing to Insect Brush Border Membranes. *Journal of Economic Entomology*, 88(5),
       1197-202.

[108] Duranti, M., & Gius, C. (1997). Legume seeds: protein content and nutritional value.
       Field Crops Research, 53(1-3), 31-45.

Chapter 6

Soybean, Nutrition and Health

Sherif M. Hassan

Additional information is available at the end of the chapter

1. Introduction

Soybean (*Glycine max* L.) is a species of legume native to East Asia, widely grown for its edible bean which has several uses. This chapter will focuses on soybean nutrition and soy food products, and describe the main bioactive compounds in the soybean and their effects on human and animal health.

2. Soybean and nutrition

Soybean is recognized as an oil seed containing several useful nutrients including protein, carbohydrate, vitamins, and minerals. Dry soybean contain 36% protein, 19% oil, 35% carbohydrate (17% of which dietary fiber), 5% minerals and several other components including vitamins [1]. Tables 1 and 2 show the different nutrients content of soybean and its by-products [2]

Soybean protein is one of the least expensive sources of dietary protein [3]. Soybean protein is considered to be a good substituent for animal protein [4], and their nutritional profile except sulfur amino acids (methionine and cysteine) is almost similar to that of animal protein because soybean proteins contain most of the essential amino acids required for animal and human nutrition. Researches on rats indicated that the biological value of soy protein is similar to many animal proteins such as casein if enriched with the sulfur-containing amino acid methionine [5]. According to the standard for measuring protein quality, Protein Digestibility Corrected Amino Acid Score, soybean protein has a biological value of 74, whole soybeans 96, soybean milk 91, and eggs 97[6]. Soybeans contain two small storage proteins known as glycinin and beta-conglycinin.

Nutrient	Soybean				
	Flour	Protein concentrate	Seed heat processed	Meal solvent extracted	Seed without hulls, meal solvent extracted
Protein%	13.3	84.1	37.0	44.0	48.5
Fat%	1.6	0.4	18.0	0.8	1.0
Linoleic acid %	-	-	8.46	0.40	0.40
Crude fiber%	33.0	0.2	5.5	7.0	3.9
Calcium%	0.37	0.0	0.25	0.29	0.27
Total phosphorus%	0.19	080	058	065	0.62
Non phytate phosphorus %	-	0.32	-	0.27	0.22
Potassium %	1.50	0.18	1.61	2.00	1.98
Chlorine%	0.02	0.02	0.03	0.05	0.05
Iron (mg/kg)	-	130	80	120	170
Magnesium %	0.12	0.01	0.28	0.27	0.30
Manganese(mg/kg)	29	1	30	29	43
Sodium %	0.25	0.07	0.03	0.01	0.02
Sulfur %	0.06	0.71	0.22	0.43	0.44
Copper (mg/kg)	-	7	16	22	15
Selenium (mg/kg)	-	0.10	0.11	0.10	0.10
Zinc (mg/kg)	-	23	25	40	55
Biotein(mg/kg)	0.22	0.3	0.27	0.32	0.32
Choline(mg/kg)	640	2	2.860	2794	2731
Folacin (mg/kg)	0.30	2.5	4.2	1.3	1.3
Niacin(mg/kg)	24	6	22	29	22
Pantothenic acid (mg/kg)	13.0	4.2	11.0	16.0	15.0
Pyridoxine(mg/kg)	2.2	5.4	10.8	6.0	5.0
Riboflavin(mg/kg)	3.5	1.2	2.6	2.9	2.9
Thiamin(mg/kg)	2.2	0.2	11.0	4.5	3.2
Vitamin B12 (µg/kg)	-	-	-	-	-
Vitamin E (mg/kg)	-	-	40	2	3

Table 1. Shows composition of soybean and some soybean by-product.

Nutrient	Soybean				
	flour	Protein concentrate	Seed heat processed	Meal solvent extracted	Seed without hulls, meal solvent extracted
Arginine%	0.94	6.70	2.59	3.14	3.48
Glycine %	0.40	3.30	1.55	1.90	2.05
Serine%	-	5.30	1.87	2.29	2.48
Histidine%	0.18	2.10	0.99	1.17	1.28
Isoleucine%	0.40	4.60	1.56	1.96	2.12
Leucine %	0.57	6.60	2.75	3.39	3.74
Lysine%	0.48	5.50	2.25	2.69	2.96
Methionine%	0.10	0.81	0.53	0.62	0.67
Cystine%	0.21	0.49	0.54	0.66	0.72
Phenylalanine%	0.37	4.30	1.78	2.16	2.34
Tyrosine%	0.23	3.10	1.34	1.91	1.95
Threonine%	0.30	3.30	1.41	1.72	1.87
Tryptophan%	0.10	0.81	0.51	0.74	0.74
Valine%	0.37	4.40	1.65	2.07	2.22

Table 2. Shows amino acids contain of soybean and some soybean by-product.

On the other hand, Soy vegetable oil is another product of processing the soybean crop used in many industrial applications. Soybean oil contains about 15.65% saturated fatty acids, 22.78% monounsaturated fatty acids, and 57.74% polyunsaturated fatty acids (7% linolenic acid and 54% linoleic acid) [7]. Furthermore, soybeans contain several bioactive compounds such as isoflavones among other, which possess many beneficial effects on animal and human health [8].

Soybean is very important for vegetarians and vegans because of its rich in several beneficial nutrients. In addition, it can be prepared into a different type of fermented and non-fermented soy foods. Asians consume about 20–80 g daily of customary soy foods in many forms including soybean sprouts, toasted soy protein flours, soy milk, tofu and many more. Also fermented soy food products consumed include tempeh, miso, natto, soybean paste and soy sauce among other [9, 10]. This quantity intake of soy foods is equivalent daily to 25 and 100 mg total isoflavones [11] and between 8 and 50 g soy protein [12]. On the other hand, western people consume only about 1–3 g daily soy foods mostly as soy drinks, breakfast cereals, and soy burgers among other processed soy food forms [10].

Soybean is used as the raw material for oil milling, and the residue (soybean meal) can be mainly used as source of protein feedstuff for domestic animals including pig, chicken, cattle, horse, sheep, and fish feed and many prepackaged meals as well [1]. It is widely used as a filler and source of protein in animal diets, including pig, chicken, cattle, horse, sheep, and fish feed [13]. In general, soybean meal is a great source of protein ranged from 44-49%, but methionine is usually the only limiting amino acid and contains some anti-nutritional factors such as trypsin inhibitor and hemaglutinins (lectins) which can be destroyed by heating and fermenting the soybean meal before use. Textured vegetable protein (TVP) is another soybean byproduct has been used for more than 50 years as inexpensively and safely extending ground beef up to 30% for hamburgers or veggie burgers, without reducing its nutritional value and in many poultry and dairy products (soy milk, margarine, soy ice cream, soy yogurt, soy cheese, and soy cream cheese). as well [1, 13, 14, 15]. The total estimates of feed consumed for broilers, turkeys, layers and associated breeders production over the world in 2006 was about 452 million tones [16]. This estimated value is calculated depending on poultry feeds containing about 30% soybean meal on average. Therefore, 136 million tones of soybean meal are used annually in poultry feeds. As a generalization, the numbers shown can be multiplied by 0.3 for an estimate of the needs of soybean meal. Soy-based infant formula (SBIF) is another soybean product that can be used for infants who are allergic to pasteurized cow milk proteins. It is sold in powdered, ready-to-feed, and concentrated liquid forms without side effects on human growth, development, or reproduction [17, 18, 19].

There are several types commercially available of non fermented soy foods, including soy milk, infant formulas, tofu (soybean curd), soy sauce, soybean cake, tempeh, su-jae, and many more. However, fermented foods include soy sauce, fermented bean paste, natto, and tempeh, among others. Fermented soybean paste is native to the East and Southeast Asia countries such as Korea, China, Japan, Indonesia, and Vietnam [20]. Korean soy foods including kochujang (fermented red pepper paste with soybean flour) and long-term fermented soybean pastes (doenjang, chungkukjang, and chungkookjang) are now internationally accepted foods [20]. Furthermore, natto and miso are originally Japanese soy food types of chungkukjang and doenjang, respectively. China also has different fermented soybean products including doubanjiang, douche (sweet noodle sauce), tauchu (yellow soybean paste), and dajiang. Chungkukjang is a short-term fermented soy food similar to Japanese natto, whereas doenjang, kochujang, and kanjang (fermented soy sauce) undergo long term fermentation as do Chinese tauchu and Japanese miso.

In general, this fermentation of soy foods changes the physical and chemical properties of soy food products including the color, flavor and bioactive compounds content. These changes differ according to different production methods such as the conditions of fermentation, the additives, and the organisms used such as bacteria or yeasts during their manufacture. These changes differ as well as whether the soybeans are roasted as in chunjang or aged as in tauchu before being ground. In addition to physicochemical properties, the fermentation of these soybean products changes the bioactive components, such as isoflavonoids and peptides, in ways which may alter their nutritional and health effects.

Also, the nutritional value of cooked soybean depends on the pre-processing and the method of cooking such as boiling, frying, roasting, baking, and many more. The quality and quantity of soybean components is considerably changed by physical and chemical or enzymatic processes during the producing of soy-based foods [21, 22, 23, 24, 25, 26]. Fermentation is a great processing method for improving nutritional and functional properties of soybeans due to the increased content of many bioactive compounds. On the other hand, the conformation of soy protein (glycinin) is easily altered by heat (steaming) and salt [27]. Many large molecules in raw soybean are broken down by enzymatic hydrolysis during fermentation to small molecules, which are responsible for producing new functional properties for the final products. For example, isoflavones, which are mostly present as 6-O-malonylglucoside and β-glycoside conjugates and associated with soybean proteins, are broken down by heat treatment and fermentation [28]. In general, the chemical profiles of various minor components related to health benefits and nutritional quality of products are also affected by fermentation [29]. It is usual to heat-treat legume components to denature the high levels of trypsin inhibitors soybean [30]. The digestibility of some soy foods are as follows: steamed soybeans 65.3%, tofu 92.7%, soy milk 92.6%, and soy protein isolate 93–97% [1].

3. Bioactive compounds of soybean

Many bioactive compounds are isolated from soybean and soy food products including isoflavones, peptides, flavonoids, phytic acid, soy lipids, soy phytoalexins, soyasaponins, lectins, hemagglutinin, soy toxins, and vitamins and more [31]. Flavonoids are low-molecular-weight polyphenolic compounds classified according to their chemical structure into flavonols, flavones, flavanones, isoflavones, catechins, anthocyanidins and chalcones [32]. Typical flavonoids are kaempferol, quercetin and rutin (the common glycoside of quercetin), belonging to the class of flavonols. Isoflavones (soy phytoestrogens) is a subgroup of flavonoids. The major isoflavones in soybean are genistein, daidzein, and glycitein, representing about 50, 40, and 10% of total isoflavone profiles, respectively. Soy isoflavones, daidzein and genistein, are present at high concentrations as a glycoside in many soybeans and soy food products such as miso, tofu, and soy milk. Soybeans contain 0.1 to 5 mg total isoflavones per gram, primarily genistein, daidzein, and glycitein, the three major isoflavonoids found in soybean and soy products [33]. These compounds are naturally present as the β-glucosides genistin, daidzin, and glycitin, representing 50% to 55%, 40% to 45%, and 5% to 10% of the total isoflavone content, respectively depending on the soy products [8]. Formononetin is another form of isoflavone found in soybeans and can be converted in the rumen (in sheep and cow) into a potent phytoestrogen called equol [34].

Recently, there has been increased interest in the potential health benefits of other bioactive polypeptides and proteins from soybean, including lectins (soy lectins are glycoprotein) and lunasin. Lunasin is a novel peptide originally isolated from soybean foods [35]. Lunasin concentration is ranged from 0.1 to 1.3 g/100 g flour [36, 37], and from 3.3 to 16.7 ng/mg seed [38]. Soybean phytosterols usually include four major or types: β-sitosterol, stigmasterol, campesterol, and brassicasterol, all of which make good raw materials for the production of steroid hormones. Triterpenoid saponins in the mature soybean are divided into two groups; group A soy saponins have undesirable astringent taste, and group B soy saponins have the health promoting properties [39, 40]. Group A soy saponins are found only in soybean hypocotyls, while group B soy saponins are widely distributed in legume seeds in both hypocotyls (germ) and cotyledons [39]. Saponin concentrations in soybean seed are ranged from 0.5 to 6.5% [41, 42].

Soybeans also contain isoflavones called genistein and daidzein, which are one source of phytoestrogens in the human diet. Soybeans are a significant source of mammalian lignan precursor secoisolariciresinol containing 13–273 µg/100 g dry weight [43]. Another phytoestrogen in the human diet with estrogen activity is coumestans, which are found in soybean sprouts. Coumestrol, an isoflavone coumarin derivative is the only coumestan in foods [44, 45]. Soybeans and processed soy foods are among the richest foods in total phytoestrogens present primarily in the form of the isoflavones daidzein and genistein [46].

4. Soybean and health

4.1. Beneficial effects of soybean

Recent research of the health effects of soy foods and soybean containing several bioactive compounds received significant attention to support the health improvements or health risks observed clinically or *in vitro* experiments in animal and human.

4.1.1. Effects on cancer

Recent studies suggested that soy food (soy milk) and soybean protein containing flavonoid genistein, Biochanin A, phytoestrogens (isoflavones) consumption is associated with lowered risks for several cancers including breast [11,47,48,49,50,51,52], prostate [53,54], endometrial [52,55], lung [56], colon [57], liver [58], and bladder [59] cancers.

Isoflavones (genistein) use both hormonal and non-hormonal action in the prevention of cancer, the hormonal action of isoflavones has been postulated to be through a number of pathways, which include the ability to inhibit many tyrosine kinases involved in regulation of cell growth, to enhance transformation growth factor-β which inhibits the cell cycle progression, as well as to influence the transcription factors that are involved in the expression of stress response-related genes involved in programmed cell death [60,61]. Other nonhormonal mechanisms by which isoflavones are believed to increase their anticarcinogenic effects are via their anti-oxidant, anti-proliferative, anti-angiogenic and anti-inflammatory properties [62].

On the other hand, soy proteins and peptides showed potential results in preventing the different stages of cancer including initiation, promotion, and progression [63]. They noted that Kunitz trypsin inhibitor (KTI), a protease inhibitor originally isolated from soybean, inhibited carcinogenesis due to its ability to suppress invasion and metastasis of cancer cells. Also, [64] found that soybean lectins and lunasin were able to possess cancer chemopreventive activity *in vitro*, *in vivo* (in human).

Cell culture experiments have demonstrated that a novel soybean seed peptide (lunasin) prevented mammalian cells transformation induced by chemical carcinogens without affecting morphology and proliferation of normal cells [65]. Lunasin purified from defatted soybean flour showed potent activity against human metastatic colon cancer cells. Lunasin caused cytotoxicity in four different human colon cancer cell lines [66]. It has been also demonstrated that lunasin causes a dose-dependent inhibition of the growth of estrogen independent for human breast cancer [67].

4.1.2. Effect on hypercholesterolemia and cardiovascular diseases

Soy food and soybean protein containing isoflavones consumption lowered hypercholesterolemia [68, 69, 70, 71]. Many studies reported that soybean protein consumption lowered incidence of cardiovascular diseases [68]. Soy isoflavone suppress excessive stress-induced hyperactivity of the sympatho-adrenal system and thereby protect the cardiovascular system [72].

Several studies reported a relation between soybean protein consumption and the reduction in cardiovascular risk in laboratory animal's models by reducing plasma cholesterol levels [68, 69]. Reduction in the incidence of hypercholesterolemia and cardiovascular diseases in Asian countries depending on their diets rich in soy protein was reported [73]. Another study found that the substitution of the animal protein with soybean protein resulted in a significantly decrease in plasma cholesterol levels, mainly LDL (low-density lipoprotein) cholesterol [74]. In the same way, [69] showed that after replacing animal protein with soybean protein consumption for hypercholesterolemia persons resulted in a significant decrease of 9.3% of total plasma cholesterol, mainly 12.9% of LDL cholesterol level and 10.5% of triglycerides. The health beneficial effect for replacing animal protein with soy protein consumption showed the most effective in the highest hypercholesterolemic depend on the initial plasma cholesterol levels [70, 71] without or with the lowest effects in normocholesterolemic persons.

Several research attentions have been paid to the high dietary intake of isoflavones because of their potentially beneficial effects associated with a reduction in the risk of developing cardiovascular diseases. On the other hand, other studies conducted out to establish whether soybean protein and/or isoflavones could be responsible for the hypocholesterolemic effects of soybean diets and therefore their beneficial effects on cardiovascular disease. By studying the effect of soy bean protein and isoflavones, [75] reported that these major components of soybean flour (soybean proteins and soybean isoflavones) independently decreased serum cholesterol. Recent study reported that soybean protein containing isoflavones significantly reduced serum total cholesterol, LDL cholesterol, and triacylglycerol and significantly increased HDL (high-density lipoprotein) cholesterol, but the changes were related to the level and duration of intake, and gender and initial serum lipid concentrations of the persons [76].

Some studies have shown that soybean oil effective in lowering the serum cholesterol and LDL levels, and likely can be used as potential hypocholesterolemic agent if used as a dietary fat and ultimately help prevent atherosclerosis and heart diseases [77]. Soybean oil is a rich source of vitamin E, which is essential to protect the body fat from oxidation and to scavenge the free radicals and therefore helps to prevent their potential effect upon chronic diseases such as coronary heart diseases and cancer [78]. The FDA granted the following health claim for soy: "25 grams of soy protein a day, as part of a diet low in saturated fat and cholesterol, may reduce the risk of heart disease [79].

4.1.3. Effect on osteoporosis and menopause

Soy food and soybean isoflavones consumption lowered osteoporosis, improved bone health and other bone health problems [80, 81, 82]. In addition, consumption of soy foods may reduce the risk of osteoporosis and help alleviate hot flashes associated with menopausal symptoms which are major health concerns for women [83].

4.1.4. Hypotensive activity

Soy food kochujang extract consumption lowered hypertension [84]. The angiotensin I converting enzyme inhibitory peptide isolated from soybean hydrolysate and Korean soybean paste enhanced anti-hypertensive activity *in vivo* [85], causing a fall in blood pressure compared with thiazide diuretics or beta-blockers for mild essential hypertension [86].

4.1.5. Effect on insulin secretion and energy metabolism

Increasing the insulin secretion followed by glucose challenge was recorded when male monkeys fed soybean protein and isoflavones [87, 88]. Flavonoid genistein, tyrosine kinase inhibitor, inhibited insulin signaling pathways [60]. Dietary isoflavones induced alteration in energy metabolism in human [89]. They also noted an inhibition of glycolysis and a general shift in energy metabolism from carbohydrate to lipid metabolism due to isoflavone interference.

4.1.6. Effect on blood pressure and endothelial function

Reduction in the blood pressure via renin-angiotensin system activity (one of the most important blood pressure control systems in mammals) was recorded by feeding rats on diet containing commercial purified soybean saponin [90]. They found that soybean saponin inhibited renin activity *in vitro* and that oral administration of soybean saponin at 80 mg/kg of body weight daily to spontaneously hypertensive rats for 8 weeks significantly reduced blood pressure. In addition, [91] studied the effects of dietary intake of soybean protein and isoflavones on cardiovascular disease risk factors in high risk, 61 middle aged men in Scotland. For five weeks, half the men fed diets containing at least 20 g of soybean protein and 80 mg of isoflavones daily. The effects of isoflavones on blood pressure, cholesterol levels, and urinary excretion were measured, and then compared to those of the remaining men who were fed placebo diet containing olive oil. Men that fed soybean in their diet showed significant decrease in both diastolic and systolic blood pressure. In addition, [92] found that feeding soy nut significantly decreased systolic and diastolic pressure in hypertensive postmenopausal women. On the other hand, [93] found no effect of soybean protein with isoflavones on blood pressure in hypertensive persons. Soy protein and soy isoflavones intake improved endothelial function and the flow-induced dilatation in postmenopausal hyper-cholesterolemia women by raising the levels of endothelial nitric oxide synthase (eNOS), a regulator of the cardiovascular function [94, 95, 96, 97]. Furthermore, chronic administration of genistein increased the levels of NOS in spontaneously hypertensive rats [98, 99].

4.1.7. Effects on platelet aggregation and fibrinolytic activity

The effect of genistein, a protein tyrosine kinase inhibitor on platelet aggregation was exhib-ited [100,101]. Nattokinase, a strong fibrinolytic enzyme, in the vegetable cheese natto (a popular soybean fermented Japanese food) showed approximately fourtimes stronger activ-ity than plasmin in the clot lysis assay [102]. However, intraduodenal administration natto-kinase decreased fibrinogen plasma levels in rats [103,104] and in humans [105]. In addition, soybean protein and peptides exhibited anti-fatigue activity helping in performing exercise and delaying fatigue [106], antioxidant [107,108], anti-aging, skin moisturizing, anti-solar, cleansing, and hair-promoting agent [109].

The beneficial effect of Soy isoflavonne (daidzein) on human health extends to the preven-tion of cancer [110], cardiovascular disease [111]. Also, soybean isoflavones (genistein, daid-zein, and their beta glycoside conjugates) showed antitumor [112], estrogenic [113], antifungal activities [114]. Soy isoflavonne (daidzein) stimulates catecholamine synthesis at low concentrations [115]. However, daidzein at high concentrations (1-100 μM) inhibited catecholamine synthesis and secretion induced by stress or emotional excitation. Recent studies recoded an improvement in cognitive function, particularly verbal memory [116] and in frontal lobe function [117] with the use of soy supplements. Glyceollins molecules are also found in the soybean and exhibited an antifungal activity against *Aspergillus sojae*, the fungal ferment used to produce soy sauce [118]. They are phytoalexins with an antiestrogen-ic activity [119].

4.2. Harmful effects

Despite the several beneficial effects documented of soybean consumption, there are some controversial effects claimed in recent studies on animal and human health. Soybean con-tains several naturally occurring compounds that are toxic to humans and animals such as the trypsin (a serine protease found in the digestive system) inhibitors, phytic acid, toxic components such as lectins and hemagglutinins, some metalloprotein such as soyatoxin and many more other biological of soyatoxin. Some studies reported high levels of protease or trypsin inhibitors (1-5% of total protein) in legume seeds such as soybean [120]. *In vivo* stud-ies using rat, high levels of exposure to trypsin inhibitors isolated from raw soy flour cause pancreatic cancer whereas moderate levels cause the rat pancreas to be more susceptible to cancer-causing agents. However, the US FDA concluded that low levels of soybean protease (trypsin) inhibitors cause no threat to human health. For human consumption, soybeans must be cooked with "wet" heat to destroy the trypsin inhibitors (serine protease inhibitors). Raw soybeans, including the immature green form, are toxic to humans, swine, chickens, and in fact, all monogastric animals [121]. Tofu intake was associated with worse memory, but tempeh (a fermented soy product) intake was associated with better memory [122]. Isoflavones might increase breast cancer risk in healthy women or worsen the prognosis of breast cancer patients [123].

Soy compounds	Biological properties	Selected references
Genistein, daidzein, lectins and lunasin	Anticancer	[11, 47, 48,49,50, 51,52, 53,54, 55,56,57,58,59,62,63, 65,66,67,110,113]
Isoflavones and oil	Hypercholesterolemia	[68,69,70,71,73,74,75,76,77]
Daidzein and oil	Cardiovascular diseases	[68,72,73,77,78,79,91, 111]
Isoflavones	Osteoporosis and menopause	[80,81,82,83]
Genistein	Hypertensive	[84,85,86,98,99]
Genistein	Insulin secretion and energy metabolism	[60,87,88,89]
Saponin and genistein	Blood pressure and endothelial function	[90,91,92,94,95,96,97,98,99]
Genistein	Plate aggregation and fibrinolyic activity	[100,101,102,103,104,105]
Genistein	Antioxidant	[107,108]
Protein and peptides	Anti aging	[109]
Protein and peptides	Anti-fatigue	[106]
Genistein	Anti-flammation	[62]
Genistein, daidzein and Glyceollins	Antifungal	[114,118]
Genistein, daidzein and Glyceollins	Estrogenic activity	[113,119]
Daidzein	Catecholamine synthesis	[115]
Genistein	Anti-angiogenic	[62]

Table 3. Summarizes some beneficial effects of some soybean compounds on animal and human health

Phytic acid is also criticized for reducing vital minerals due to its chelating effect, especially for diets already low in minerals [124]. Phytic acid present in soybean seeds binds to minerals and metals to form phytate (chelated forms of phytic acid with magnesium, calcium, iron, and zinc). Phytate is not digestible and impermeable molecules through cell membranes for humans or nonruminant animals. In addition, phytic acid prevents the body to use many essential minerals such as magnesium, calcium, iron and especially zinc. Unfermented soy products contain high levels of lectins/hemagglutinins. Hemagglutinin makes red blood cells unable to absorb oxygen. However, the soybean fermentation process deactivates soybean hemagglutinins and makes the amounts of lectins present in soybeans inconsiderable. However, some dried soybean products may still contain a large amount of active or toxic lectins. These lectins are believed to cause allergic in a human body. Recently, a metalloprotein named soyatoxin exhibiting toxicity to mice (LD_{50} 7-8 mg/kg mouse upon intraperitoneal injection) was identified. Regardless of the beneficial effect of genistein, there are some controversies about safety and harmful effect of soybean food supplementation rich in genistein. Some studies reported that genistein is not safe and has harmful effects on human health. Consumption of genistein-rich soy food and supplements during pregnancy has been suggested to raise the risk of infant leukemias [125]. In addition, some researches showing stimulatory effect of genistein on proliferation of some breast cancer cells lines increase the concerning problem about the safety of genistein intake for breast cancer women [126]. Recent study reported that administration 56g soy protein powder daily caused a reduction in serum testosterone up to 4% in four weeks in a test group of 12 healthy males [127]. Finally, allergy to soy is common, and the food is listed with other foods. Only a few reported studies have attempted to confirm allergy to soy by direct challenge with the food under controlled conditions [127]. Table (3) shows several beneficial effects reported of some soybean compounds on animal and human health.

Acknowledgements

This work was supported by a grant from the Deanship of Scientific Research, King Faisal University, Kingdom of Arabia Saudi. The authors have no conflict of interest for the information presented in this review.

Author details

Sherif M. Hassan

College of Agricultural and Food Sciences, King Faisal University, Kingdom of Saudi Arabia

References

[1] Liu K.S. Chemistry and Nutritional Value of Soybean Components. In: Soybean: Chemistry, Technology, and Utilization. New York: Chapman & Hall; 1997.p25-113.

[2] Composition of Feedstuffs Used in Poultry Diets. In: National Research Council (NRC). Washington, DC: National Academy of Sciences; 1994.p61-68.

[3] Derbyshire, E, Wright DJ, Boulter D. Review: Legumin and vicilin, storage proteins of legume seeds. Phytochemistry 1976:15:3.

[4] Sacks FM, Lichtenstein A, Van Horn L, Harris W, Kris-Etherton P, Winston M. "Soy Protein, Isoflavones, and Cardiovascular Health. An American Heart Association Sci-

ence Advisory for Professionals from the Nutrition Committee". Circulation 2006;113(7):1034–1044.

[5]   Hajos G, Gelencsér E, Grant G, Bardocz S, Sakhri M, Duguid TJ, Newman AM, Pusztai A. Effects of Proteolytic Modification and Methionine Enrichment On the Nutritional Value of Soya Albumins For Rats. The Journal of Nutritional Biochemistry 1996;7:481-487.

[6]   FAO/WHO (1989). Protein Quality Evaluation: Report of the Joint FAO/WHO Expert Consultation. Bethesda, MD (USA): Food and Agriculture Organization of the United Nations (Food and Nutrition Paper) 1989:51.

[7]   Wolke RL"Where There's Smoke, There's a Fryer". The Washington Post.2007.

[8]   Young VR. Soy Protein in Relation to Human Protein and Amino Acid Nutrition. Journal of American Dietetic Association 1991;91:828-835.

[9]   Wang H, Murphy, P. Isoflavone content in commercial soybean foods. Journal of Agricultural and Food Chemistry 1994;42(8):1666-1673.

[10] Fournier DB, Erdman JW, Gordon GB. Soy, its components, and cancer prevention: a review of the in vitro, animal, and human data. Cancer Epidemiology, Biomarkers and Prevention 1998;7(11):1055–1065.

[11] Messina M, McCaskill-Stevens W, Lampe JW. Addressing the soy and breast cancer relationship: review, commentary, and workshop proceedings. Journal of the National Cancer Institute 2006;98:1275-1284.

[12] Erdman JrJ, Jadger T, Lampe J, Setchell KDR, Messina M. Not all soy products are created equal: caution needed in interpretation of research results. The Journal of Nutrition 2004;134(5):S1229–S1233.

[13] Riaz, MN. Soy Applications in Food. Boca Raton. Florida: CRC Press;2006.

[14] Hoogenkamp HW. Soy Protein and Formulated Meat Products. Wallingford. Oxon, UK: CABI Publishing; 2005:14.

[15] Joseph GE. Soy Protein Products. AOCS Publishing. 2001.

[16] Leesons S., Summers J. Commercial Poultry Nutrition. Guelph, Ontario, Canada: Nottingham University Press; 2005.

[17] Giampietro PG, Bruno G, Furcolo G, Casati A, Brunetti E, Spadoni GL, Galli E. Soy Protein Formulas in Children: No Hormonal Effects in Long-term Feeding. Journal of Pediatric Endocrinology and Metabolism 2004;17(2):191–196.

[18] Strom BL. "Exposure to Soy-Based Formula in Infancy and Endocrinological and Reproductive Outcomes in Young Adulthood". The Journal of the American Medical Association 2001;286(7):807–814.

[19] Merritt RJ, Jenks BH. Safety of Soy-Based Infant Formulas Containing Isoflavones: The Clinical Evidence. The Journal of Nutrition2004;134(5):1220S–1224S.

[20] Kwon DY, Daily JW, Kim HJ, Park S. Antidiabetic Effects of Fermented Soybean Products on Type 2 Diabetes. Nutrition Research 2010;30:1-13.

[21] Park JS, Lee MY, Kim JS, Lee TS. Compositions of Nitrogen Compound and Amino acid in Soybean Paste (Doenjang) Prepared with Different Microbial Sources. Korean Journal of Food Science and Technology 1994; 26:609-615.

[22] Garcia MC, Torre M, Marina ML, Laborda F. Composition and Characterization of Soybean and Related Products. Critical Reviews in Food Science and Nutrition 1997;37:361-391

[23] Nakajima N, Nozaki N, Ishihara K, Ishikawa A, Tsuji H. Analysis of Isoflavone Content in Tempeh, a Fermented Soybean, and Preparation of a New Isoflavone enriched Tempeh. Journal of Bioscience and Bioengineering 2005;100: 685-687.

[24] Yamabe, S, Kobayashi-Hattori, K, Kaneko, K, Endo, K, Takita, T. Effect of Soybean Varieties on the Content and Composition of Isoflavone in Rice-koji Miso. Food Chemistry 2007;100:369-374.

[25] Jang CH, Park CS, Lim JK, Kim JH, Kwon DY, Kim YS, Shin DH, Kim JS. Metabolism of Isoflavone Derivatives during Manufacturing of Traditional Meju and Doenjang. Food Science and Biotechnology 2008; 17:442-445.

[26] Baek LM, Park LY, Park KS, Lee SH. Effect of Starter Cultures on the Fermentative Characteristics of Cheonggukjang. Korean Journal of Food Science and Technology 2008;.40:400-405.

[27] Kim KS, Kim S, Yang HJ, Kwon DY. Changes of Glycinin Conformation due to pH, Heat and Salt Determined by Differential Scanning Calorimetry and Circular Dichroism. International Journal of Food Science and Technology 2004;39:385-393.

[28] Choi HK, Yoon JH, Kim YS, Kwon DY. Metabolomic Profiling of Cheonggukjang during Fermentation by 1H NMR Spectrometry and Principal Components Analysis. Process Biochemistry 2007;42:263-266

[29] Kim, NY, Song EJ, Kwon DY, Kim HP, Heo MY. Antioxidant and Antigenotoxic Activities of Korean Fermented Soybean. Food and Chemical Toxicology 2008; 46:1184-1189.

[30] Teakle R., Jensen J. Heliothis punctiger. In: Singh R., Moore R. (ed.) Handbook of Insect Rearing. Amsterdam: Elsevier; 1985.p312 – 322.

[31] Davis J, Iqbal MJ, Steinle J, Oitker J, Higginbotham DA, Peterson RG. Soy Protein Influences the Development of the Metabolic Syndrome in Male Obese ZDFxSHHF Rats. Hormone and Metabolic Research 2007;37:316-325.

[32] Rice-Evans C. Flavonoid antioxidants. Current Medicinal Chemistry 2001; 8(7): 797-807.

[33] Park OJ, Surh Y-H. Chemopreventive potential of epigallocatechin gallate and genistein: evidence from epidemiological and laboratory studies. Toxicology Letters 2004; 150(1):43-56.

[34] Tolleson WH, Doerge DR, Churchwell MI, Marques MM, Roberts DW. Metabolism of Biochanin A and Formononetin by Human Liver Microsomes in Vitro. Journal of Agricultural and Food Chemistry 2002;50(17):4783-4790.

[35] Galvez AF, Revilleza MJR, de Lumen BO. A novel methionine-rich protein from soybean cotyledon: cloning and characterization of cDNA (accession No. AF005030). Plant Register #PGR97-103. Plant Physiology 1997;114:1567-1569.

[36] de Mejia EG, Vasconez M, de Lumen BO, Nelson R. Lunasin concentration in different soybean genotypes, commercial soy protein, and isoflavone products. Journal of Agricultural and Food Chemistry 2004;52(19):5882-5887.

[37] Wang W, Dia VP, Vasconez M, Nelson RL, de Mejia EG. Analysis of soybean protein-derived peptides and the effect of cultivar, environmental conditions, and processing on lunasin concentration in soybean and soy products. Journal of AOAC International 2008;91(4):936-946.

[38] de Mejia EG., Dia VP. Chemistry and biological properties of soybean peptides and proteins. In: Cadwallader K. et al. (ed.) Chemistry, Texture and Flavor of Soy. USA:American Chemical Society; 2010b.p133-154. Available from http://pubs.acs.org.

[39] Shiraiwa M, Harada K, Okubo K. Composition and structure of group B saponin in soybean seed. Agricultural and biological chemistry 1991;55:911-917.

[40] Kuduo S, Tonomura M, Tsukamoto C. Isolation and structure elucidation of DDMP-conjugated soyasaponins as genuine saponins from soybean seeds. Bioscience, Biotechnology, and Biochemistry. 1993;57:546-550.

[41] Ireland, PA, Dziedzic SZ, Kearsley MW. Saponin content of soya and some commercial soya products by means of high-performance liquid chromatography of the sapogenins. Journal of the Science of Food and Agriculture 1986;37:694-698.

[42] Berhow MA, Kong SB, Vermillion KE, Duval SM. Complete quantification of group A and group B soyasaponins in soybeans. Journal of Agricultural and Food Chemistry 2006;54:2035-2044.

[43] Adlercreutz H, Mazur W, Bartels P, Elomaa V, Watanabe S, Wähälä K, Landström M, Lundin E, Bergh A, Damber JE, Aman P, Widmark A, Johansson A, Zhang JX, Hallmans G. "Phytoestrogens and Prostate Disease". The Journal of Nutrition 2000;130(3): 658S-659S.

[44] De Kleijn MJ, Van Der Schouw YT, Wilson PW, Grobbee DE, Jacques PF. Dietary Intake of Phytoestrogens is Associated With a Favorable Metabolic Cardiovascular Risk Profile in Postmenopausal U.S. Women: The Framingham Study. The Journal of Nutrition 2002;132(2):276-282.

[45] Valsta LM, Kilkkinen A, Mazur W, Nurmi T, Lampi A-M, Ovaskainen M-L, Korho-nen T, Adlercreutz H, Pietinen P. Phyto-oestrogen Database of Foods and Average Intake in Finland. British Journal of Nutrition 2003;89 (5):S31–S38.

[46] Thompson, Lilian U.; Boucher, Beatrice A.; Liu, Zhen; Cotterchio, Michelle; Kreiger N. Phytoestrogen Content of Foods Consumed in Canada, Including Isoflavones, Li-gnans, and Coumestan". Nutrition and Cancer 2006;54 (2): 184–201.

[47] Peterson G, Barnes S Genistein and Biochanin A. Inhibit the Growth of Human Pros-tate Cancer Cells but not Epidermal Growth Factor Receptor Autophosphorylation. Prostate 1993;22:335-345.

[48] Wu AH, Ziegler RG, Horn-Ross PL, Nomura AMY, West DW, Kolonel LN, Rosenthal JF, Hoover RN, Pike MC. Tofu and risk of breast cancer in Asian-Americans. Cancer Epidemiological Biomarkers Preview 1996;5(11):901–906.

[49] Wu AH, Ziegler RG, Nomura AM, West DW, Kolonel LN, Horn-Ross PL, Hoover RN, Pike MC. Soy intake and risk of breast cancer in Asians and Asian Americans. The American Journal of Clinical Nutrition 1998;68:1437S-1443S.

[50] Zheng W, Dai Q, Custer LJ, Shu XO, Wen WQ, Jin F, Franke AA. Urinary excretion of isoflavonoids and the risk of breast cancer. Cancer Epidemiology Biomarkers & Pre-vention 1999;8(1):35-40.

[51] Boyapati SM, Shu XO, Ruan ZX, Dai Q, Cai Q, Gao YT, Zheng W. Soyfood intake and breast cancer survival: a followup of the Shanghai Breast Cancer Study. Breast Can-cer Res Treat 2005;92:11-17.

[52] Lof M, Weiderpass E. Epidemiological evidence suggests that dietary phytoestrogens intake is associated with reduced risk of breast, endometrial and prostate cancers. Nutrition Research 2006; 26(12):609-619.

[53] Peterson G, Barnes S. Genistein Inhibition of the Growth of Human Breast Cancer Cells: Independence from Estrogen Receptors and the Multi-drug Resistance Gene. Biochemical and Biophysical Research Communications 1991; 179: 661-667.

[54] Jacobsen BK, Knutsen SF, Fraser GE. Does high soy milk intake reduce prostate can-cer incidence? The Adventist health study (United States). Cancer Causes Control 1998; 9(6):553-557.

[55] Goodman MT, Wilkens LR, Hankin JH, Lyu LC, Wu AH, Kolonel LN. Association of soy and fiber consumption with the risk of endometrial cancer. American Journal of Epidemiology 1997;146(4):294-306.

[56] Swanson CA, Mao BL, Li JY, Lubin JH, Yao SX, Wang JZ, Cai SK, Hou Y, Luo QS, Blot WJ. Dietary determinants of lung-cancer risk - results from a case-control study in Yunnan province, China. International Journal of Cancer 1992;50(6):876-880.

[57] Azuma N, Machida K, Saeki T, Kanamoto R, Iwami K. Preventive effect of soybean resistant proteins against experimental tumorigenesis in rat colon. Journal of Nutritional Science and Vitaminology 2000;46(1):23–29.

[58] Kanamoto R, Azuma N, Miyamoto T, Saeki T, Tsuchihashi Y, Iwami K. Soybean resistant proteins interrupt an enterohepatic circulation of bile acids and suppress liver tumorigenesis induced by azoxymethane and dietary deoxycholate in rats. Bioscience, Biotechnology, and Biochemistry 2001; 65(4):999–1002.

[59] Sun CL, Yuan JM, Arakawa K, Low SH, Lee HP, Yu MC. Dietary soy and increased risk of bladder cancer: The Singapore Chinese health study. Cancer Epidemiological Biomarkers Preview 2002;11(12):1674-1677.

[60] Akiyama T, Ishida J, Nakagawa S, Ogawara H, Watanabe S, Itoh N, Shibuya M, Fukami Y. Genistein a specific inhibitor of tyrosine-specific protein kinases. The Journal of Biological Chemistry 1987;262:5592-5595.

[61] Zhou Y, Lee AS. Mechanism for the suppression of the mammalian stress response by genistein, an anticancer phytoestrogen from soy. Journal of National Cancer Institute 1998:9(5):381–388.

[62] Gilani G.S., Anderson J.J.B., editor. Phytoestrogens and Health. IL, USA: AOCS Press; 2002.

[63] de Mejia EG, Dia VP. The role of nutraceutical proteins and peptides in apoptosis, angiogenesis, and metastasis of cancer cells. Cancer Metastasis Review 2010a; 29(3): 511-528.

[64] de Mejia EG. Bradford T, Hasler C The anticarcinogenic potential of soybean lectin and lunasin. Nutrition Reviews 2003;61(7):239-246.

[65] Galvez A.F, Chen N, Macasieb J, de Lumen BO. Chemopreventive property of a soybean peptide (Lunasin) that binds to deacetylated histones and inhibit acetylation. Cancer Research 2001; 61(20):7473-7478.

[66] Dia VP, and de Mejia EG. Lunasin induces apoptosis and modifies the expression of genes associated with extracellular matrix and cell adhesion in human metastatic colon cancer cells. Molecular Nutrition and Food Research 2011;55(4): 623-634.

[67] Hsieh C-C, Hernandez-Ledesma B, de Lumen BO. Lunasin, a novel seed peptide, sensitizes human breast cancer MDA-MB-231 cells to aspirin-arrested cell cycle and induced-apoptosis. Chemico-Biological Interactions 2010;186(2):127-134.

[68] Carroll KK. Hypercholesterolemia and atherosclerosis: effects of dietary protein. Fed Proc 1982;41:2792-2796.

[69] Anderson JW, Johnstone BM, Cook-Newell ME. Meta-analysis of the effects of soy protein intake on serum lipids. The New England Journal of Medicine 1995;333:276-282.

[70] Crouse JR III, Morgan T, Terry JG, Ellis J, Vitolins M, Burke GL. A randomized trial comparing the effect of casein with that of soy protein containing varying amounts of isoflavones on plasma concentrations of lipids and lipoproteins. Archives of Internal Medicine 1999;159:2070-2076.

[71] Jenkins DJ, Kendall CW, Jackson CJ, Connelly PW, Parker T, Faulkner D, Vidgen E, Cunnane SC, Leiter LA, Josse RG. Effects of high- and low isoflavone soyfoods on blood lipids, oxidized LDL, homocysteine, and blood pressure in hyperlipidemic men and women. The American Journal of Clinical Nutrition 2002; 76:365-372.

[72] Yanagihara N, Toyohira Y, Shinohara Y. Insights into the pharmacological potentials of estrogens and phytoestrogens on catecholamine signaling. Annals of the New York Academy of Sciences 2008;1129: 96-104.

[73] Keys A. Seven Countries: A Multivariate Analysis of Death and Coronary Heart Disease. Cambridge, Massachusetts:. Harvard University Press;1980.

[74] Descovich GC, Ceredi C, Gaddi A, Benassi MS, Mannino G, Colombo L, Cattin L, Fontana G, Senin U, Mannarino E, Caruzzo C, Bertelli E, Fragiacomo C, Noseda G, Sirtori M, Sirtori CR. Multicentre study of soybean protein diet for outpatient hypercholesterolaemic patients. Lancet 1980;2:709-712.

[75] Potter SM, Baum JA, Teng H, Stillman RJ, Shay NF, Erdman Jr JW. Soy protein and isoflavones: their effects on blood lipids and bone density in postmenopausal women, The American Journal of Clinical Nutrition 1998;68 (Suppl):1375S-1379S.

[76] Zhan S, Ho SC. Meta-analysis of the effects of soy protein containing isoflavones on the lipid profile. The American Journal of Clinical Nutrition 2005;81:397-408.

[77] Kummerow FA, Mahfouz MM, Zhou Q. Trans fatty acids in partially hydrogenated soybean oil inhibit prostacyclin release by endothelial cells in presence of high level of linoleic acid, Prostaglandins & other Lipid Mediators 2007;84(3-4):138–153.

[78] Lu C, Liu Y. Interaction of lipoic acid radical cations with vitamins C and E analogue and hydroxycinnamic acid derivatives. Archives of Biochemistry and Biophysics 2002;406:78–84.

[79] Henkel J. Soy:Health Claims for Soy Protein, Question About Other Components. FDA Consumer (Food and Drug Administration) 2000;34 (3): 18–20.

[80] Toda, T, Uesugi, T, Hirai, K, Nukaya, H, Tsuji, K, Ishida, H. New 6-O-acyl isoflavone Glycosides from Soybeans Fermented with Bacillus subtilis (natto). I. 6-OSuccinylated Isoflavone Glycosides and Their Preventive Effects on Bone Loss in Ovariectomized Rats Fed a Calcium-Deficient Diet. Biological & Pharmaceutical Bulletin 1999a; 22(11):1193-1201.

[81] Toda T, Uesugi T, Hirai K, Nukaya H, Tsuji K, Ishida H. New 6-O-acyl isoflavone glycosides from soybeans fermented with Bacillus subtilis (natto). I. 6-Osuccinylated isoflavone glycosides and their preventive effects on bone loss in ovariectomized rats fed a calcium-deficient diet, Biological & Pharmaceutical Bulletin 1999b22:1193-1201.

[82] Messina M, Messina V. Soyfoods, soybean isoflavones, and bone health: a brief overview. Journal of Renal Nutrition 2000;10:63-68.

[83] Persky VW, Turyk ME, Wang L, Freels S, Chatterton RJ, Barnes S, Erdman JJ, Sepkovic DW, Bradlow HL, Potter S. Effect of soy protein on endogenous hormones in postmenopausal women. American Society for Clinical Nutrition 2002;75(1):145–153.

[84] Kim SJ, Jung KO, Park KY. Inhibitory Effect of Kochujang Extracts on Chemically Induced Mutagenesis. Journal of Food Science and Nutrition 1999; 4:38-42.

[85] Ahn SW, Kim KM, Yu KW, Noh DO, Suh HJ. Isolation of angiotensin I converting enzyme inhibitory peptide from soybean hydrolysate. Food Science Biotechnology 2000;9(3):378–381.

[86] Pool JL, Smith SG, Nelson EB, Taylor AA, Gomez HJ. Angiotensin converting enzyme inhibitors compared with thiazide diuretics or beta-blockers as monotherapy for treatment of mild essential hypertension. Current Opinion Cardiology 1989;4:ll-15.

[87] Sites CK, Cooper BC, Toth MJ, Gastaldelli A, Arabshahi A, Barnes S. Effect of Daily Supplement of Soy Protein on Body Composition and Insulin Secretion in Postmenopausal Women. Fertility and Sterility 2007; 88:1609-1617.

[88] Wagner JD, Zhang L, Shadoan MK, Kavanagh K, Chen H, Tresnasari K, Kaplan JR, Adams MR. Effects of soy protein and isoflavones on insulin resistance and adiponectin in male monkeys. Metabolism 2008;57:S24-S31.

[89] Solanky KS, Bailey NJ, Beckwith-Hall BM, Bingham S, Davis A, Holmes E, Nicholson JK, Cassidy A. Biofluid 1H NMR-based metabonomic techniques in nutrition research - metabolic effects of dietary isoflavones in humans. The Journal of Nutritional Biochemistry 2005;16:236-244.

[90] Hiwatashi K, Shirakawa H, Hori K, Yoshiki Y, Suzuki N, Hokari M, Komai M, Takahashi S. Reduction of blood pressure by soybean saponins, rennin inhibitors from soybean, in spontaneously hypertensive rats. Bioscience, Biotechnology, and Biochemistry 2010;74:2310-2312.

[91] Sagara M, Kanda T, NJelekera M, Teramoto T, Armitage L, Birt N, Birt C, Yamori Y. Effects of dietary intake of soy protein and isoflavones on cardiovascular disease risk factors in high risk, middle-aged men in Scotland. Journal of the American College of Nutrition 2004; 23:85-91.

[92] Welty, FK, Lee, KS, Lew, NS, Zhou, JR. Effect of soy nuts on blood pressure and lipid levels in hypertensive, prehypertensive, and normotensive postmenopausal women. Archives of Internal Medicine 2007;167:1060-1067.

[93] Teede HJ, Giannopoulos D, Dalais, FS, Hodgson, J, McGrath, BP. Randomised, controlled, cross-over trial of soy protein with isoflavones on blood pressure and arterial function in hypertensive subjects. Journal of the American College of Nutrition 2006;25:533-540.

[94] Cuevas AM, Irribarra VL, Castillo OA, Yanez MD, Germain AM. Isolated soy protein improves endothelial function in postmenopausal hypercholesterolemic women. European Journal of Clinical Nutrition 2003;57:889-894.

[95] Lissin LW, Oka R, Lakshmi S, Cooke JP. Isoflavones improve vascular reactivity in post-menopausal women with hypercholesterolemia. Vascular Medicine 2004;9:26-30.

[96] Colacurci N, Chiantera A, Fornaro F, de NV, Manzella D, Arciello A, Chiantera V, Improta L, Paolisso G. Effects of soy isoflavones on endothelial function in healthy postmenopausal women. Menopause 2005;12:299-307.

[97] Hall WL, Formanuik NL, Harnpanich D, Cheung M, Talbot D, Chowienczyk PJ, Sanders TA. A meal enriched with soy isoflavones increases nitric oxidemediated vasodilation in healthy postmenopausal women. Journal of Nutrition 2008;138:1288-1292.

[98] Vera R, Sanchez M, Galisteo M, Villar IC, Jimenez R, Zarzuelo A, Perez-Vizcaino F, Duarte J. Chronic administration of genistein improves endothelial dysfunction in spontaneously hypertensive rats: involvement of eNOS, caveolin and calmodulin expression and NADPH oxidase activity. Clinical Science 2007;112:183-191.

[99] Si H, Liu, D. Genistein, a soy phytoestrogen, upregulates the expression of human endothelial nitric oxide synthase and lowers blood pressure in spontaneously hypertensive rats. Journal of Nutrition 2008;138:297-304.

[100] Nakashima S, Koike T, Nozawa Y. Genistein, a protein tyrosine kinase inhibitor, inhibits thromboxane A2-mediated human platelet responses. Molecular Pharmacology 1991;39:475-480.

[101] McNicol A. The effects of genistein on platelet function are due to thromboxane receptor antagonism rather than inhibition of tyrosine kinase. Prostaglandins Leukot Essent Fatty Acids 1993;48:379-384.

[102] Fujita M, Nomura K, Hong K, Ito Y, Asada A, Nishimuro S. Purification and characterization of a strong fibrinolytic enzyme (nattokinase) in the vegetable cheese natto, a popular soybean fermented food in Japan. Biochemical and Biophysical Research Communications 1993;197:1340-1347.

[103] Fujita M, Hong K, Ito Y, Fujii R, Kariya K. Nishimuro Shrombolytic effect of nattokinase on a chemically induced thrombosis model in rat. Biological & Pharmaceutical Bulletin 1995;18:1387-1391.

[104] Suzuki Y, Kondo K, Matsumoto Y, Zhao BQ, Otsuguro K, Maeda T, Tsukamoto Y, Urano T, Umemura K. Dietary supplementation of fermented soybean, natto, suppresses intimal thickening and modulates the lysis of mural thrombi after endothelial injury in rat femoral artery. Life Sciences 2003;73:1289-1298.

[105] Hsia CH, Shen MC, Lin JS, Wen YK, Hwang KL, Cham TM, Yang NC. Nattokinase decreases plasma levels of fibrinogen, factor VII, and factor VIII in human subjects. Nutrition Research 2009;29:190-196.

[106] Marquezi ML, Roschel HA, Costa ADS, Sawada LA, Lancha AH. Effect of aspartate and asparagine supplementation on fatigue determinants in intense exercise. International Journal of Sport Nutrition and Exercise Metabolism 2003;13(1):65–67.

[107] Chen HM, Muramoto K, Yamauchi F, Fujimoto K, Nokihara K. Antioxidative properties of histidine-containing peptides designed from peptide fragments found in the digests of a soybean protein. Journal of Agricultural and Food Chemistry 1998; 46(1): 49–53.

[108] Tsai P, Huang P. Effects of isoflavones containing soy protein isolate compared with fish protein on serum lipids and susceptibility of low density lipoprotein and liver lipids to in vitro oxidation in hamsters. The Journal of Nutritional Biochemistry 1999;10(11):631–637.

[109] Sudel KM, Venzke K, Mielke H, Breitenbach U, Mundt C, Jaspers S, Koop U, Sauermann K, Knussman-Hartig E, Moll I, Gercken G, Young AR, Stab F, Wenck H, Gallinat S. Novel aspects of intrinsic and extrinsic aging of human skin: beneficial effects of soy extract. Photochemistry and Photobiology 2005;81(3), 581–587.

[110] Birt DF, Hendrich S, Wang W. Dietary agents in cancer prevention: flavonoids and isoflavonoids. Pharmacology & therapeutics 2001;.90(2-3):157-177.

[111] Rimbach G, Boesch-Saadatmandi C, Frank J, Fuchs D, Wenzel U, Daniel H, Hall WL, Weinberg PD. Dietary isoflavones in the prevention of cardiovascular disease – A molecular perspective. Food and Chemical Toxicology 2008;46(4):1308–1319.

[112] Coward L, Barnes N, Setchell K, Barnes S. Genistein, daidzein, and their beta glycoside conjugates: antitumor isoflavones in soybean food from American and Asian diets. Journal of Agricultural Food Chemistry 1993; 41(11):1961-1967.

[113] Doerge DR, Sheehan DM. Goitrogenic and Estrogenic Activity of Soy Isoflavones, Environmental Health Perspectives Supplements 2002;110(3):349-353.

[114] Weidenboerner, M, Hindorf, H, Jha, HC, Tsotsonos, P, Egge, H. Antifungal Activity of Isoflavonoids in Different Reduced Stages on Rhizoctonia solani and Sclerotium rolfsii. Phytochemistry 1990;29(3):801-803.

[115] Liu M, Yanagihara N, Toyohira Y, Tsutsui M, Ueno, S, Shinohara Y. Dual effects of daidzein, a soy isoflavone, on catecholamine synthesis and secretion in cultured bovine adrenal medullary cells. Endocrinol 2007;148:5348-5354.

[116] Kritz-Silverstein D, Von Mühlen D, Barrett-Connor E, Bressel MA. Isoflavones and Cognitive Function in Older Women: The Soy and Postmenopausal Health in Aging (SOPHIA) Study. Menopause 2003;10(3):196–202.

[117] File SE, Hartley DE, Elsabagh S, Duffy R, Wiseman H. Cognitive Improvement After 6 Weeks of Soy Supplements in Postmenopausal Women is Limited to Frontal Lobe Function. Menopause 2005;12(2):193–201.

[118] Kim HJ, Suh H-J, Lee CH, Kim JH, Kang SC, Park S, Kim J-S. Antifungal Activity of Glyceollins Isolated From Soybean Elicited with Aspergillus Sojae. Journal of Agricultural and Food Chemistry 2010;58 (17): 9483–9487.

[119] Tilghman SL, Boué SM, Burow ME. Glyceollins, a Novel Class of Antiestrogenic Phytoalexins. Molecular and Cellular Pharmacology 2010;2(4):155–160.

[120] Sastry M, Murray D. The contribution of trypsin inhibitors to the nutritional value of chickpea seed protein. Journal of Science Food and Agriculture 1987;40: 253 – 261.

[121] Circle SJ, Smith AH. Soybeans: chemistry and technology. Westport, Conn: Avi Pub. Co.; 1972.

[122] Hogervorst E, Sadjimim T, Yesufu A, Kreager P, Rahardjo TB. High Tofu Intake is Associated with Worse Memory in Elderly Indonesian Men and Women". Dementia and Geriatric Cognitive Disorders 2008;26 (1): 50–57.

[123] Messina MJ, Loprinzi CL. Soy for breast cancer survivors: a critical review of the literature. The Journal of Nutrition 2001;131(11):3095–3108.

[124] National Research Council (NRC), Committee on Food Protection, Food and Nutrition Board "Phytates". Toxicants Occurring Naturally in Foods. Washington, DC: National Academy of Sciences; 1973.p363–371.

[125] Hengstler JG, Heimerdinger CK, Schiffer IB, Gebhard S, Sagemuller J, Tanner B, Bolt HM, Oesch F Dietary topoisomerase II-poisons: contribution of soy products to infant leukemia? Experimental and Clinical Sciences International online journal for advances in science Journal 2002;1:8-14.

[126] Lavigne JA, Takahashi Y, Chandramouli GVR, Liu H, Perkins SN, Hursting SD, Wang TTY Concentration-dependent effects of genistein on global gene expression in MCF-7 breast cancer cells: an oligo microarray study. Breast Cancer Research and Treatment 2008;110(1):85-98.

[127] Goodin S, Shen F, Shih WJ, Dave N, Kane MP, Medina P, Lambert GH, Aisner J, Gallo M, DiPaola RS Clinical and Biological Activity of Soy Protein Powder Supplementation in Healthy Male Volunteers. Cancer Epidemiology Biomarkers & Prevention 2007;16 (4): 829–833.

[128] Cantani A, Lucenti P Natural History of Soy Allergy and/or Intolerance in Children, and Clinical Use of Soy-protein Formulas. Pediatric Journal of Allergy and Clinical Immunology 1997;8 (2):59–74.

Chapter 7

Effects of Soybean Trypsin Inhibitor on Hemostasis

Eugene A. Borodin, Igor E. Pamirsky,
Mikhail A. Shtarberg, Vladimir A. Dorovskikh,
Alexander V. Korotkikh, Chie Tarumizu,
Kiyoharu Takamatsu and Shigeru Yamamoto

Additional information is available at the end of the chapter

1. Introduction

Soy bean trypsin inhibitor (SBTI) belongs to the family of serpins – serine protease inhibitors widely distributed in the nature [Silverman G.A. et al., 2004; 21]. Serpins participate in the regulation of proteopytic reactions underling very important physiological and pathological processes such as digestion [16], blood clotting [3, 14, 17], immunity [44] apoptosis [36, 42], inflammation [10], dystrophy [31], carcinogenesis [22, 11] and so on. Despite the apparent importance of the correction of imbalances of the proteolysis in pathology, in fact only the pancreatic trypsin inhibitor (aprotinin) is used more or less widely as a protease inhibitor drug (Trasisol, Contrical, Gordox etc) in the treatment of some diseases [24, 13, 40]. Being the animal protein aprotinin possesses substantial disadvantages [23, 12] and attempts to develop new drug on the base of plant or recombinant proteins or peptidomimetics are undertaken [35, 27]. SBTI is one of the candidates for such a role [34].

Among proteolytic processes hemostasis is of special importance because of the high frequency of it's disturbances, accompanied many pathological processes and diseases [20, 18]. At first sight it looks strange to expect that plant protease inhibitor SBTI should influence hemostasis in humans because blood clotting represents the cascade of proteolytic reactions catalyzed by highly specific proteases [43]. However, results of the very early studies support such possibility [26, 41]. The present study was aimed to investigate the influence of SBTI and aprotinin on hemostasis using modern bioinformatics approach and classical in vitro methods. The concrete purposes of the study included:

1. To establish the extent of structural homology, compare functional activities and evaluate potential targets of SBTI and aprotinin among the human proteases by bioinformatics (*in silico*) methods.

2. To investigate the influence of trypsin, SBTI and aprotinin on some indexes of blood clotting, fibrilolysis and platelet aggregation as well as on the hemolytic activity of the complement *in vitro*.

3. To elucidate the possibility of the regulation of the total proteolytic and tripsin-inhibiting activity of blood plasma *in vivo* by the consumption of soy foods enriched with soy protein isolate (SPI) possessing thermo-stable fraction of SBTI.

2. The study of aprotinin and SBTI *in silico*

In silico methods (bioinformatics) entails the creation and advancement of databases, algorithms, computational techniques to solve formal and practical problems arising from the management and analysis of biological data. One of the potential areas for exploiting these methods is computational drug design. We used *in silico* methods in our study for the comparison of structural homology, functional activities and evaluation of the potential targets of SBTI and aprotinin among the human proteases.

From fifty databases we had browsed in the INTRNET nine contained information on aprotinin and SBTI (Table 1). From these PDB (Protein Data Bank) and UniProt/Swiss-Prot were the most informative and we used these databases in our study.

Database	ID in the Database	
	Aprotinin	SBTI
UniProt-Swiss-Prot http://www.expasy.org	BPT1_BOVIN (P00974)	ITRA_SOYBN (P01070)
Blocks - most highly conserved regions of proteins http://www.ebi.ac.uk	P00974	IPR002160
COG - the database of Clusters of Ortologous Groups of proteins http://www.ncbi.nlm.nih.gov	Precursor P00974; NP_001001554;	-
GTOP - Genomes TO Protein structures and functions http://spock.genes.nig.ac.jp	btau0:ENSBTAG00000017328	?atha0:At1g17860.1
iProClass - an integrated, comprehensive and annotated Protein Classification database http://pir.georgetown.edu	P00974/BPT1_BOVIN; PIRSF001621	-
LIGAND - LIGAND chemical database for enzyme reactions	50059016; 3809839; bta:616039; 100156830; Bt.32343;BPT1_BOVIN	-

Database	ID in the Database	
	Aprotinin	SBTI
http://www.pasteur.fr		
MMDB - Molecular Modeling Data Base http://www.ncbi.nlm.nih.gov	P00974 (Precursor); 1QLQ0	1AVU
PDB - Protein Data Bank http://www.rcsb.org	1OA6	1AVU; 1BA7
MEROPS (peptidases) http://merops.sanger.ac.uk	I02.001	I03.001

Table 1. Databases containing the information on aprotinin and SBTI.

2.1. The homology of the primary structures of SBTI and aprotinin

To investigate the homology of the primary structures of SBTI and aprotinin we used BLAST (Basic Local Alignment Search Tool) algorithm, sequence alignment editor Bio Edit 5.0.9 and got information on these proteins from Protein Data Bank and Uni Prot using FAS-TA (Table 2), MOL and PDB format-files. The extent of homology of the total sequences of SBTI and aprotinin, calculated by the method of the multiple paired alignment, makes up 10% (Figures 1 and 2). The low homology of the total sequences of should be attributed to the different lengths of polypeptide chains these proteins – 58 amino acids in aprotinin [45] and 181 in SBTI [39].

Traditional abbreviation	Ala	Asn	Cys	Asp	Glu	Phe	Gly	His	Ile	Lis	Met
FASTA fromat	A	B	C	D	E	F	G	H	I	K	M
Traditional abbreviation	Asn	Pro	Gln	Arg	Ser	Tre	Val	Trp	Tyr	Glu	*
FASTA fromat	N	P	Q	R	S	T	V	W	Y	Z	-

Table 2. Designation of aminoacids in FASTA format. *- symbol in FASTA format corresponds to the gap of unlimited length

Comparison of the separate domains of SBTI and aprotinin reveals greater sequence similarity. Thus, the extent of homology of the N-terminal region of SBTI (5-60 amino acid residues) and the sequence of aprotonin without last two amino acids makes up 21%. The central region of SBTI (85-116 amino acid residues) and the region of aprotinin within 25-56 amino acids show 24% similarity. The highest extent of homology up to 35% is characteristic for C-terminal region of SBTI (132-181 amino acid residues) and aprotenin sequence without six C-terminal amino acids.

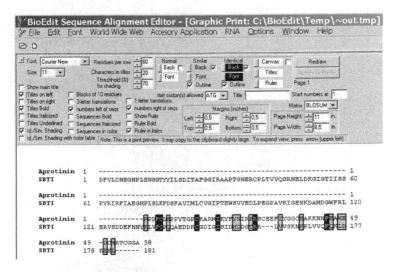

Figure 1. The screen of the Bio Edit 5.0.9 with the results of calculation of homology of Aprotinin and SBTI. The identical amino acids are marked with black and similar with gray. Amino acid sequence are represented in FASTA format.

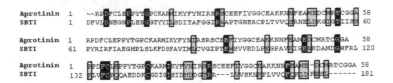

Figure 2. Homological regions of Aprotinin and SBTI (according to Bio Edit 5.0.9). The identical amino acids are marked with black and similar with gray.

2.2. Tertiary structures of SBTI and aprotinin

For the visualization of the 3D-structures of SBTI and aprotinin we used Chem Office 5.0 and Yasara 6.2.5 software. The files in MOL format, containing the information on the sequences of these protease inhibitors, were imputed into Chem Office 5.0 software and converted in PDB format. The obtained PDB files were imputed into Yasara 6.2.5 software. The results of the generation of the electronic 3D-structures of aprotinin and SBTI are presented in Figure 3. Apparently there there is similarity of 3D-structures of these proteins.

We failed to find molecular targets of SBTI and aprotinin because there was no necessary software in free excess.

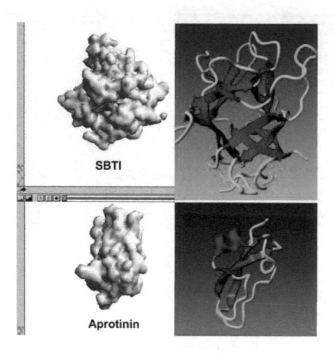

Figure 3. Electronic 3D-structures of SBTI and aprotinin. Left – model with the single surface (Chem Office 5.0). Right-ribbon diagram (Yasara 6.2.5).

2.3. Spectra of biological activity of SBTI and aprotinin

The exploiting the PASS software shows four potential activities of aprotinin and three for SBTI. Both SBTI and aprotinin may inhibit rennin as well as angiotensin- and endothelein-converting enzymes according to the revealing of the possible biological activities of these protease inhibitors by ISIS/Draw 2.4 and PASS Professional chemical structure drawing programs (Figure 4). The possibility of exerting of such effects (drug-likeness) is nearly the same in both protease inhibitors. Aprotinin may inhibit neutral endopeptidase, also. For all the abovementioned activities probability of the activity (P_a) is higher than the probability of the lack of activity (P_i). SBTI and aprotinin should not reveal serious toxicity, mutagenic, carcinogenic and teratogenic effects according to in silico, in vitro and in vivo studies.

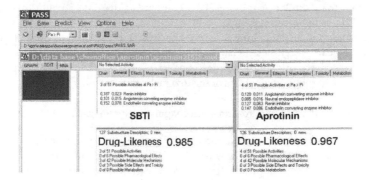

Figure 4. Spectra of the biological activities of aprotinin and SBTI according to PASS software. Drug-Likeness – probability of exerting the effect. Probability of the activity (P_a) and the lack of activity (P_i). The average accuracy of the prognosis makes up more than 85%.

3. The study of aprotinin and SBTI *in vitro*

In silico methods testify the common features in the primary structure, 3D-structures, functional activities of t SBTI and aprotinin and allow to propose that SBTI should influence processes of hemostasis similar to aprotinin. To prove this assumption we studied the influence of trypsin, SBTI and aprotinin on blood clottting, fibrilolysis and platelets aggregation *in vitro*.

3.1. Trypsin-inhibitory activity of SBTI and aprotinin

We measured tyipsin-inhibitory activity of aprotinin, SBTI and low molecular weight protease inhibitors (ε-aminocapronic acid and D-lysine) by their ability to inhibit BAEE-esterase activity of the trypsin [33]. The results are presented in the table 3.

Trypsin inhibitor	Aprotinin	SBTI	ε-aminocapronic acid	D-lysine
Trypsin-inhibitory activity (IU/mg)	141±13	61±0,9	1,36±0,04	1,31±0,02
Trypsin-inhibitory activity (IU/nmole)	0,922±0,085	1,23±0,02	-	-

Table 3. Trypsin-inhibitory activity of aprotinin, SBTI, ε-aminocapronic acid and D-lysine.

Trypsin-inhibitory activity aprotinin, SBTI, ε-aminocapronic acid and D-lysine was measured as their ability to inhibit BAEE-esterase activity of trypsin.

Tyipsin-inhibitory activity, calculated per mg of inhibitor, decrease down in the order: aprotinin > SBTI > aminocapronic acid > D-lysine. Thus, tyipsin-inhibitory activity of aprotinin is

2-times higher than similar activity of SBTI. However, taking into account that molecular weight of SBTI (20100) is 3-times higher than molecular weight of aprotinin (6514), it should be concluded that the trypsine-inhibitory activity of SBTI, calculated per mole of the inhibitor, should be nearly 1,5-times higher than trypsine-inhibitory activity of aptotinin. Trypsine-inhibitory activity of ε-aminocapronic acid and D-lysine when calculated per mole of the inhibitor is negligibly low because of the low molecular weight.

3.2. The influence of SBTI and aprotinin on coagulation hemostasis

The following indexes, characterizing mainly the first and the second phases of blood clotting, were measured: prothrombin time (reflects activity of coagulation factors YII, V, X and II), activated partial thromboplastin time (reflects activity of coagulation factors YII and VIII), activated clotting time (a measure of the anticoagulation affects of heparin) and thrombin time (time of the formation of fibrin clot). Results are summarized in the table 4.

Indexes	Time,sec			
	Plasma (control) (1)	Plasma + trypsin (2)	Plasma + aprotinin (3)	Plasma + SBTI (4)
Prothrombin time	20±0,9	13±0,9 $P_{1-2}<0,0001$	21±0,9 $P_{1-3}"/0,05$	31±0,9 $P_{1-4}<0,0001$
Activated partial thromboplastin time	36±1,7	3±0,9 $P_{1-2}<0,0001$	113±1,8 $P_{1-3}<0,0001$	105±2,7 $P_{1-4}<0,0001$
Thrombine time	16±0,9	2,5±0,5 $P_{1-2}<0,0001$	24±0,9 $P_{1-3}<0,0001$	∞
Ativated clotting time	88±9,7	58±4,7 $P_{1-2}<0.002$	157±12 $P_{1-4}<0.0001$	258±17 $P_{1-4}<0.001$
Euglobulin lysis time	400±34,7	182±7 $P_{1-2}<0,0001$	No lysis of blood clot	No lysis of blood clot

Table 4. The influence of trypsin, SBTI and aprotinin on the indexes of hemostasis *in vitro*. The final concentration of trypsin in the incubation medium consisted of 0,01%, SBTI and aprotinin - 0,1%..

Prothrombin time decreases 33-35% in the presence of trypsin from 20±0,9 to 13±0,9 sec, increases insignificantly in the presence of aprotinin and increases 1,5-times up to 31±0,9 sec in the presence of SBTI. Activated partial thromboplastin time sharply decreases by trypsin from 36±1,7 sec in the control samples to 3,0+0,9 sec and increases 3-times by aprotinin (113±1,8 sec) and SBTI (105±2,7 sec). The changes of the activated clotting time under the influence of trypsin, aprotinin and SBTI are similar. Thrombin time decreases nearly 7-fold in the presence of trypsin from 16±0,9 to 2,5±0.5 sec and increases on 50% by aprotinin to 24±0,9 sec. SBTI completely blocked the formation of fibrin clot at least within 10 min. Similar effects trypsin and it's inhibitors exerts on fibrinolysis. Euglobulin lysis time (factor XII–

callicrein-dependent fibrinolysis) decreases 60% by trypsin from 400±35 to 182±7 sec. Aptotinin and SBTI in the abovementioned concentrations entirely inhibit the lysis of fibrin clot (Table 4).

3.3 The influence of SBTI and aprotinin on platelet aggregation

Results of the study of the influence of SBTI and aprotinin on platelets aggregation are presented in the table 5 and Figures 5-8.

	Plasma (control) (1)	Plasma + aprotinin (2)	Plasma + SBTI (3)	Plasma + trypsin (4)
Reversible (ADP)	22±1,8	14±0,9 $P_{1-2}<0,01$	15±0,9 $P_{1-3}<0,01$	34±0,9 $P_{1-4}<0,01$
T_{MA}	55±4,7	55±4,7	55±4,7	65±15
Two-phase (ADP):				
Fist phase				
MA	48±1,8	35±0,9 $P_{1-2}<0,01$	34±1 $P_{1-3}<0,01$	57±3 $P_{1-4}<0,01$
TMA	85±4,7	68±10,5	70±9	80±10
Second phase				
MA	70±2	56±2 $P_{1-2}<0,02$	58±1 $P_{1-3}<0,05$	99±1 $P_{1-4}<0,02$
TMA	350±9,2	325±9,2	320±9,3	310±9,2
Irreversible (ADP)				
MA	49±0,9	33±0,9 $P_{1-2}<0,02$	29±0,9 $P_{1-3}<0,02$	100 $P_{1-4}<0,001$
T_{MA}	400±18	410±9,3	410±9,3	210±26,4of
Two-phase (adrenalin):				
Fist phase				
MA	23±0,9	16±1,8 $P_{1-2}<0,02$	16±1,8 $P_{1-3}<0,02$	32±0,9 $P_{1-4}<0,01$
TMA	110±9,3	125±13,1	125±13,1	130±9,3
Second phase				
MA	47±2,4	44±2 $P_{1-2}"/0,05$	44±1,8 $P_{1-3}"/0,05$	55±0,9 $P_{1-4}<0,02$
TMA	390±9,3	430±18	430±18	435±16,2

Table 5. The effects of aprotinin, SBTI and trypsin on the aggregation of platelets initiated by ADP (reversible, two-phase and irreversible aggregation) and adrenalin (two-phase aggregation). MA – maximum aggregation (transmittance, %), T_{MA} – time of accomplishment of maximum aggregation (sec).

The effects of trypsin, SBTI and aprotinin on the aggregation of platelets initiated by ADP (reversible, two-phase and irreversible aggregation) and adrenalin (two-phase aggregation) are similar to their effects on blood coagulation and fibrinolysis. Trypsin increases platelet aggregation (the increase of the maximum of all types of aggregation and decrease of maximum aggregation time of irreversible ADP-initiated aggregation). SBTI and aprotinin reveals anti-aggregation effect manifested by the decrease of maximum aggregation.

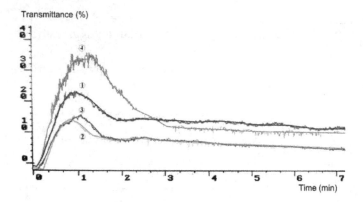

Figure 5. ADP-initiated reversible aggregation of platelets in the presence of aptotinin, SBTI and trypsin in vitro. The platelet aggregation was initiated by the addition of the 5 mkM ADP. 1 – controle; 2 – aprotinin - 1%; 3 – SBTI - 1%; 4 – trypsin – 0,1%.

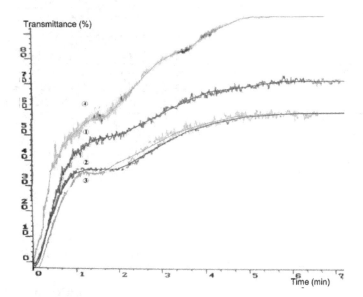

Figure 6. ADP-initiated two-phase aggregation of platelets in the presence of aptotinin, SBTI and trypsin in vitro. The platelet aggregation was initiated by the addition of the 5 mkM ADP. 1 – controle; 2 – aprotinin - 1%; 3 – SBTI - 1%; 4 – trypsin – 0,1%.

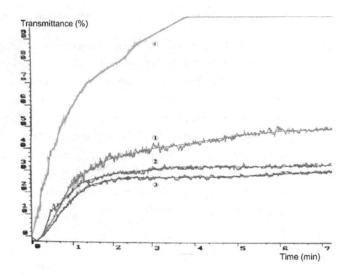

Figure 7. ADP-initiated irreversible aggregation of platelets in the presence of aptotinin, SBTI and trypsin in vitro. The platelet aggregation was initiated by the addition of the 25 mkM ADP. 1 – controle; 2 – aprotinin - 1%; 3 – SBTI - 1%; 4 – trypsin – 0,1%.

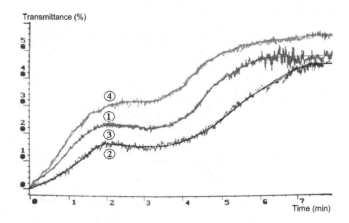

Figure 8. Adrenalin-initiated two-phase aggregation of platelets in the presence of trypsin, aptotinin and SBTI in vitro. The platelet aggregation was initiated by the addition of the 25 mkM adrenalin. 1 – controle; 2 – trypsin - 0,1%; 3 – aprotinin - 1%; 4 – SBTI - 1%.

So, the results of *in vitro* experiments proves the ability of SBTI to influence the hemostatsis predicted by the *in silico* studies.

3.4. The influence of SBTI and aprotinin on the hemolytic activity of the complement

Limited cascade proteolysis underlies the activation of the complement system, which includes nearly 20 individual proteins [30, 37]. Majority of these proteins belongs to the serine protease family (subunits C1r and C1s, C3/C5-cinvertase of the classical way of activation, factors I and D) [15]. While complement system does not have direct relation to hemostasis and targeted on alien cells and microorganisms, in some conditions complement system may play a role in the disruption of intrinsic cells of the organism and among them blood cells [15, 37]. Because of this we investigated the influence of trypsin, SBTI and aprotinin on the hemolytic activity of the complement assuming that SBTI and aprotinin may retard hemolysis of red blood cells by inhibiting activation of some components of the complement system. Results are presented in the table 6 and Figures 9-10.

	Lag-phase (min)	Total time of hemolysis, including lag-phase (min)
Control (1)	3,5±0,5	8,5±1,5
Aprotinin (1%) (2)	3,2±0,15 P_{1-2}"/0,05	7,25±0,25 P_{1-2}"/0,05
SBNI (1%) (3)	3,2±0,15 P_{1-3}"/0,05	7,25±0,25 P_{1-3}"/0,05
Trypsin(0,01%) (4)	4,25±0,25 P_{1-4}<0,02	15,5±0,5 P_{1-4}<0,02

Table 6. The influence of aprotinin, SBTI and trypsin on the hemolytic activity of complement *in vitro*

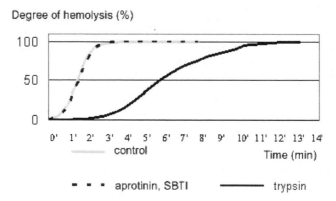

Figure 9. The hemolytic activity of the complement in the presence of aprotinin, SBTI and trypsin in vitro. The lag-phase is not shown. 1 – controle; 2 – aprotinin - 1%; 3 – SBTI - 1%; 4 – trypsin – 0,1%.

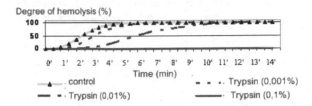

Figure 10. The hemolytic activity of the complement in the presence of different concentrations of trypsin in vitro. The lag-phase is not shown.

SBTI and aprotinin in the final concentrations 0,001-0,1% do not influence hemolysis time and the only statistically insignificant effect consists of small decrease of lag-phase from 3,5±0,3 to 3,2±0,15 min (p>0,05) and total time of hemolysis from 8,5±1,5 to 7,25±0,15 min (p>0,05). Trypsin suppresses the hemolytic activity of the complement in the concentration-dependent manner. In the presence of 0.0001% and 0,001% trypsin the lag-phase of the hemolysis increases to 4,25±0,25 min (p<0,02) and total time of hemolysis to 11,5±0,5 min (p<0,02) and 15,5±1,6 min (p<0,01), respectively. 0,01% trypsin entirely blocks the hemolysis. The inhibiting effect of trypsin should be attributed to the disruption of C3 component of the complement system.

Thus, the results obtained show that aprotinin and SBTI in the abovementiioned concentrations do not influence the hemolytic activity of the complement.

4. The influence of the consumption of soy priotein isolate by healthy persons on the indexes of proteolysis in the blood serum

Soya is cultivated as a foodstuff, having favorable effect on human health, more than 3000 years [28]. Within the last decades soy foods have started to be used in Europe, USA and in Russia as dietotherapy means at a number of diseases [6, 29]. In particular, we have shown the cholesterol-lowering effect of soy protein, resulting from the consumption of SPI enriched cookies by persons with modern hypelipidemia [7, 9]. The consumption of soy foods is followed by the antioxidant effect [Santana et al., 2008, 8, 9] due to the high antioxidant content in these foods [2, 9]. Antioxidant effect is important for the treatment of diseases followed by the development of the oxidative stress [1, 38].

The other possible use of soy foods in dietotherapy may be the correction of disturbances of proteolytic processes. Activation of the tissue proteolysis accompanies different diseases and pathological processes [34]. Soy protein foods contain Bouman-Birk type trypsin inhibitor – SBTI [4]. This protease inhibitor has been officially recognized as a component of foodstuff [25]. There are data showing the possibility of absorbtion of SBTI in the intestin after the consumption of soy protein foods [22]. So, it seams reasonable to assume that the consumption of soy foods will follow anti-proteolytic effect. Because of this it was interesting to elucidate whether prolonged consumption of the cookies enriched with soy protein isolate (SPI) will influence total proteolytic and trypsin-inhibiting activity of the blood plasma in humans or not?

4.1. Tripsin-inhibiting activity of SPI

According to the characteristics, specified by manufacturer, protein content of SPI consists of 90%. From this 10% belongs to protease inhibitors [32]. Within the processing of soy beans into SPI the former are expoused to high temperatures, which inactivates the major part of soy proteins. However, about 5-20 % of SBTI are represented by the thermostable fraction [19, 22]. The measurement of the trypsin-inhibiting activity of SPI shows that tryp-sin-inhibiting activity of SPI consists of 1,4±0,1 IU/mg of SPI. For example 1 mg of pure SBTI possesses activity equal nearby 60 IU. Thus, the active SBTI makes up 2,7% from SPI mass.

4.2. The influence of the consumption of SPI by healthy persons on the total proteolytic and trypsin inhibiting activity of blood serum

30 adult people aged 35-67 years without the expressed signs of chronic diseases consumed cookies, enriched with SPI (30% protein content), for two months [7]. Fasting blood samples were drawn before and after the dietary treatment. Serum samples were frozen and ana-lyzed for total proteolytic (BAEE-esterase activity) and tripsin-inhibiting activity [33]. Daily intake of 100g of cookies corresponds to consumption of 30g of SPI and about 0,85g of the active SBTI (nearly 52 00 IU). The total consumption of SPI for two months consists of 1,8 kg or 50 g of an active SBTI. Twenty-eight participants (19 females and 9 males) could complete the trial. Results are presented in the table 7.

	Total proteolytic activity (relative units)	Trypsin-inhibitory activity (IU/ml)
Before the consumption	0,343±0,010	113±3,6
After the consumption	0,282±0,008 p<0,05	137±5,3 p<0,01

Table 7. Total proteolytic and trypsin-inhibitory activity of blood serum of healthy persons before and after two months consumption of soy protein isolate.

Total proteolytic activity of blood serum was measures as serum BAEE-esterase activity and trypsin-inhibitory activity by inhibiting by serum of BAEE-esterase activity of trypsin.

Two months long SPI consumption was followed by 18% decrease of total proteolytic activi-ty of blood serum from 0,343±0,010 to 0,282±0,008 relative units (p<0,05) and by 21% increase of trypsin-inhibiting activity from 113±3,6 to 137±5,3 IU/ml (p<0,01). The results obtained show possibility of the regulation of proteolysis *in vivo* by the consumption of soy foods.

5. Discussion

The disturbances of the hemostasis accompany many pathological processes and diseases [20, 18]. Both blood clotting and fibrinolysis represents the cascade of proteolytic reactions catalyzed by highly specific proteases [43]. To correct disturbances of the fibrinolysis as well as imbalances of the other proteolytic processes the animal trypsin inhibitor aprotinin is used as a drug (Trasisol, Contrical, Gordox etc) [24, 13, 40]. The effectiveness of the aprotinin in the treatment of some diseases has been questioned recently because of it's substantial disadvantages [23, 12]. Following consultation with the German Federal Institute for Drugs and Medical Devices, the U.S. Food and Drug Administration, Health Canada, and other health authorities, the producer of Trasilol - Bayer announced in 2007 that it has elected to temporarily suspend worldwide marketing of Trasylol® (aprotinin injection) until final results from the Canadian BARTtrial can be compiled, received and evaluated. Information regarding the decision has been posted to Bayer's websites [5]. Because of this attempts to develop new protease inhibitor drug are undertaken [35, 27]. One of the candidates for such a role is plant protease inhibitor - SBTI.

In the present study we exploited the advantages of the modern bioinformatics for establishing the extent of structural homology, comparing functional activities and evaluating potential targets of SBTI and aprotinin among the human proteases. The results of *in silico* study testifies apparent homology of SBTI and aprotinin manifested by the common features in the primary structure, 3D-structures, functional activities of these proteins and allow us to propose that SBTI should influence processes of hemostasis similar to aprotinin. The investigation of the influence of SBTI and aprotinin on coagulation and thrombocyte hemostasis by *in vitro* methods prove this assumption and show that both proteins inhibit blood clotting, fibrinolusis and platelet aggregation which is evident from the increase of prothrombin time, activated partial thromboplastin time, activated clotting time, thrombin time and inhibition of fibrinolysis. We investigated the influence of SBTI and aprotinin on the hemolytic activity of the complement assuming that SBTI and aprotinin may retard hemolysis of red blood cells by inhibiting activation of some components of the complement system. However, both inhibitors do not influence hemolysis time.

While the major part of SBTI is disrupted by heating, nearly 20% of inhibitor are thermostable and remains active in soy foods [19]. Part of the consumed SBTI are absorbed in the intestine after the consumption of soy foods [22]. Because of this we elucidated whether prolonged consumption of the cookies, enriched with SPI, will influence total proteolytic and trypsin-inhibiting activity of the blood plasma in humans or not? Daily intake of 100g of cookies corresponds to consumption of about 0,85g of the active SBTI. The total consumption of SBTI for the study period consists of 50 g of an active inhibitor. The consumption of 100 g of cookies daily for two months was followed by 18% decrease of total proteolytic activity of blood serum and by 21% increase of trypsin-inhibiting activity. The results obtained testify possibility of the soft regulation of proteolysis *in vivo* by the consumption of soy foods.

At first, the ability of SBTI to inhibit blood clotting was shown more than half a centaury ago [26, 41]. However, these results had no any practical consequences and in fact were forgotten for a long time. Our study represent new attempt to revive interest to SBTI as possible protease inhibiting drug.

Author details

Eugene A. Borodin[1*], Igor E. Pamirsky[2], Mikhail A. Shtarberg[1], Vladimir A. Dorovskikh[1], Alexander V. Korotkikh[1], Chie Tarumizu[3], Kiyoharu Takamatsu[4] and Shigeru Yamamoto[3]

*Address all correspondence to: borodin@amur.ru

1 Amur State Medical Academy, Blagoveshchensk, Russian Federation

2 Institute of Geology and Natural Management of the Far Easteren Branch of Russia Academy of Sciences, Russian Federation

3 Asian Nutrition and Food Culture Research Center, Jumonji University, Saitama, Japan

4 Fuji Oil Co., Ltd Osaka, Japan

References

[1] Aivatidi, C., Vourliotakis, G., Georgopoulos, S., Sigala, F., Bastounis, E., & Papalambros, E. (2011, March). Oxidative stress during abdominal aortic aneurysm repair--biomarkers and antioxidant's protective effect: a review. *Eur. Rev. Med. Pharmacol. Sci.*, 15(3), 245-252, 1128-3602.

[2] Anderson, R. L., & Wolf, W. J. (1995, March). Compositional changes in trypsin inhibitors, phytic acid, saponins and isoflavones related to soybean processing. *J Nutr.*, 125(3), 581S-588S, 0022-3166.

[3] Bernstein, J. A. (2008). Hereditary angioedema: a current state-of-the-art review, VIII: current status of emerging therapies. *Ann. Allergy Asthma Immunol* [1, 2], S41-S46, 1081-1206.

[4] Birk, Y. (1985). The Bowman-Birk inhibitor. Trypsin- and chymotrypsin-inhibitor from soybeans. *Int J Pept Protein Res.*, 25(2), 113-31, 0367-8377.

[5] Bayer (2007). Available from http://www.trasylol.com http://www.pharma.bayer.com http://www.bayerscheringpharma.de/trasylol/en http://www.bayerhealthcare.com/trasylol/en

[6] Borodin, E. A., Aksyonova, T. V., & Anishchenko, N. I. (2000). Soy foods. New role. *Vestnik DVO RAN* [3], 72-85.

[7] Borodin, E. A., Menshikova, I. G., Dorovskikh, V. A., Feoktistova, N. A., Shtarberg, M. A., Yamamoto, T., Takamatsu, K., Mori, H., & Yamamoto, S. (2009, December). Ef-

fects of soy protein Isolate and casein on blood lipids and glucose in Russian adults with moderate hyperlipidemia. *J. Nutr. Sci. Vit*, 50(6), 492-497, 0301-4800.

[8] Borodin, E.A., Dorovskikh, V. A., Aksyonova, T. V., & Shtarberg, M. A. (2001, October-December). Lipid composition and the antioxidant properties of soy milk in vitro and in vivo. *Far Eastern Medical Journal (Russian)* [4], 26-30, 1994-5191.

[9] Borodin, E.A., Menshikova, I. G., Dorovskikh, V. A., Feoktistova, N. A., Shtarberg, M. A., Aksenova, T. V., Yamamoto, T., Takamatsu, K., Mori, H, & Yamamoto, S. (2011). Antioxidant and hypocholesterolemic effects of soy foods and cardiovascular disease. *In: Soybean and Health Ed. by Hany A. El-Shemy. INTECH*, 407-424, 978-9-53307-535-8.

[10] Correale, M., Brunetti, N. D., De Gennaro, L., & Di Biase, M. (2008). Acute phase proteins in atherosclerosis (acute coronary syndrome) Cardiovasc. *Hematol. Agents Med. Chem.*, 6(4), 272-277, 1871-5257.

[11] Clemente, A., Moreno, F. J., Marнn-Manzano, Mdel. C., Jimйnez, E., & Domoney, C. (2010). The cytotoxic effect of Bowman-Birk isoinhibitors, IBB1 and IBBD2, from soybean (Glycine max) on HT29 human colorectal cancer cells is related to their intrinsic ability to inhibit serine proteases. *Mol. Nutr. Food. Res*, 54(3), 396-405, 1613-4125.

[12] Deanda, A. Jr, & Spiess, B. D. (2012, July). Aprotinin revisited. *J Thorac Cardiovasc Surg*, 13, 1344-4964.

[13] Dietrich, W. (2009, Feb). Aprotinin: 1 year on. *Curr Opin Anaesthesiol*, 22(1), 121-127, 0952-7907.

[14] Dupont, D. M., Madsen, J. B., Kristensen, T., Bodker, J. S., Blouse, G. E., Wind, T., & Andreasen, P. A. (2009). Biochemical properties of plasminogen activator inhibitor-1. *Front Biosci*, 1(14), 1337-1361, 1093-4715.

[15] Ehrnthaller, C., Ignatius, A., Gebhard, F., & Huber-Lang, M. (2011, March-April). New insights of an old defense system: structure, function, and clinical relevance of the complement system. *Mol Med.*, 17(3-4), 317-329, 1076-1551.

[16] Erickson, R. H., & Kim, Y. S. (1990). Digestion and absorption of dietary protein. *Annu Rev Med*, 41, 133-139, 0066-4219.

[17] Gray, E., Hogwood, J., & Mulloy, B. (2012). The anticoagulant and antithrombotic mechanisms of heparin. *Handb. Exp. Pharmacol.*, 207, 43-61, 0171-2004.

[18] Grabowski, E. F., Yam, K., & Gerace, M. (2012). Evaluation of hemostasis in flowing blood. *Am. J. Hematol.*, 87(1), 51-55, 0361-8609.

[19] Hathcock, J. N. (1991). Residue trypsin inhibitor: data needs for risk assessment. *Adv Exp Med Biol.*, 289, 273-279, 0065-2598.

[20] Hook, K. M., & Abrams, C. S. (2012, February). The loss of homeostasis in hemostasis: new approaches in treating and understanding acute disseminated intravascular coagulation in critically ill patients. *Clin Transl Sci.*, 5(1), 85-92, 1579-2242.

[21] Huntington, J. A. (2011, July). Serpin structure, function and dysfunction. *J Thromb Haemost.*, 9, 26-34, 1538-7933.

[22] Kennedy, A. R. (1998). Bowman-Birk inhibitor from soybeans as anticarcinogenic agent. *J. Clin. Nutr.*, 68(6), 1406-1412, 0002-9165.

[23] Kristeller, J. L., Roslund, B. P., & Stahl, R. F. (2008). Benefits and risks of aprotinin use during cardiac surgery. *Pharmacotherapy.*, 28(1), 112-124, 0277-0008, 1875-9114.

[24] Levy, J. H., & Sypniewski, E. (2004). Aprotinin: a pharmacologic overview. *Orthopedics.*, 27(6), 653-658, 0147-7447.

[25] Losso, J. N. (2008). The biochemical and functional food properties of the bowman-birkinhibitor. *Crit. Rev. Food Sci. Nutr.*, 48(1), 94-118, 1040-8398.

[26] Macfarlane, R.G., & Pilling, J. (1946). Anticoagulant action of soya-bean trypsin inhibitor. *Lancet* [2], 888, 0140-6736.

[27] Mazza, F., Tronconi, E., Valerio, A., Groettrup, M., Kremer, M., Tossi, A., Benedetti, F., Cargnel, A., & Atzori, C. (2006). The non-peptidic HIV protease inhibitor tipranavir and two synthetic peptidomimetics (TS98 and TS102) modulate Pneumocystis carinii growth and proteasome activity of HEL299 cell line. *J. Eukaryot. Microbiol.*, 53(1), S144-S146, 1066-5234.

[28] Messina, M., & Messina, V. (2003, May-June). Provisional Recommended Soy Protein and Isoflavone Intakes for Healthy Adults: Rationale. *Nutr Today*, 38(3), 100-109, 0002-9666X.

[29] Messina, M. (2010, July). A brief historical overview of the past two decades of soy and isoflavone research. *J. Nutr.*, 140(7), 1350S-1354S, 0022-3166.

[30] Mizuno, M., & Matsuo, S. (2010, January). Recent knowledge of complement system and the protective roles. *Nihon Rinsho.*, 68(6), 49-52, 0047-1852.

[31] Morris, C. A., Selsby, J. T., Morris, L. D., Pendrak, K., & Sweeney, H. L. (2010). Bowman-Birkinhibitor attenuates dystrophic pathology in mdx mice. *J. Appl. Physiol*, 109(5), 1492-1499, 8750-7587.

[32] Mosse, D., & Pernolle, D. (1986). *In: Chemisry and Biochemistry of legumes. Moscow, "Agropromizdat", Russian*, 248, 0369-8629.

[33] Nartikova, V. F., & Paskhina, T. S. (1977). Determination of anti-trypsine activity in human blood plasma. *In: Modern Methods in Biochemistry. Ed. By V.N. Orekhovich. Moskow, "Medicine" (Russian)*, 188-191, 0042-8787.

[34] Pamirsky, I., Borodin, E., & Shraberg, M. (2012). Regulation of proteolysis by plant and animal inhibitors. *LAP Lambert Academic Publishing.*, 105, 0000-9783.

[35] Pogue, G. P., Vojdani, F., Palmer, K. E., Hiatt, E., Hume, S., Phelps, J., Long, L., Bohorova, N., Kim, D., Pauly, M., Velasco, J., Whaley, K., Zeitlin, L., Garger, S. J., White, E., Bai, Y., Haydon, H., & Bratcher, B. (2010). Production of pharmaceutical-grade re-

combinant aprotinin and a monoclonal antibody product using plant-based transient expression systems. *Plant Biotechnol J.*, 8(5), 638-654, 1467-7644.

[36] Radović, N., Cucić, S., & Altarac, S. (2008). Molecular aspects of apoptosis. *Acta Med. Croatica.*, 62(3), 249-256, 1330-0164.

[37] Sarma, J. V., & Ward, P. A. (2011, January). The complement system. *Cell Tissue Res.*, 343(1), 227-235, 0030-2766X.

[38] Shi, Y. C., & Pan, T. M. (2012, April). Red mold, diabetes, and oxidative stress: a review. *Appl Microbiol Biotechnol.*, 94(1), 47-55, 0175-7598.

[39] Sigma. (2006). Catalog reagents ang labware. Sigma-Aldrich, Inc. 0013-7227 , 1440.

[40] Smith, M., Kocher, H. M., & Hunt, B. J. (2010, January). Aprotinin in severe acute pancreatitis. *J. Clin. Pract.*, 4(1), 84-92, 1368-5031.

[41] Tagnon, H. J., & Soulier, J. P. (1946). Anticoagulant activity of the trypsin inhibitor from soy bean flour. *Proc. Soc. Exper. Biol. & Med*, 61, 440, 0003-4819.

[42] Tang, M., Asamoto, M., Ogawa, K., Naiki-Ito, A., Sato, S., Takahashi, S., & Shirai, T. (2009). Induction of apoptosis in the LNCaP human prostate carcinoma cell line and prostate adenocarcinomas of SV40T antigen transgenic rats by the Bowman-Birk inhibitor. *Pathol Int.*, 59(11), 790-796, 13205-4630.

[43] Walsh, P. N., & Ahmad, S. S. (2002). Proteases in blood clotting. *Essays Biochem.*, 38, 95-111, 0071-1365.

[44] Wouters, D., Wagenaar-Bos, I., van Ham, M., & Zeerleder, S. (2008). C1 inhibitor: just a serine protease inhibitor? New and old considerations on therapeutic applications of C1 inhibitor. *Expert. Opin. Biol. Ther*, 8(8), 1225-1240, 1744-7682.

[45] Zhirnov, O. P., Klenk, H. D., & Wright, P. F. (2011, July). Aprotinin and similar protease inhibitors as drugs against influenza. *Antiviral Res.*, 92(1), 27-36, 16635429-9999.

Chapter 8

Approach for Dispersing a Hydrophilic Compound as Nanoparticles Into Soybean Oil Using Evaporation Technique

Kenjiro Koga

Additional information is available at the end of the chapter

1. Introduction

Most popular formulation dispersing a hydrophilic compound into oil phase such as soybean oil is emulsions. An emulsion is a dispersed system that consists of water, oil, and surfactant. In general, apparatuses of an emulsifier, a homogenizer, etc. are used for the preparation. As the pharmaceutical trial to disperse water-soluble compounds in an oil phase, the form of the emulsion is very important. Namely, for pharmaceutical preparations containing a hydrophilic drug dispersed uniformly into the oil phase, water-in-oil (w/o) and water-in-oil-in-water (w/o/w) emulsions are preferred. In these cases, hydrophilic drug molecules must retain a high-density in the dispersed water phase of the emulsion; doing so depends on the oil-to-water partition coefficient of the drug. Furthermore, decreasing the particle size in the dispersed water phase is necessary. Much pharmaceutical technical information about adjusting the size of particles is now available: for example, rotating membrane emulsification [1, 2], shirasu porous glass membrane emulsification [3, 4], electrocapillary emulsification [5, 6]. These methods adjust particle size on the basis of membrane pore size and shearing force, which depends on the flow of dispersion medium or on contact-surface dielectric constant differences between the dispersion medium and the dispersion phase. Therefore, these technologies are advantageous in that they can produce uniform particle sizes. In this chapter, a simple method of preparing w/o emulsions with a narrow range of polydispersity is described. In this method, a Polytron homogenizer and an evaporator are used as apparatuses. Namely, specific and expensive apparatuses were not used. Glycyrrhizin monoammonium (GZ) and indocyanine green (ICG) were used as a hydrophilic compound. Here, the phase behavior, stability in terms of particle size of w/o emulsions prepared using the novel method and the sustained release characteristics drug from nano-sized w/o emulsions were investigated [7, 8].

2. Selection of emulsifier for the preparation of stable w/o emulsion

The choice of ideal emulsifier is an important to prepare physicochemically stable w/o emulsion. Furthermore, the emulsifier must be safe in human. Therefore, mainly non-ionic surfactants added in foods or medicines were chosen. The list used in this experiment was shown in Table 1.

No.	Surfactants	Product name	HLB
1	condensed ricinoleic acid tetraglycerin ester	CR-310[1]	2.5
2	polyethyleneglycol distearate ester	CDS-400[2]	8.5
3	hexaglycerin sesquistearate ester	SS-500[1]	10.1
4	tetraglycerin monostearate ester	MS-310[1]	10.2
5	tetraglycerin monooleate ester	MO-310[1]	10.2
6	tetraglycerin monolaurate ester	ML-310[1]	10.3
7	polyethyleneglycol(10EO) monostearate ester	MYS-10[2]	11
8	hexaglycerin monostearate ester	MS-500[1]	12.2
9	hexaglycerin monooleate ester	MO-500[1]	12.2
10	polyethyleneglycol(10EO) monolaurate ester	MYL-10[2]	12.5
11	stearyl macrogol glycerides	GELUCIRE 50/13[3]	13
12	hexaglycerin monolaurate ester	ML-500[1]	13.5
13	lauroyl macrogol-32 glycerides	GELUCIRE 44/14[3]	14
14	decaglycerin monooleate ester	MO-750[1]	14.5
15	Polyethyleneglycol (25EO) monostearate ester	MYS-25[2]	15
16	decaglycerin monolaurate ester	DECAGLYN 1-L[2]	15.5
17	decaglycerin monolaurate ester	ML-750[1]	15.7
18	polyoxyethylene(20EO) oleylether	BO-20[2]	17
19	polyethyleneglycol(40EO) monostearate ester	MYS-40[2]	17.5
20	polyoxyethylene(30EO) phytosterol	BPS-30[2]	18
21	polyoxyethylene(21EO) laurylether	BL-21[2]	19
22	polyoxyethylene(18EO)nonylphenylether	NP-18TX[2]	19
23	polyoxyethylene(25EO)laurylether	BL-25[2]	19.5
24	polyoxyethylene(30EO)octylphenylether	OP-30[2]	20
25	polyoxyethylene(20EO)nonylphenylether	NP-20[2]	20

Number 1), 2), and 3) indicated in product name were gifts from Sakamoto Yakuhin Kogyo Co.Ltd., Nikko Chemicals Co. Ltd., and Gattefossé, respectively.

Table 1. Surfactants used for the preparation of w/o emulsions.

The following examination was carried out in order to estimate the stability of w/o emulsion. Each surfactant (0.75 g) and soybean oil (6.75 g) were put in a glass tube. The mixture was dissolved or dispersed uniformly at 60°C for 15 min. GZ solution (2.25 mL of 40 mg/mL in dissolved with 100 mM phosphate buffered solution, pH7.4) was added into the glass tube, then immediately, the solution with oil, water phase, and surfactant was emulsified using a Polytron homogenizer (PT-MR 3100, Kinematica AG, Littau/Luzern, Switzerland) at 20,000 rpm for 3 min. The state of the w/o emulsions was observed at 24 h after the preparation. The stability of the w/o emulsions was estimated by objective evaluation scale (OES; 0, 1, 2, 3, 4, and 5). Namely, OES 0 is a completely discrete state without emulsifying. OES 1 is a biphasic separate state with wispy cloud in bottom phase, OES 2 is a biphasic separate state with weakly white turbidity in bottom phase, OES 3 is a biphasic separate state with moderately white turbidity in bottom phase, OES 4 is a biphasic separate state with strongly white turbidity in bottom phase, and OES 5 is a stable state without phase separation.

Figure 1. Observed properties of emulsions with 24 kinds of emulsifiers at 24 h after the preparation using a Polytron homogenizer. The photo of emulsion with CR-310 was referred in Figure 3A. The number under each photo is a product number shown in Table 1.

The results in the observed state after preparation of w/o emulsions are shown in Figure 1. Moreover, Figure 2 shows the relationship between HLB number of emulsifiers and OES.

CR-310 and CDS-400 among 25 kinds of surfactants were convenient for the preparation of stable w/o emulsion. The results were identified with the theory that surfactant with low HLB is suitable for the preparation of w/o emulsions. A difference of viscosity as physico-chemical properties was observed between CR-310 and CDS-400. Namely, the viscosity of the w/o emulsion with CDS-400 was high like cream, that of the emulsion with CR-310 was the same as the viscosity of soybean oil. Therefore, it was clear that CR-310 among used sur-factants was most convenient emulsifier for the preparation of w/o emulsions when soybean oil was used as an oil phase.

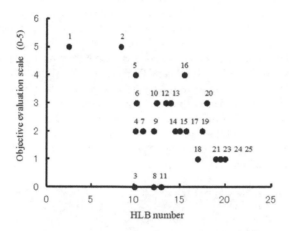

Figure 2. The relationship between HLB number of surfactants and objective evaluation scale. The number of each symbol is a product number shown in Table 1.

3. Preparation of nano-sized w/o emulsions

GZ solution (400 mg/mL) was prepared by dissolving GZ powder at 60°C in 100 mM phos-phate-buffered solution (pH7.4) containing 8.0% (w/v) L-arginine. L-arginine was used to in-hibit the gelation of GZ [9]. Soybean oil (4.50 g) and CR-310 (0.50 g) were mixed in a glass tube, and then the mixture was heated for 15 min at 60°C in order to blend uniformly. The GZ solution and the mixture of oil and emulsifier were cooled down at room temperature, and 400 mg/mL GZ solution (2.2, 3.3, or 4.4 g) was added to glass tubes containing the mix-ture. First, the w/o emulsions were prepared by agitating the mixture with a Polytron ho-mogenizer (PT-MR 3100, Kinematica AG) at 20,000 rpm for 3 min. Second, the w/o emulsions were placed into a 50-mL round-bottom flask, which was then placed into a rota-ry evaporator (R-210, Buchi Labortechnik AG, Flawil, Switzerland) equipped with a vacuum controller (V-850) and a vacuum pump (V-700). The vacuum was initially set to 120 hPa, and then decreased at a rate of 10 hPa per minute until 20 hPa was reached. The mixture was then subjected to these vacuum conditions at 40°C for 90 min. The prepared emulsions were separated into either glass vials or 10-mL centrifuge tubes, depending on the analyses to be done. For phase behavior comparisons, w/o emulsions without GZ was also prepared using distilled water (3.3 g) instead of GZ solution [7].

4. Phase behavior during the preparation of w/o emulsions

The w/o emulsions prepared by adding GZ solution slowly changed in turbidity to pale white, and after 23-24, 26-27, and 31-32 min for GZ solutions of 2.2, 3.3, and 4.4 g, respectively, to clear or slightly turbidity. The samples remained clear only for approximately 2 min. When subjected to prolonged evaporation, the emulsions rapidly changed in turbidity to white as solid dispersion. Figure 3A shows photographs of the emulsions prepared with GZ before evaporation (0 min) and 10, 15, 26, and 90 min after evaporation. To confirm the relationship between evaporation time and phase behavior in w/o emulsions with or without GZ, similar experiments on w/o emulsions lacking GZ were carried out (Figure 3B). The turbidity of w/o emulsions without GZ remained white up to 15 min after evaporation, before gradually changing to pale white. The w/o emulsions finally became transparent 90 min after evaporation. These results suggest that the phase behavior may depend on the water content of the w/o emulsions.

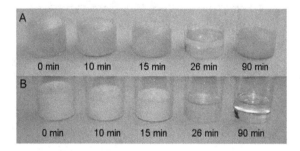

Figure 3. Photographs of w/o emulsions at the indicated times after evaporation. (A) Emulsion prepared with a GZ solution containing 3.3 g GZ. (B) Emulsion prepared without GZ (water phase is distilled water). The times shown below each vial represent the length of time samples underwent evaporation.

5. Particle size of water phase in w/o emulsions

GZ solution (400 mg/mL) was prepared by dissolving GZ powder at 60°C in 100 mM phosphate-buffered solution (pH 7.4) containing 8.0% (w/v) L-arginine and 10 µg/mL fluorescein sodium. Fluorescein sodium was used as a marker for the fluorescent observation of the dispersion state of water phase in the emulsions. After the GZ solution (3.3 g), soybean oil (4.50 g), and CR-310 (0.50 g) were mixed, w/o emulsions were prepared according to the method

described above. The w/o emulsion before dehydration was analyzed with a confocal laser
scanning microscope LSM510 (Carl Zeiss GmbH, Jena, Germany) and LMS Image Browser
Software (Carl Zeiss GmbH). The excitation and fluorescein wavelengths were set to 405
and 488 nm, respectively.

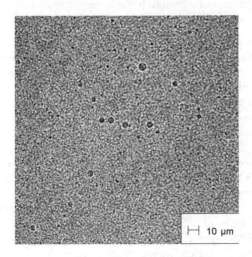

Figure 4. Fluorescence photomicrograph of a w/o emulsion following the addition of fluorescein sodium to the water
phase. The dark gray areas represent the aqueous phase. The position of water phase containing GZ is indicated as
green fluorescent particles.

Figure 4 shows the photo by a confocal laser scanning microscope. The water phase was uniformly distributed as small droplets (< 5 μm), indicating that before evaporation the particle
size distribution in the water phase of GZ sample distributed widely after dispersal with a
Polytron homogenizer.

Emulsions	Particle size in relative frequency (nm)				
	10%	**25%**	**50%**	**75%**	**90%**
2.2 g GZ sample	135	170	225	287	341
3.3 g GZ sample	219	253	299	349	394
4.4 g GZ sample	310	504	1105	1381	1610

Table 2. Particle size and size distribution of clear or slightly turbid w/o emulsions.

The particle sizes of w/o emulsions at 23, 26, and 31 min for GZ solutions of 2.2, 3.3, and 4.4
g after evaporation were analyzed using a Particle Size Analyzer (LS 13 320; Beckman Coulter, Inc., Fullerton, CA, United States). Table 2 presents the average particle sizes and size
distributions of the clear or slightly turbid w/o emulsions after evaporation. The size distribution of the three kinds of emulsions was narrow, with relative frequency values of 10, 25,

75, and 90%. These results suggest that dehydration proceeded in a way that caused the dispersed phase to approach the narrow distribution. In particular, the particle size distribution of the 2.2 and 3.3 g GZ samples converged toward the nano-size range. This was consistent with our prediction that large water droplets would efficiently evaporate from the surface of the w/o emulsions as their round bottom flasks were rotated and that particle size distribution would narrow as a function of evaporation time.

During the preparation of transparent w/o emulsions containing GZ, the only component removed by evaporation was water. Thus, the components dissolved in the water phase (GZ, L-arginine, and phosphate salts) were gradually concentrated as the water content decreased. The observation that a simple evaporation process changed the turbidity of the w/o emulsions from white to clear within a short time suggests that the particles comprising the dispersed phase became extremely small in size. In fact, the particle size range of sample which prepared with 3.3 g GZ solution was 219-394 nm by dynamic light scattering assay (LS 13 320; Beckman Coulter, Inc.). The state of the nano-sized droplets was observed in transmission electron microscopy (TEM). Namely, a clear w/o emulsion was prepared by adding GZ solution (3.3 g) to the mixture of soybean oil (4.50 g) and CR-310 (0.50 g), and the resulting clear w/o emulsion was passed through a quantitative filter (No. 5B, Advantec; Toyo Roshi Kaisha, Ltd., Tokyo, Japan). The filter was then hardened with cured acryl resin. After embedding the filter, ultrathin sections were obtained by cutting the surface of the block containing the filter on an ultramicrotome equipped with a diamond knife (EM-Ultracut-s; Leica Microsystems Vertrieb GmbH, Wetzlar, Germany). The ultrathin sections were mounted onto freezing support grids and then stained with ruthenium tetrachloride. Next, the emulsified particles were observed with a transmission electron microscope (TEM; JEM-2100; Jeol Ltd., Tokyo, Japan).

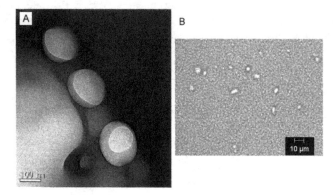

Figure 5. A) TEM photograph of dispersed particles in a clear w/o emulsion prepared by adding GZ solution. (B) Solid GZ particles after 90 min evaporation.

To determine whether the nano-sized droplets exist as a liquid state in the water phase, the shape of particles in the clear w/o emulsion (3.3 g GZ sample) was observed with TEM. As the water content of the emulsion decreases with evaporation, part of the dissolved GZ may precipitate as a solid state from the dispersed phase. If this hypothesis is accurate, then the particles may be not spherical. However, TEM analysis revealed that the particles were spheres of approximately 200 nm in diameter and were maintained a uniform globe (Figure 5A), strongly suggesting that the nanoparticles in the clear w/o emulsions existed as a liquid phase.

Following additional evaporation, the turbidity of the w/o emulsions again changed to white, indicating a change in phase behavior. This suggests that the hydrophilic components dissolved in the water phase separated as solid states. In fact, microscopic analysis demonstrated that the w/o emulsions containing GZ contained 1-10 μm-diameter solid particles >90 min after evaporation (Figure. 5B) [8].

6. Water contents in w/o emulsions

The water content (%, w/w) of the w/o emulsions was analyzed using a Karl Fischer titration apparatus (870, KF Titrino Plus; Metrohm Shibata Co., Ltd., Tokyo, Japan). The water content (%, w/w) of the 2.2, 3.3, and 4.4 g GZ samples before evaporation was 16.8, 21.9, and 25.4% (w/w), respectively. On the other hand, the water content of the clear or slightly turbid 2.2, 3.3, and 4.4 g GZ samples was 7.8 ± 0.1, 9.4 ± 0.3, and $11.9 \pm 0.3\%$ (w/w), respectively. Furthermore after 90 min of evaporation, the water content of the 2.2, 3.3, and 4.4 g GZ samples was all in the range of 1.3-1.8% (w/w). The water content of w/o emulsion lacking GZ was 9.3 ± 0.2 and $0.18 \pm 0.02\%$ (w/w), respectively, at 26 and 90 min after evaporation. These results indicate that the phase behavior of the w/o emulsions changed from white turbid to translucent or transparent when the water content reached approximately 8-12% (w/w). Since the water content of the clear or slightly turbid w/o emulsions correlated well with GZ content (0.8, 1.2, and 1.6 g for 2.2, 3.3, and 4.4 g GZ samples; correlation coefficient was 0.992), it is plausible that the precipitation of hydrophilic components, such as GZ, L-arginine and phosphate salts, was responsible for the increased turbidity resulting after prolonged evaporation.

From these results, it is concluded that the water contents of w/o emulsions changed the phase behavior in the emulsions, that is, from white turbid phase to clear phase when the water content reached to be approximately 9%, and then from clear phase to white turbid phase. The decreasing rate of the water content is affected by the setting of a vacuum controller. In the experiments, although the relationship between the decreasing rate of the water content and maintained interval with clear phase in the w/o emulsions was not investigated, the pressure condition in the evaporation process will be a major problem for the simple preparation of clear w/o emulsions.

7. Stability of w/o emulsions

The stability of the w/o emulsions were evaluated according to two criteria: i) the uniform dispersal of the water phase containing GZ in the emulsion; and ii) the size distribution at steady-state conditions. Clear or slightly turbid w/o emulsions (2.2, 3.3, and 4.4 g GZ solutions) were transferred to 10-mL centrifuge tubes (Figure 6), which remained undisturbed at 20 ± 2°C for 10 days. The w/o emulsion prepared with 3.3 g GZ solution remained under undisturbed conditions continuously for 65 days. Next, a pipette was inserted 1 cm from the top or bottom of each centrifuge tube and a small amount (50 mg) of emulsion was removed and transferred into two screw vials (one vial for the "top" sample, the other vial for the "bottom" sample). Methanol (30 mL) was added to each vial, and the vials were shaken for 15 min on a vortex mixer. The methanol-containing GZ samples were adequately diluted with 100 mM phosphate-buffered solution (pH 7.4) and were injected into an HPLC system in order to determine the GZ concentration in the samples [10]. The water content (%, w/w) in the w/o emulsions obtained from both samples was analyzed by Karl Fischer titration.

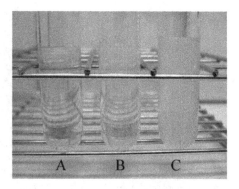

Figure 6. Photograph of clear and slightly turbid w/o emulsions. (A) 2.2 g GZ sample. (B) 3.3 g GZ sample. (C) 4.4 g GZ sample.

7.1. Uniformity of GZ concentration and water content in w/o emulsions

The GZ concentrations of 2.2, 3.3, and 4.4 g GZ samples that were clear or slightly turbid were 11.1, 14.5, and 17.0% (w/w), respectively. The GZ concentrations and water content of clear or slightly turbid w/o emulsions 10 days and 65 days after preparation are shown in Table 3. As the phase behavior changes with time, the aggregation or coalescence of the dispersed phase occurs more frequently after emulsification is achieved, because the emulsion is thermodynamically unstable. If the dispersed stability is not maintained, then the GZ concentration and water content in the emulsion will differ in the parts of the emulsion near the top and bottom of the sample. This is because the dispersed phase containing GZ moves toward the bottom due to specific gravity differences. GZ concentration in the samples were analyzed in the present study, however, the GZ concentrations in the top and bottom parts of the emulsions were identical in the 2.2 and 3.3 g GZ samples but not in the 4.4 g GZ sample. Furthermore, similar results were obtained in our comparative analysis of water content. These results suggest that the dispersed stability of the 2.2 and 3.3 g GZ w/o emulsions was extremely high. On the other hand, the difference in phase behavior in the top and bottom layers of the 4.4 g GZ sample suggests dispersed instability. Specifically, after 10 days the turbidity of the bottom layer became clearer than that of the top layer, indicating that the dispersed phase containing 4.4 g GZ gradually moved toward the bottom of the 10-mL centrifuge tube. The specific gravity of the dispersed phase in the w/o emulsions increased with increasing GZ concentration.

w/o emulsions	GZ concentration (%, w/w)		Water content (%, w/w)	
	upper position	bottom position	upper position	bottom position
2.2 g GZ sample	11.5 ± 0.7	11.2 ± 0.4	8.2 ± 0.2	8.4 ± 0.2
3.3 g GZ sample	14.9 ± 0.5	14.7 ± 0.5	10.1 ± 0.2	10.2 ± 0.3
3.3 g GZ sample**	15.0 ± 0.4	14.8 ± 0.4	9.8 ± 0.3	9.6 ± 0.2
4.4 g GZ sample	10.2 ± 0.8	17.4 ± 0.7	7.4 ± 0.1	12.0 ± 0.3

GZ concentration and water content were determined after the sample was stored undisturbed for 10 days at 20 ± 2°C. *Samples were obtained 1 cm from the top and bottom of the emulsions contained within a 10-mL centrifuge tube. **The sample was stored undisturbed for 65 days at 20 ± 2°C.

Table 3. GZ concentration and water content in different parts of the w/o emulsions*.

7.2. Uniformity of particle size of dispersed phase in w/o emulsions

The particle size distribution of the GZ sample containing 3.3 g GZ solution at 65 days after preparation was 226-421 nm. This range was similar to that measured during immediate intervals after preparation. In general, refinement of emulsion particle size lowers thermodynamic stability, because the phase behavior of an emulsion-which is a very complicated system of oil, water, and surfactant-is affected by decreases in particle size [11]. Nano-sized emulsions, in particular, are not as stable as micron-sized emulsions. Therefore, nano-sized emulsions must be stabilized with polymers and excessive amounts of surfactants [12]. In GZ sample, the concentration of surfactant in the oil phase was 10% (w/w); the emulsions did not contain other stabilizers. However, the GZ sample (nano-sized emulsions) was stable at least for 2 months. Moreover, although the particle sizes of the top and bottom layers in the 4.4 g GZ sample were not determined, the aggregation of dispersed particles suggests that the size distribution in this w/o emulsion tended to be on the large side. Taken together, these observations suggest that dispersed stability decreases with increasing GZ content.

8. Sustained release effect by nano-sized w/o emulsions

From the viewpoint of medical treatment, drug release from w/o emulsions is important for the efficiency of controlled release. The pharmacokinetics of GZ by nano-size w/o emulsion, aqueous formulation, o/w emulsion, and w/o emulsion with solid GZ was investigated in order to clarify the degree of the controlled release. Furthermore, the release characteristics of a hydrophilic compound, indocyanine green (ICG), from administered subcutaneous site in rats was observed using a near-infrared fluorescent camera (Photo Dynamic Eye, PDE).

8.1. In vivo experiments in GZ pharmacokinetics

Pharmacokinetic studies of GZ were investigated in detail in human [13, 14], in rat [15, 16], and other species. The elimination half-life of GZ in rats after the intravenous administration (20-50 mg/kg) is approximately 2-4 h in plasma [17, 18]. GZ is rapidly excreted into bile via multidrug resistance-associated protein 2 (MRP2) ATP-binding cassette transporter C2 (ABCC2) transporter [19]. Therefore, the release of GZ from w/o emulsions was estimated as GZ elimination into bile.

The protocol of this study was approved by the Committee of Animal Use of Hokuriku University. All animal experiments were conducted in accordance with the Institutional Guidelines of Care and Use of Laboratory Animals. Male Sprague-Dawley rats (180-200 g) were housed for at least 10 days in a clean room. The rats were given free access to commercial chow and water and were maintained according to the Hokuriku University Animal Guidelines. For in vivo experiments in GZ formulations, the rats (250-280 g) were randomly divided into four treatment groups as four rats per group.

A GZ stock solution (400 mg/mL) was prepared at 60°C in 100 mM phosphate-buffered solution, pH7.4, containing 8.0% (w/v) L-arginine. The GZ stock solution was stored in a refrigerator. An aqueous formulation of GZ (150 mg/mL; Rp-II) was prepared by adding 100 mM phosphate-buffered solution (pH7.4) to the GZ stock solution. Preparation of an oil-in-water (o/w) emulsion of GZ was as follows: soybean oil (1.00 g), HCO-60 (0.12 g), and egg yolk lecithin (0.12 g) were blended uniformly by heating at 90°C for 15 min on a block heater. The mixture was then cooled at room temperature. The o/w emulsion of GZ (150 mg/mL; Rp-III) was prepared by combining the soybean oil mixture (1.0 mL), GZ stock solution (1.16 mL), and 100 mM phosphate-buffered solution, pH7.4 (0.84 mL) and by using a Polytron homogenizer (PT-MR 3100) at 20000 rpm for 3 min for emulsification.

Preparation of an w/o emulsion of GZ was described above. The GZ stock solution (3.3 g) was then added to the lukewarm mixture, which was emulsified using a Polytron homogenizer (PT-MR 3100) at 20000 rpm for 3 min. The w/o emulsion was placed into a 50-mL round-bottom flask, which was then set in a rotary evaporator (R-210, Buchi Labortechnik AG) equipped with a vacuum controller (V-850). The vacuum was initially set to 120 hPa at 40°C; thereafter, the pressure was decreased at a rate of 10 hPa per min until 20 hPa was reached. To prepare Rp-I and Rp-IV, the dehydration was continued for 27 min and 120 min, respectively, and then adjusted the GZ concentration to 150 mg/mL by adding the soybean oil/CR-310 (9 : 1, w/w) mixture. Adminidtration method of four kind formulations and sampling of bile in rats were described in detail in reference [8].

As the characteristics of used formulations, the average particle sizes in the Rp-I and Rp-III formulations were 299 nm and 376 nm, respectively. The 10-90% ranges of size distribution in Rp-I and Rp-III were 208-402 nm and 255-512 nm, respectively. Two peaks in size distribution were observed in the Rp-IV formulation: 312 nm and 5000 nm. Microscopic observations revealed that large-size GZ particles were solid GZ, because they were not spherical. The water content in Rp-IV (1.5%, w/w) was very low compared with that in Rp-I (9.4%, w/w). Moreover, the small- and large-size particles were not observed after evaporation for 120 min in a w/o emulsion lacking GZ.

Almost all GZ transported to hepatocytes via the blood is eliminated into bile as unchanged GZ (i.e., not metabolized) [20, 21]. Therefore, the elimination rate of GZ in bile reflects the bioavailability of GZ in hepatocytes. Figure 7 shows the cumulative elimination (%) of GZ over time after subcutaneous administration of Rp-I, Rp-II, Rp-III, and Rp-IV in rats. After the administration of Rp-I, cumulative elimination at 8 h, 24 h, and 72 h as a function of administered GZ dose (50 mg/kg) in bile was 11%, 20%, and 47%, respectively. These results indicate that the Rp-I formulation resulted in a sustained release of apparent zero-order kinetics. The average elimination rate of GZ up to 72 h was 84.2 ± 14.2 μg/h.

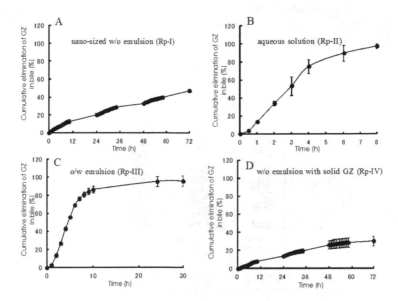

Figure 7. Cumulative elimination of GZ in bile after subcutaneous administration of GZ formulations in rats. (A) Nano-sized w/o emulsion encapsulating GZ, (B) aqueous solution of GZ, (C) o/w emulsion containing GZ, and (D) w/o emulsion with solid GZ. GZ concentrations were all adjusted to 150 mg/mL. The GZ dose administered to all rats was all 50 mg/kg. Data represent means ± S.D. of four experiments.

With the Rp-II formulation, the cumulative elimination at 4 h and 8 h as a function of administered GZ dose was 75% and 97%, respectively. In intravenous and subcutaneous administration models, no difference in the elimination rate of GZ has been observed after intravenous administration of GZ [10]. These results suggest that GZ dissolved in phosphate buffered solution rapidly diffused in the hyperdermis, before being transferred into the general circulation. With the Rp-III formulation, the cumulative elimination of GZ at 8 h and 30 h was 81% and 96%, respectively. The elimination rate of GZ in Rp-III was faster than that in Rp-I but slower than that in Rp-II, suggesting that the oil phase of the o/w emulsion inhibited the diffusion of the water phase containing GZ in subcutaneous regions, even though GZ was dissolved in the outer water phase of o/w emulsion. On the other hand, the cumulative elimination of GZ in Rp-IV was the lowest among the four formulations: The GZ elimination in bile at 8 h, 24 h, and 72 h was 7.1%, 14%, and 31%, respectively. As with the elimination kinetics of the Rp-I formulation, the elimination kinetics of GZ in Rp-IV showed that GZ was released in a sustained fashion for up to 72 h. Since Rp-IV contained solid GZ in w/o emulsions, it was speculated that the elimination rate of GZ in bile after the administration of Rp-IV would be slower than that of GZ in Rp-I. In fact, the eliminated amount of GZ in bile after the administration of Rp-IV was 0.64-fold compared to that of Rp-I. These results suggest that reduced water content in w/o emulsions delays hydration in the subcutaneous region and that much time is required to dissolve the dispersed solid GZ in Rp-IV.

To determine more precisely the characteristics of sustained GZ release from w/o emulsions, the rates of GZ elimination in bile were recalculated every 24 h after the administration of Rp-I and Rp-IV (Figure 8).

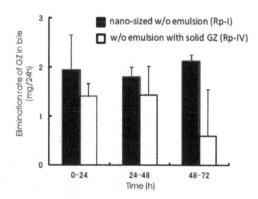

Figure 8. Elimination rate per 24 h of GZ in bile after subcutaneous administration of Rp-I and Rp-IV in Rats. Administration dose of GZ was 50 mg/kg in both formulations. Data represent means ± S.D. of four experiments.

The elimination of GZ in Rp-I occurred at a constant rate for 72 h, i.e., the rate ranged from 1.80 to 2.12 mg/day. On the other hand, elimination of GZ in Rp-IV decreased from 1.40-1.41 mg/day at 48 h to 0.60 mg/day at 48-72 h. As the reason for the decrease in GZ elimination rate at 48-72 h in Rp-IV, it was predicted that the presence of solid dispersed GZ may be involved deeply the transfer rate of GZ from subcutaneous site to liver. Actually, dissolved-state GZ and solid-state GZ exist in Rp-IV. Although it was considered that the dissolved GZ in Rp-IV was transferred to liver as similar to GZ in Rp-I, solid GZ particles must be dissolved to some extent which can be passed vascular system such as vein and lymph capillary in order to transfer GZ from subcutaneous site to liver. Therefore, after 48 h, the proportion of solid GZ for residual GZ in the subcutaneous site will increase certainly. As a result, it was guessed that GZ elimination into bile decreased based on the decrease of transfer rate from subcutaneous site to liver. These results indicate that Rp-I was a substantially superior formulation compared to Rp-IV in

terms of sustained release in bile. It was hypothesized that the small and narrowrange polydispersity (208-402 nm in Rp-I) of the dispersed phase in w/o emulsions may be important for sta-
bilizing the release rate of GZ from these emulsions.

8.2. Tissue distribution of ICG from subcutaneous site in rats

For the experiments of ICG administration, two male Sprague-Dawley rats (250, 255 g) were
used. A w/o emulsion encapsulating ICG was prepared as follows: 5 mg/mL ICG solution (3.3
g) was added to the mixture of soybean oil (4.50 g) and CR-310 (0.50 g) heated at 60°C. Next
procedure was the same with the preparation step of Rp-I described above. This ICG solution
and w/o emulsion encapsulating ICG were used to observe the tissue distribution of drugs
from the subdermal injection site. To monitor over time the delayed drug distribution and the
diffusion of hydrophobic formulation in the subdermal site, a w/o emulsion encapsulating
ICG instead of GZ was prepared. ICG is a hydrophilic fluorescent dye and biocompatibility
marker with excitation and emission spectra in the near-infrared wavelength range of 600 to
900 nm, and the maximum emission wavelength of ICG in vivo is 845 nm [22, 23]. The kinetics
of ICG in vivo was observed using a non-invasive, near-infrared fluorescent camera (Photo
Dynamic Eye; PDE), because the near-infrared-wavelength monitoring barely affects biological molecules such as water and hemoglobin. For the experiments of the microscopic image using a fluorescent probe, ICG, rats were anesthetized with an intraperitoneal administration of
sodium pentobarbital (50 mg/kg), and then were kept in a supine position on an operation
plate. A PDE was set on the upper position of 20 cm from the rat and the photo image was
monitored using a personal computer. Two samples, 5 mg/mL ICG solution (0.05 mL) and w/o
emulsion encapsulating ICG (0.05 mL) were administered to right and left hind legs of rat, respectively. The diffusion of ICG was observed for 60 min.

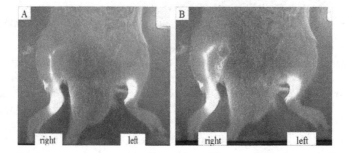

Figure 9. Photographs taken at (A) 15 min and (B) 60 min after the subcutaneous injection of ICG solution into the right
leg and w/o emulsion encapsulating ICG into the left leg injection volume of ICG was 50 mL in both formulations.

The purpose of the PDE experiments was to determine whether the diffusion rate of w/o emulsions at the subdermal site is remarkably slower than that of solutions like Rp-II of GZ formulation. Figure 9 shows photographs of rats' hind legs 15 min and 60 min after injecting ICG (right leg) and after injecting w/o emulsion encapsulating ICG (left leg). The ICG solution was rapidly absorbed into capillary blood vessels within 15 min. Furthermore, ICG also reached peripheral lymphatics. On the other hand, ICG in w/o emulsions remained in the vicinity of the administration site for 60 min. These results suggested that the outer oil phase inhibited the diffusion of w/o emulsion encapsulating ICG and/or the release of ICG from the w/o emulsion. The viscosity of lipophilic formulations is generally high. Therefore, one would expect the release of ICG from a w/o emulsion to be delayed compared to that from aqueous formulations of ICG. The expectation will be correspondent with the decrease of diffusion and/or release of GZ from Rp-I of GZ formulation.

9. Conclusion

It was clarified that GZ, a hydrophilic compound, was dispersed as nanoparticles into soybean oil by using the evaporation technique. The prepared oil phase containing GZ was nano-emulsions with low polydispersity, and was stable at least 2 months. This ideal dispersion method had to make water content approximately 9%, and it was clear that the further dehydration became solid dispersion. It is concluded that a hydrophilic compound can be dispersed easily into an oil phase such as soybean oil by utilizing this method. Concretely, the w/o emulsions containing GZ (2.2 g of GZ [11.1%, w/w] and 3.3 g of GZ [14.5%, w/w]) with narrow-ranged polydispersity and high-dispersed stability were easily prepared by the measurable removal of water using a Polytron homogenizer and a rotary evaporator. The water content in 2.2 and 3.3 g GZ samples had to be 7.8 and 9.4% (w/w), respectively, because decreasing the water content beyond these levels caused the phase behavior to change (e.g., white turbid). The particle size distribution (relative frequency values ranging from 10 to 90%) of the clear w/o emulsions was in the range of 135 to 421 nm as the samples remained undisturbed for 65 days at $20 \pm 2°C$. The w/o emulsion preparation method described in the present study provides useful information on the lipophilic formulations of GZ.

A nano-sized w/o emulsion of GZ (Rp-I) showed sustained elimination of GZ in bile at a relatively constant rate for 72 h. The sustained GZ elimination in bile was strongly affected by diffusion of the w/o emulsion and by the release of GZ from the emulsion to the perimeter of the subdermal site, based on the PDE observations with ICG. Indeed, the average elimination rate of GZ in bile was 0.084 mg/h over 72 h, when Rp-I (50 mg/kg as GZ) was administered subcutaneously. If GZ release from Rp-I will be maintained as zeroorder elimination (0.084 mg/h), 6-7 day are needed until the GZ release finishes in the rats. Namely, GZ in Rp-I will slowly transfer from subcutaneous tissue to liver 20-fold periods as compared with GZ in Rp-II, the elimination of almost all GZ finished 8 h. These results indicate that the nano-sized w/o emulsion encapsulating GZ, which can be subcutaneously administered, will be useful as a new sustained-release formulation.

Acknowledgements

Author thanks Cokey Co. Ltd. for the gift of glycyrrhizin monoammonium, and thanks Sakamoto Yakuhin Kogyo Co. Ltd. and Nikko Chemicals Co. Ltd. for the gifts of several kinds of surfactants.

Author details

Kenjiro Koga*

Address all correspondence to: k-koga@hokuriku-u.ac.jp

Faculty of Pharmaceutical Sciences, Hokuriku University, Japan

References

[1] Aryanti, N., Hou, R., & Williams, R. A. (2009). Performance of a rotating membrane emulsifier for production of coarse droplets. *Journal of Membrane Science*, 326(1), 9-18.

[2] Vladisavljevic, G. T., & Williams, R. A. (2006). Manufacture of large uniform droplets using rotating membrane emulsification. *Journal of Colloid and Interface Science*, 299(1), 396-402.

[3] Vladisavljevic, G. T., Surh, J., & Mc Clements, J. D. (2006). Effect of emulsifier type on droplet disruption in repeated shirasu porous glass membrane homogenization. *Langmuir: the ACS journal of surfaces and colloids*, 22(10), 4526-4533.

[4] Hosoya, K., Bendo, M., Tanaka, N., Watabe, Y., Ikegami, T., & Minakuchi, H. (2005). An application of silica-based monolithic membrane emulsification technique for easy and efficient preparation of uniformly sized polymer particles. *Macromolecular Materials and Engineering*, 290(8), 753-758.

[5] Sakai, H., Tanaka, K., Fukushima, H., Tsuchiya, K., Sakai, K., Kondo, T., & Abe, M. (2008). Preparation of polyurea capsules using electrocapillary emulsification. *Colloids and Surfaces, B: Biointerfaces*, 66(2), 287-290.

[6] Abe, M., Nakayama, A., Kondo, T., Morishita, H., Tsuchiya, K., Utsumi, S., Ohkubo, T., & Sakai, H. (2007). Preparation of tiny biodegradable capsules using electrocapillary emulsification. *Journal of Microencapsulation*, 24(8), 777-786.

[7] Koga, K., Liu, H., & Takada, K. (2009). A simple preparation method for dispersing glycyrrhizin solution into oil phase using an evaporation technique. *Journal of Drug Delivery Science and Technology*, 19(6), 431-435.

[8] Koga, K., Nishimon, Y., Ueta, H., Matsuno, K., & Takada, K. (2011). Utility of Nano-sized, water-in-oil emulsion as a sustained release formulation of glycyrrhizin. *Biological & Pharmaceutical Bulletin*, 34(2), 300-305.

[9] Koga, K., Kawashima, S., Shibata, N., & Takada, K. (2007). Novel formulations of a liver protection drug glycyrrhizin. *Yakugaku Zasshi*, 127(7), 1103-1114.

[10] Koga, K., Tomoyama, M., Ohyanagi, K., & Takada, K. (2008). Pharmacokinetics of glycyrrhizin in normal and albumin-deficient rats. *Biopharmaceutics & Drug Disposition*, 29(7), 373-381.

[11] Porras, M., Solans, C., Gonzalez, C., & Gutierrez, J. M. (2008). Properties of water-in-oil (W/O) nano-emulsions prepared by a low-energy emulsification method. *Colloids and Surfaces, A: Physicochemical and Engineering Aspects*, 324(1-3), 181-188.

[12] Sood, A. (2008). Modeling of the particle size distribution in emulsion polymerization. *Journal of Applied Polymer Science*, 109(3), 1403-1419.

[13] Yamamura, Y., Kawakami, J., Santa, T., Kotaki, H., Uchino, K., Sawada, Y., Tanaka, N., & Iga, T. (1992). Pharmacokinetic profile of glycyrrhizin in healthy volunteers by a new high-performance liquid chromatographic method. *Journal of Pharmaceutical Sciences*, 81(10), 1042-1046.

[14] Ploeger, B., Mensinga, T., Sips, A., Seinen, W., Meulenbelt, J., & De Jongh, J. (2001). The pharmacokinetics of glycyrrhizic acid evaluated by physiologically based pharmacokinetic modeling. *Drug Metabolism Reviews*, 33(2), 125-147.

[15] Yamamura, Y., Santa, T., Kotaki, H., Uchino, K., Sawada, Y., & Iga, T. (1995). Administration-route dependency of absorption of glycyrrhizin in rats: intraperitoneal administration dramatically enhanced bioavailability. *Biological & Pharmaceutical Bulletin*, 18(2), 337-341.

[16] Takahashi, M., Nakano, S., Takeda, I. , Kumada, T. , Sugiyama, K., Osada, T. , Kiriyama, S., Toyoda, H. , Shimada, S. , & Samori, T. (1995). The pharmacokinetics of the glycyrrhizin and glycyrrhetic acid after intravenous administration of glycyrrhizin for the patients with chronic liver disease caused by type C hepatitis virus. *The Japanese journal of gastro-enterology*, 92(12), 1929-1936.

[17] Ishida, S. , Sakiya, Y. , Ichikawa, T., & Taira, Z. (1992). Dose-dependent pharmacokinetics of glycyrrhizin in rats. *Chemical & Pharmaceutical Bulletin*, 40(7), 1917-1920.

[18] Koga, K., Takekoshi, K., Kaji, A., Hata, Y., Ogura, T., & Fujishita, O. (2006). Basic study on development of glycyrrhizin formulations for self administration. *Iryo Yakugaku*, 32(5), 414-419.

[19] Horikawa, M., Kato, Y., Tyson, C. A., & Sugiyama, Y. (2002). The potential for an interaction between MRP2 (ABCC2) and various therapeutic agents: probenecid as a candidate inhibitor of the biliary excretion of irinotecan metabolites. *Drug Metabolism and Pharmacokinetics*, 17(1), 23-33.

[20] Ishida, S., Sakiya, Y., Ichikawa, T., Taira, Z., & Awazu, S. (1990). Prediction of glycyr-rhizin disposition in rat and man by a physiologically based pharmacokinetic model. *Chemical & Pharmaceutical Bulletin*, 38(1), 212-18.

[21] Ichikawa, T., Ishida, S., Sakiya, Y., Sawada, Y., & Hanano, M. (1986). Biliary excretion and enterohepatic cycling of glycyrrhizin in rats. *Journal of Pharmaceutical Sciences*, 75(7), 672-675.

[22] Baker, K. J. (1966). Binding of sulfobromophthalein sodium and indocyanine green by plasma (sub α1)-lipoproteins. *Proceedings of the Society for Experimental Biology and Medicine*, 122(4), 957-963.

[23] Benson, R. C., & Kues, H. A. (1978). Fluorescence properties of indocyanine green as related to angiography. *Physics in Medicine & Biology*, 23(1), 159-163.

Chapter 9

Recent Advances on Soybean Isoflavone Extraction and Enzymatic Modification of Soybean Oil

Masakazu Naya and Masanao Imai

Additional information is available at the end of the chapter

1. Extraction of soybean isoflavone

1.1. Soybean isoflavones and attractive potential of supercritical carbon dioxide (SCCO$_2$)

Isoflavones produced from bioresources are gaining attention as attractive components in food supplements. Isoflavones are heterocyclic phenols with a structure very similar to that of estrogens. Isoflavone displays like estrogens and has anti estrogen activity; it influences sex hormone metabolism and related biological activity [1,2] and prevents osteoporosis [3,4], arteriosclerosis [5], dementia [2], and cancer [6,7].

Soybeans contain 12 different isoflavones classified into two components, glycosides and aglycons. Glycoside isoflavone has a glucose chain in its molecular structure; aglycon isoflavone does not have a glucose structure.

Ninety-three percent of isoflavones are produced and stored as glycoside. Therefore, in practical separation processes, glycoside isoflavones were the major fraction and were recognized as the main target group rather than aglycons. This article focuses on daidzin, genistin and glycitin as typical glycosides. Their aglycons (i.e., daidzein, genistein and glycitein) were examined for comparison. The aglycons have no glycoside chain; their chemical structure is depicted in Fig. 1.

Methods of extracting isoflavones from soybean have been previously examined by using organic solvent [8], pressurized liquid [9], ultrasound [10,11], and supercritical carbon dioxide [12-16]. Supercritical carbon dioxide has been the favorite extraction medium for many food functional components, i.e. caffeine [17-20], capsaicin [21,22], carotenoids [23-26], polyphenol [27-30], aspirin [31], and coenzyme Q10 [32].

In general, the solubility of polar components in the $SCCO_2$-only system was very low because carbon dioxide has non-polar characteristics. The solubility of polar components has been well enhanced by adding polar components to the $SCCO_2$ system. The added component was referred to as an entrainer. Ethanol was effectively employed as an entrainer for extraction and applied to caffeine [17,19], capsaicin [21], catechin [27], epicatechin [28], aspirin[31], and coenzyme Q10 [32]. Rostagno et al. (2002) successfully extracted large amounts of isoflavones from soybean flour by using methanol aqueous solution as an entrainer [14]. Zuo et al. (2008) also extracted isoflavones from soybean meal by using methanol [16].

To design practical separation processes using $SCCO_2$, it is necessary to establish a reliable database of the entrainer's enhancement effects. This would facilitate both the choice of a suitable entrainer for an objective component and the quantitative evaluation of separation yield of a target component in actual processes.

In this chapter, we demonstrate the solubility of isoflavones in $SCCO_2$ with ethanol added. The solubility in an $SCCO_2$-only system was also measured for comparison. The effect of the entrainer on solubility is discussed with the hydrophobicity of guest components evaluated from their molecular structure. The thermodynamic relationship between the solubility and the parameter indicated a non-ideal state in $SCCO_2$ [33].

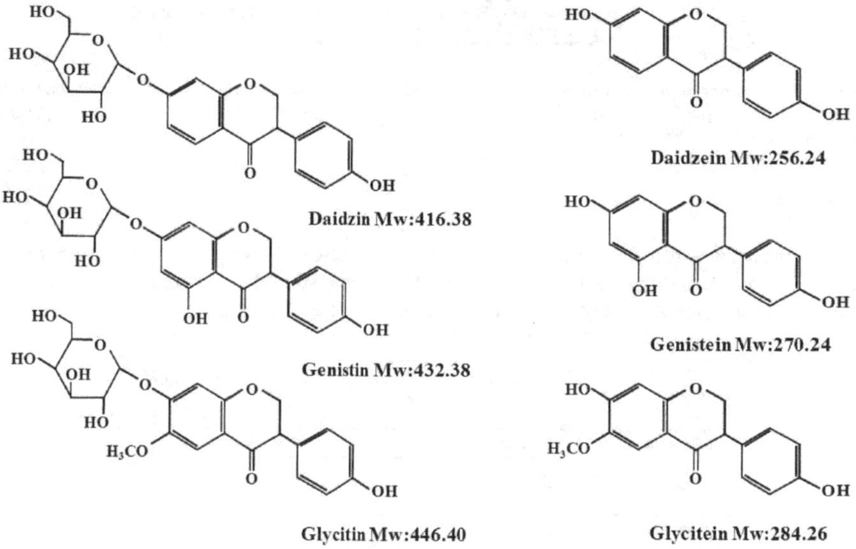

Daidzein Mw:256.24

Daidzin Mw:416.38

Genistin Mw:432.38

Genistein Mw:270.24

Glycitin Mw:446.40

Glycitein Mw:284.26

Figure 1. Chemical structure of isoflavones in soybean

1.2. Solubility of isoflavones and effect of entrainer

1.2.1. Experimental

A circulation flow of $SCCO_2$ was employed for the experimental extraction system (JASCO Co., Ltd., Tokyo) as presented in Fig. 2. The 1.0mL stainless-steel extraction vessel was installed in an extraction line with a total volume of 19.8mL. The extraction temperature was set at 313K. The pressure range was from 15 to 25MPa. The CO_2 volumetric flow rate in the extraction line was adjusted to a constant 5mL/min at 15MPa and 25MPa.

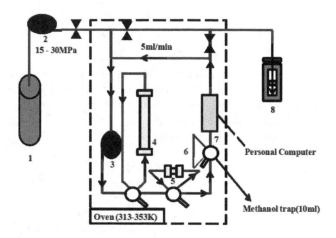

Figure 2. Schematic of experimental apparatus used in measuring the solubility in $SCCO_2$. (1) CO_2 gas cylinder, (2) compressor, (3) circulating pomp, (4) diffusion column, (5) extraction vessel, (6) sample loop with a methanol trap, (7) UV-detector, and (8) exhaust regulator

1.2.2. Solubility of isoflavones

Table 1 summarizes the solubility of isoflavones in the $SCCO_2$-single system. In general, the isoflavones were hardly extracted by the $SCCO_2$-single system. In particular, the solubility of glycoside isoflavones was very low it could not be detected by HPLC.

	Isoflavone	Solubility
		[mol-isoflavone/mol-$SCCO_2$]
Glycoside	Daidzin	not detected
	Genistin	not detected
	Glycitin	not detected
Aglycon	Daidzein	5.14×10^{-10}
	Genistein	6.38×10^{-10}
	Glycitein	not detected

Table 1. The solubility of isoflavones in pure $SCCO_2$ without ethanol at 313K, 25MPa

1.2.3. Effect of entrainer (ethanol) on solubility of isoflavones

Figure 3 presents the solubility of daidzin (as glycoside) and daidzein (as aglycon) in the $SCCO_2$ and ethanol binary system. Solubility S was increased remarkably by increasing the molar fraction of ethanol, M. This trend was also obtained at 25MPa. The solubility of genistin (as glycoside) and genistein (as aglycon) presented in Fig. 4 also exhibited the same trend. This remarkable influence of the molar fraction of ethanol also seemed to be similar between glycitin (as glycoside) and glycitein (as aglycon) (Fig. 5). The results indicated that the solubility of hydrophilic glycoside isoflavones (daidzin, genistin, and glycitin) depended more strongly on the molar fraction of ethanol.

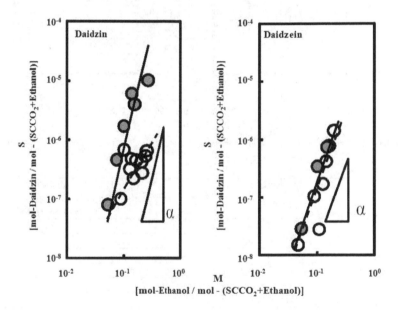

Figure 3. Solubility of daidzin and daidzein in $SCCO_2$ and ethanol binary system at 15 and 25 MPa and at 313K, ○ 15 MPa, ● 25 MPa.

As seen in Fig. 3, the solubility of daidzin at 25MPa was far greater than that at 15MPa. Ethanol depended more heavily on the molar fraction at 25MPa than at 15MPa. In contrast to genistin and genistein, the solubility was almost the same in spite of the increased pressure (Fig. 4). The dependency on the molar fraction of ethanol was similar for 25MPa and 15MPa.

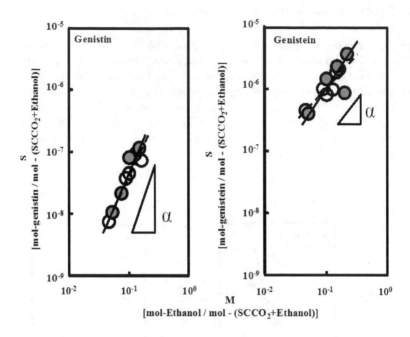

Figure 4. Solubility of genistein into SCCO₂ and ethanol binary system at 15 and 25 MPa and at 313K ○ 15 MPa, ● 25 MPa.

	Isoflavones	Solubility ratio $\equiv \dfrac{\text{The solubility } at\ 25MPa}{\text{The solubility } at\ 15MPa}[-]$
Glycoside	Daidzin	6.3
	Genistin	1.8
	Glycitin	-
Aglycon	Daidzein	1.3
	Genistein	1.8
	Glycitein	-
The solubility were referred from Fig. 3 and Fig. 4.		
The molar fraction of ethanol M was set at 0.1.		

Table 2. The solubility ratio of isoflavones in SCCO₂ with ethanol at 313K. The solubility were referred from Fig. 3 and Fig. 4. The molar fraction of ethanol M was set at 0.1.

The solubility ratio was defined as the solubility of 25 MPa divided by that of 15 MPa. The molar fraction of ethanol was set at 0.10, as evaluated from Figs. 3 and 4, and summarized in Table 2. In the case of daidzin, the solubility ratio was calculated as 6.3 fold. It was especially high among the tested isoflavones, i.e. 1.8 (Genistin), 1.3 (Daidzein), and 1.8 (Genistein). The solubility of daidzin was strongly affected by extraction pressure in four tested isoflavones.

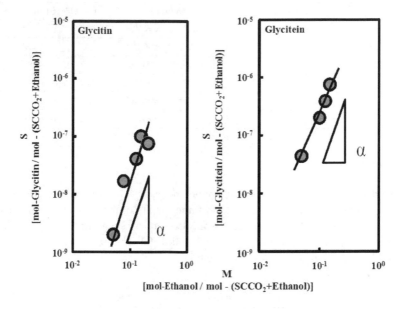

Figure 5. Solubility of glycitin and glycitein in SCCO$_2$ and ethanol binary system at 25 MPa and 313K.

The solubility of daidzin (glycoside) exceeded that of daidzein (aglycon). This trend appeared especially strong in daidzin, in contrast to that of other isoflavones. For other isoflavones, glycosides (genistin and glycitin) were less soluble than the corresponding aglycons (genistein and glycitein) due to their hydrophilic nature and the glycoside chain in their molecular structure. The detailed reasons for the special behavior of daidzin and daidzein are not clear at present.

The enhanced solubility after adding ethanol was preliminarily evaluated by the logarithmic dependency of α on the molar fraction of ethanol M. The solubility S was proportional to the $\alpha^{th.}$ power of M as indicated in empirical equation Eq. (1).

$$S \propto M^{\alpha} \qquad (1)$$

The power term α is summarized in Table 3. The power term α of glycoside isoflavones (daidzin, genistin, glycitin) at both 15MPa and 25MPa often exceeded 3.0. As presented in Table 4, glycoside isoflavones are commonly more hydrophilic than their corresponding aglycons. Additive ethanol concentration in SCCO$_2$ strongly affected the solubility of hydrophilic isoflavones. The power term α of daidzein (aglycon) exceptionally exceeded 3.0 in spite of its hydrophobic nature. The detailed mechanism of solubilization must be investigated further. It may be related to a slight difference of molecular structure.

	Isoflavones	α(15MPa) [-]		α(25MPa) [-]	
Glycoside	Daidzin	3.42	R^2=0.708	3.69	R^2=0.907
	Genistin	3.08	R^2=0.996	3.41	R^2=0.932
	Glycitin	No data		3.27	R^2=0.852
Aglycon	Daidzein	3.17	$R^2$0.967	3.02	R^2=0.975
	Genistein	0.728	R^2=0.985	1.72	R^2=0.999
	Glycitein	No data		1.85	R^2=0.988
	R^2 : Correlation coefficient				

Table 3. Power term α of Eq.(1) on the solubility enhancement

The dependency on the molar fraction of ethanol increased at higher pressures. The solubilities of genistin and genistein are almost the same in spite of the pressure change. The dependency on molar fraction of ethanol was also similar, suggesting that the solubility depended heavily on the amount of ethanol added. Power term α became large under $SCCO_2$ at higher pressures, except for daidzein.

	Isoflavones	Log P [-]
Glycoside	Daidzin	0.232
	Genistin	0.837
	Glycitin	0.230
Aglycon	Daidzein	1.29
	Genistein	2.09
	Glycitein	1.85

Table 4. The evaluated Log P of isoflavones

1.3. Conclusion

Solubilities of six different isoflavones were measured in an $SCCO_2$ system with ethanol added. Ethanol effectively increased the solubility of isoflavones. It served as an attractive entrainer with $SCCO_2$. The power term in the molar fraction of ethanol exceeded 3.0. The enhancement was remarkable in more hydrophilic isoflavones (daidzin, genistin, and glycitin). We experimentally determined the hydrophobicity (Log P) [34] of isoflavones from the equilibrium constant between 1-octanol and water. The hydrophobicity of daidzin was lowest among the tested isoflavones, and the enhancement due to adding ethanol was the highest.

Soybean and other natural bioresources are abundant sources of various glycoside isoflavones. Isoflavones will be successfully extracted from these sources for practical application by $SCCO_2$ with ethanol added.

2. Enzymatic modification of soybean lipid by lipase and immobilized lipase

2.1. Introduction

Soybean is beneficial in food applications and is attractive as a bioresource for functional components. Soybean contains many proteins and much oil. Furthermore, many functional components, isoflavone [35], lecithin [36], saponin [35,37], and oligosaccharide, [38,39] are desirable for promoting human health.

Soybean oil generally contains 52% linoleic acid, 22% oleic acid, 10% palmitic acid, and 8% linolenic acid. Soybean oil can be readily hydrolyzed by lipase like other vegetable oils. The produced fatty acids have several applications such as in manufacturing soaps, surfactants, and detergents, and in food.

Lipases have received attention for lipid modification [40-42]. They are used in fields such as food engineering, detergents, beverages, cosmetics, biomedical uses, and the chemical industry. They catalyze hydrolysis, alcoholysis, acidolysis, amidolysis, and esterification in the food and pharmaceutical industries [43-48]. Lipid modifications (hydrolysis, esterification, etc.) often lead to better quality products due to high specificity and selectivity of the lipase. Immobilized lipases have been applied in various hydrophobic reactions [42,49-51]. Reactivity of immobilized lipase was affected by physicochemical factors in reaction media [52,53]. A hydrophobic material is especially favorable for quick initiation of hydrophobic enzymatic reaction due to the easy diffusion of the substrate in the inner pores of the carrier. Previously, hydrophilic gels and solid porous carriers were often employed even for hydrophobic substrate reactions. Detailed technical data focused on carriers to quickly initiate hydrophobic enzymatic reactions, and high yield repeated-use immobilized enzymes are necessary in industrial design of hydrophobic enzyme reactions [54-56].

2.2. Process chemistry of soybean oil modification

Vegetable oils (olive oil [40,42]) can be hydrolyzed to produce monoglyceride, diglyceride, free fatty acids, and glycerol. Free fatty acids are value-added products because of their wide applications in surfactants, soap manufacturing, the food industry, and biomedical uses. The conventional and industrial method of oil hydrolysis has been carried out using a chemical catalyst at high temperatures and pressure. However, successful enzymatic hydrolysis reactions are possible without high temperatures and pressure.

Dalla, R. C. et al. investigated the continuous production of fatty acid ethyl esters from soybean oil in compressed fluids, namely carbon dioxide, propane, and n-butane, using immobilized Novozym 435 as a catalyst [57]. Their work evaluated the effects of some process variables on the production of fatty acid ethyl esters from soybean oil in compressed propane using Novozym 435 as a catalyst in a packed-bed reactor. In contrast to using carbon dioxide and n-butane, their results indicated that lipase-catalyzed alcoholysis was achieved in a continuous tubular reactor in compressed propane with high reaction yields at mild temperatures (70°C) and pressures (60 bar) and with short reaction times. The results demonstrated that lipase-catalyzed alcoholysis in a packed-bed reactor using compressed propane as solvent was promising as a potential alternative to conventional processes. It may be possible to manipulate process variables as well as reactor configurations to achieve acceptable yields.

Guan, F. et al. investigated the transesterification of a combination of two lipases [58]. A combination of two lipases was employed to catalyze methanolysis of soybean oil in an aqueous medium during production process. The aqueous medium was a mixture of 7 g soybean oil, methanol in various molar ratios (3:1, 4:1, 5:1, 6:1, and 9:1; methanol : oil) and 2mL (550U per mL) *P. pastoris-Rhizomucor miehei* lipase supernatant of fermentation broth (a water content of 28.6wt%, implies the total H_2O/weight of oil). The two lipase genes were cloned from fungal strains *Rhizomucor miehei* and *Penicillium cyclopium*, and each was expressed successfully in *Pichia pastoris*. Activities of the 1,3-specific lipase from *R. miehei* and the non-specific mono- and diacylglycerol lipase from *P. cyclopium* were 550U and 1545U per mL respectively. Enzymatic properties of these supernatants of fermentation broth (liquid lipase) were continuously stable at 4°C for more than 3 months. Under optimized conditions, the ratio of production conversion after 12h at 30°C, using *R. miehei* alone, was 68.5%. When *R. miehei* was assisted by adding *P. cyclopium*, the production conversion ratio increased to 95.1% under the same reaction conditions. The results suggested that combination of lipases with different specificity, for enzymatic conversion of more complex lipid substrates, is a potentially useful strategy to realize high conversion.

2.2.1. Hydrolysis

In hydrolysis, water is used to break the bonds of certain substances. In biotechnology and living organisms, these substances are often polymers. In hydrolysis involving an ester link between two amino acids in a protein, the products include the hydroxyl (OH) group, which becomes carboxylic acid with the addition of the remaining proton.

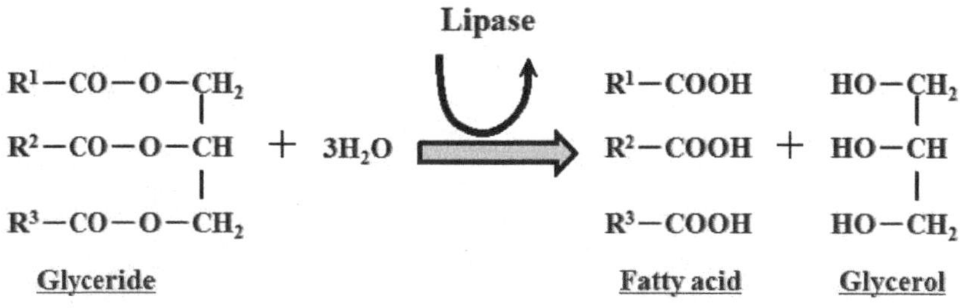

$$R^1-CO-O-CH_2$$
$$R^2-CO-O-CH \quad + \quad 3H_2O \quad \xrightarrow{\text{Lipase}} \quad R^2-COOH \quad + \quad HO-CH$$
$$R^3-CO-O-CH_2 \qquad\qquad\qquad\qquad R^3-COOH \qquad HO-CH_2$$

$$R^1-COOH \qquad HO-CH_2$$

Glyceride   **Fatty acid**   **Glycerol**

Figure 6. Hydrolysis of triglyceride catalyzed by lipase

Hydrolysis reactions in living organisms are performed with the help of catalysis by a class of enzymes known as hydrolases. The biochemical reactions that break down polymers such as proteins (peptide bonds between amino acids), nucleotides, complex sugars and starch, and fats are catalyzed by hydrolases. Within this class, lipases, amylases, and proteinases hydrolyze fats, sugars and proteins, respectively (Fig. 6).

The hydrolysis of vegetable oils is also industrially important. The complete hydrolysis of triglycerides will produce fatty acids and glycerol. These fatty acids find several applications such as in manufacturing soaps, surfactants, and detergents, and in the food industry. Since there are many kinds of natural substrates, the high specificity and selectivity of the enzymes used in the hydrolysis reaction will lead to products of better quality. Lipase has been used in the hydrolysis of different oils and fats to produce free fatty acids.

Ting, W-J. et al. investigated soybean hydrolysis by immobilized lipase in chitosan beads [59]. Their work is the culmination of their research efforts to develop an enzymatic/acid-catalyzed hybrid process for production with a view to utilizing edible and off-quality soybean oils as feedstock. They achieved a higher degree of hydrolysis. The reaction was carried out at 40°C for 12 h using binary immobilized *Candida rugosa* lipase. The conversion of free fatty acid increased rapidly from 0 to 5 h. After 5 h, the conversion of free fatty acid did not increase significantly. Almost 88% of the oil was hydrolyzed after 5 h, indicating that the feedstock for the acid-catalyzed synthesis was easily obtained by the hydrolysis of soybean oil using the binary immobilized lipase. The feedstock for acid-catalyzed production obtained after 5 h of enzymatic hydrolysis of oil contained 12% triglyceride and 88% monoglyceride, diglyceride, and free fatty acid. Problems linked to higher free fatty acid contents can be overcome by using the enzymatic/acid-catalyzed hybrid process proposed in their study. Therefore, any unrefined oil that contains different levels of free fatty acid can be used.

2.2.2. Esterification

Esterification is the chemical process of making esters, which are compounds of the chemical structure R-COOR', where R and R' are either alkyl or aryl groups (Fig. 7). The esterification process has a broad spectrum of uses from preparing highly specialized esters in chemical laboratories to producing millions of tons of commercial ester products. These commercial compounds are manufactured by either a batch or a continuous synthetic process. The batch procedure involves a single pot reactor that is filled with the acid and alcohol reactants.

Sugar fatty acid esters are widely used as non-ionic surfactants in cosmetic and food applications. Current chemical production is based on high-temperature esterification of sugars and fatty acids, using an alkaline catalyst leading to a mixture of products. Alternatively, sugar fatty acid esters can be obtained by fermentation as so-called biosurfactants. The direct esterification of sugar and fatty acid using isolated enzymes (mainly lipases) is hampered by the low solubility of sugars in most organic solvents. Good conversions can be achieved in pyridine, but this solvent is incompatible with food applications. Other solutions are based on the use of alkylglycosides or protected sugars like isopropylidene or phenylboronic acid derivatives, which require additional synthesis steps.

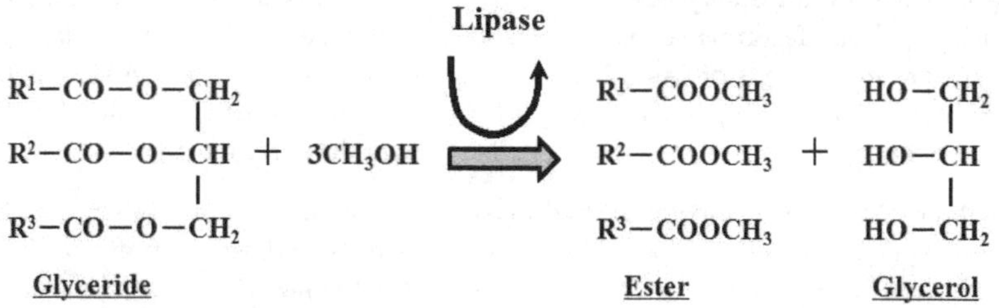

Lipase

$$R^1-CO-O-CH_2$$
$$R^2-CO-O-CH \quad + \quad 3CH_3OH \quad \Longrightarrow \quad R^2-COOCH_3 \quad + \quad HO-CH$$
$$R^3-CO-O-CH_2 \qquad\qquad\qquad\qquad R^3-COOCH_3 \qquad\qquad HO-CH_2$$

$R^1-COOCH_3$ $\quad$ $HO-CH_2$

<u>Glyceride</u> $\qquad\qquad\qquad\qquad\qquad$ <u>Ester</u> $\qquad\qquad$ <u>Glycerol</u>

Figure 7. Esterification of triglyceride catalyzed by lipase

Nagayama, K. et al. investigated lecithin microemulsion-based organogels as immobilization carriers for the esterification of lauric acid with butyl alcohol catalyzed by *Candida rugosa* lipase [60]. Gelatin was used as the gelling component of the microemulsion-based organogels. The maximum reaction rate was obtained at a G_{LW} (volume fraction of water in microemulsion-based organogel) of 75% v/v, a gelatin content of 18.5% w/v, and a lecithin concentration of 18 mM. The reaction proceeded under a reaction-controlled regime, and the reaction rate was influenced by microemulsion-based organogel compositional changes. The effective diffusion coefficient of lauric acid varied with the microemulsion-based organogel composition, while that of butyl alcohol remained constant. The partition coefficient of both substrates was affected by the microemulsion-based organogel composition. Immobilized lipase was reused in a batch-reaction system, and its activity was successfully maintained for 720h. During repeated batch reactions, lipase activity was enhanced, while the ester concentration at 48h was between 30 and 40 mM.

2.3. Immobilized enzymatic reaction of soybean lipid modification

Immobilization of lipase has been investigated to improve the stability and reusability of lipase in oil hydrolysis. For practical applications, a systematic strategy is necessary to select suitable support and organic solvents. Authors investigated a key factor of suitable support to improve enzyme activity and stability of immobilized lipase [61].

Author	Year	Enzyme	Carrier	Immobilization	Substrate	Solvent	Surfactant	Production	Reaction	Reference Number
Ahn, K. W. et al.	2011	Pseudomonas cepacia lipase / Pseudomonas fluorescens lipase	Mesoporous silica	Stirred	Soybean oil	Methanol			Methanolysis	[63]
Cao, L. et al.	1999	Candida antarctica lipase	Polypropylene / Silica gel / PEG	Adsorption / Crosslinking	Olive oil	t-butanol		Fatty acid	Esterification / Hydrolysis	[40]
Dizge, N. et al.	2009	Thermomyces lanuginosus lipase	Microsporous polymeric matrix		Soybean oil	Methanol			Transesterification batch reaction	[54]
Huang, D. et al.	2012	Rhizomucor miehei lipase			Soybean oil	Isooctane		Methyl ester	Transesterification	[64]
Khare, S.K. and Nakajima, M.	2000	Rhizopus japonicus lipase	Celite	Adsorption	Soybean oil / Stearic acid / Soybean oil / Docosahexaenoic acid / p-nitrophenyl palmitate	Hexane		1,2-dipalmitoyl-3-stearoyl glycerol / 1,3-dilauroyl-2-palmitoyl glycerol / p-nitrophenol	Transesterification / Hydrolysis	[65]
Kiatsimkul, P-P. et al.	2006	Candida rugosa lipase / Burkholderia cepacia lipase / Pseudomonas sp. lipase / Penicillium roqueforti lipase / Penicillium camemberti lipase / Aspergillus niger lipase / Mucor javanicus lipase / Rhizomucor miehei lipase			Soybean oil / Epoxidized soybean oil			Fatty acid / Glycerides / Methyl esters	Hydrolysis	[41]
Li, S-F. and Wu, W-T.	2009	Candida rugosa lipase	Polyacrylonitrile nanofiberous membranes	Shaking	Soybean oil				Hydrolysis batch reaction	[66]
Li, S-F. et al.	2011	Pseudomonas cepacia lipase	Polyacrylonitrile nanofiberous membranes	Shaking	Soybean oil			Fatty acid	Transesterification Hydrolysis	[67]
Niee, K. et al.	2001	Rhizopus delemar lipase	W/O microemulsion		Oleic acid / Octyl alcohol	Hexane	DK-ester	Fatty acid esters	Esterification	[52]
Naya, M. and Imai, M.	2012	Candida rugosa lipase	Hydrophobic porous carrier / Polypropylene Accurel MP 100	Crosslinking glutaraldehyde	Triolein	Isooctane	DK-ester	Oleic acid	Hydrolysis	[61]
Noureddini, H. et al.	2005	Pseudomonas cepacia lipase / Penicillium roqueforti lipase / Pseudomonas sp. lipase / Mucor sp. lipase / Aspergillus niger lipase / Rhizopus oryzae lipase / Penicillium camemberti lipase / Rhizopus niveus lipase / Candida rugosa lipase	Hydrophobic sol-gel support	Sol-gel method	Soybean oil	Methanol / Ethanol		Free fatty acids / Methylesters / Ethylesters	Transesterification	[68]
Ozmen, E.Y. and Yilmaz, M.	2009	Candida rugosa lipase	β-cyclodextrin-based polymer	Crosslinking	Soybean oil			Fatty acid	Hydrolysis batch reaction	[69]
Rodrigues, R.C. and Zlachia Ayub, M.A.	2011	Thermomyces lanuginosus lipase / Rhizomucor miehei lipase	Lewatit®	Multipoint-covalently immobilized	Soybean oil	Methanol		Glycerol	Transesterification / Hydrolysis batch reaction	[70]
Ting, W-J. et al.	2008	Candida rugosa lipase	Chitosan beads	Crosslinking glutaraldehyde	Soybean oil			Free fatty acid	Hydrolysis	[59]
Uehara, A. et al.	2008	Rhizopus delemar lipase	W/O microemulsion		Triolein	Isooctane	DK-ester	Oleic acid	Hydrolysis	[53]
Virto, M.D. et al.	1994	Candida rugosa lipase	Polypropylene Accurel EP-100 (1.0-0.2 mm)	Adsorption	Beef tallow / Pork lard / Olive oil	Isooctane / n-Heptane / n-pentane / Isopropanol / Ethyl ether		Free fatty acid	Hydrolysis	[42]
Wang, W. et al.	2011	Rhizomucor miehei lipase / Thermomyces lanuginosus lipase / Candida antarctica lipase	Tree commercial immobilized lipase / Lipozyme RM IM / Lipozyme TL IM / Novozym 435		Soybean oil	t-butanol		Diacylglycerol / Fatty acid	Glycerolysis	[71]
Watanabe, Y. et al.	2002	Candida antarctica lipase			Soybean oil triacylglycerols	Chloroform/methanol			Methanolysis batch reaction	[47]
Xie, W. and Ma, N.	2010	Thermomyces lanuginosus lipase	Magnetic Fe3O4 nano-particles	Mix	Soybean oil	Methanol		Fatty acid methyl esters	Transesterification batch reaction	[48]
Xie, W. and Wang, J.	2012	Candida rugosa lipase	Magnetic chitosan microspheres	Crosslinking glutaraldehyde	Soybean oil	Methanol		Fatty acid methyl esters	Transesterification batch reaction	[72]
Zhou, G. et al.	2009	Candida rugosa lipase	Mesoporous mol-like silica / Mesoporous vesicle-like silica	Physical adsorption	Butyrin	Water / Phosphate buffer saline			Hydrolysis batch reaction	[56]

Table 5. Previous investigations of enzymatic lipid modification by lipase immobilized

Immobilized enzymes have been examined for various industrial applications. In general, enzyme immobilization effectively enables separating the enzyme from products, thus facilitating their recovery and repeated use [40,42,62]. This is promising for industrial enzymatic production of various biomaterials. The main aspects of the currently investigated immobilized enzyme are as follows. First, the molecular structure of the enzyme is directly influenced by immobilization [50]. Second, enzyme reactivity is affected by the physicochemical characteristics of the enzyme carrier and the reaction media [40,51]. To quickly initiate hydrophobic enzymatic reactions, a water-in-oil (W/O) microemulsion system is desirable for achieving higher concentrations of hydrophobic substrate in the reaction media. Third, the diffusion of the substrate and the reaction products determines the rate-limiting condition in the reactivity of the immobilized enzyme [49]. Finally, repeated use of the immobilized enzyme in a practical process is a key factor in reducing costs in industrial applications.

Solid porous carriers are expected to resist compaction and deformation of carrier particles during practical use in bioreactors. Hydrophobic solid porous materials are preferred as immobilized enzymes for hydrophobic reactions. Table 5 summarizes previous hydrophobic substrate reactions using immobilized lipase. Hydrophobic materials, primarily a polypropylene porous commercial carrier called Accurel, have been employed for lipid hydrolysis and esterification. Lipase is adsorbed with strong multipoint interactions in Accurel [73]. Particle size plays a dominant role in determining the rate-limiting condition of the substrate [46,49,55,74]. The particle size as well as handling of particles was very important for both the practical design of the bioreactor and for determining reaction-rate-limiting conditions. In the Accurel EP-100 system, the effect of particle size on reaction rate was examined for a size range of 0.2 to 2.5 mm [49,55,62,74]. A higher reaction rate was obtained for a smaller immobilized carrier. Sabbani et al. reported that the reaction rate was increased six-fold by decreasing the particle size from 0.2 to 1.5 mm[55]. Montero et al. pointed out that cross-linking of lipase (*Candida rugosa*) by glutaraldehyde (GA) was promising for attaining higher reaction activity [62]. Naya and Imai investigated lipid hydrolysis using an immobilized lipase on Accurel MP100 [61]. It examined the effect of particle size on the apparent reaction rate. The technical data were expected to be used in designs for industrial application of Accurel MP100 for hydrophobic immobilized lipase reactions.

2.3.1. W/O microemulsion

W/O microemulsions are spontaneous aggregates composed of amphiphilic molecules in non-polar media. The properties of reverse micelles have been extensively investigated in the field of reverse micellar techniques. Reverse micelles enable hydrophilic proteins to be solubilized in organic solvent and are anticipated to be used as separation and enzymatic reaction media with hydrophobic substrates. When enzymes are micro-encapsulated, they are situated inside the water pool of the W/O microemulsion; whether or not they interact with the micellar interface depends on the enzyme species (Fig. 8). For example, an enzyme reaction involving lipase was observed on the interfacial layer between the hydrophobic phase containing substrates, and the hydrophilic phase containing dissolved lipase.

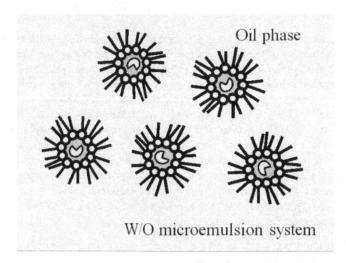

Figure 8. Schematic image of W/O microemulsion system. Micro-water pool was dispersed in bulk oil phase. ⌒ : amphiphilic molecule. ♡ :enzyme

Uehara et al. defined the reaction condition producing high reactivity over a limited range of both hydrophilicity and interfacial fluidity of the microemulsion droplet [53]. Their reaction condition was identified as the most favorable condition for sugar–ester alcohol W/O microemulsion media to perform lipid hydrolysis. The critical micelle concentration depended on the concentration of 1-butanol and was found to be inversely proportional to the second power of the 1-butanol concentration. The initial reaction rate of the hydrolysis of triolein in W/O microemulsion depended on the solubilized water content, reaching a maximum in the limited range of $2 < W_{soln} < 4$. The maximum initial reaction rate increased about 2-fold following the addition of 1-butanol. The most favorable concentration of 1-butanol for hydrolysis by *Rhizopus delemar* was identified as 3.5% v/v.

Naoe et al. investigated the esterification of oleic acid with octyl alcohol catalyzed by *Rhizopus delemar* lipase in a reverse micellar system of sugar ester DK-F-110 [52]. A high initial reaction rate was obtained by preparing a micellar organic phase with extremely low water content. The initial reaction rate decreased slightly with decreasing DK-F-110 concentration. The lipase exhibited 40% of its esterification activity after 28h incubation in the DK-F-110 reverse micellar organic phase. Sodium bis(2-ethylhexyl)sulfosuccinate (AOT) is often used as an ionic amphiphilic molecule for reverse micelle formation owing to the advantages of spontaneous aggregation, thermodynamic stability, and non-additional co-surfactant. In the work of Naoe et al. the turnover number of the DK-F-110 system was larger than that of the system using AOT.

2.3.2. Gel beads carrier

The major problem that must be solved to employ a microemulsion system in industrial processes is the recovery of the products and the repeated use of enzyme. Usual techniques such as extraction and distillation lead to poor separation because of the problems of emulsion-forming and foaming caused by the presence of surfactants. One approach to simplifying the recovery of the product and the enzyme for reuse from microemulsion based-media has been to employ gelled microemulsion systems. Interestingly, many W/O microemulsions can be gelled by adding gelatin, yielding a matrix suitable for enzyme immobilization. Cooling at room temperature causes a transparent gel with reproducible physical properties to form. These enzyme-containing, gelatin-based gels are rigid and stable in various non-polar organic solvents and may therefore be used for biotransformations in organic media. Under most conditions, the gel matrix fully retains the surfactant, gelatin, water, and enzyme components, allowing the diffusion of non-polar substrates or products between a contacting non-polar phase and the gel pellets.

Natural gelling agents such as gelatin, agar and κ-carrageenan have been tested for the formation of lecithin microemulsion-based gels as well as hydrogels presented by Stamatis, H and Xenakis, A [75]. Lipase-containing microemulsions-based organogels formulated with various biopolymers have considerable potential for their application in biotransformations. Lipase immobilized in gelatin and agar organogels exhibited good stability in catalyzing esterification reactions under mild conditions with high conversion yields. High yields (80%) were obtained with agar and κ-carrageenan organogels in isooctane. The remaining lipase activity in repeated syntheses was found to depend on the nature of the biopolymer used for forming the organogels. Gelatin and agar microemulsion-based gels had the highest operational stability. Moreover, aqueous gelatin and agar gels containing only lipase, water, and biopolymer retain their integrity in organic solvents and can also be used for the synthesis of esters.

Chitosan, poly [β-(1-4)-linked-2-amino-2-deoxy-D-glucose], is non-toxic, hydrophilic, biocompatible, biodegradable, and anti-bacterial and can be used as a material for immobilized carriers since it has a variety of functional groups that can be tailored to specific applications. Xie, W. and Wang, J. investigated the effects of various transesterification parameters on the enzymatic conversion of soybean oil [72]. In their work, magnetic chitosan microspheres were prepared by the chemical co-precipitation approach using glutaraldehyde as the cross-linking reagent for lipase immobilization. Using the immobilized lipase, the conversion of soybean oil to fatty acid methyl esters reached 87% under the optimized conditions of a methanol/oil ratio of 4:1 with the three-step addition of methanol, reaction temperature 35°C, and reaction time 30h. Moreover, the immobilized lipase could be used for four times without significant decrease of activity.

2.3.3. Polypropylene carrier

The immobilized lipase (*Candida rugosa*) using polypropylene-based hydrophobic granular porous carrier Accurel MP100 was investigated in lipid hydrolysis reactions involved in the effect of particle size on the apparent reaction rate [61]. The true shape of the original Accurel was similar to a half cylinder (Fig. 9 (a)). Macro-pores existed near the particle surfaces. Inside the particle, the micro-pores formed many branched channels (Fig. 9 (b)).

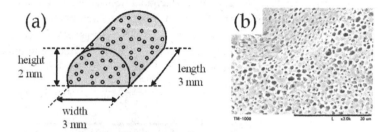

Figure 9. a). Schematic illustration of original Accurel MP100, a polypropylene-based hydrophobic granular support. The true shape of original Accurel particles seemed to be half cylinders. (b) SEM (electron microscopy) image of original Accurel particles.

The amount of immobilized lipase per unit mass of particle was increased by 19% in smaller particles (500 to 840 μm). The immobilized yield lipase based on the adsorbed amount was high (over 98%) in every class of particle size (Fig. 10). Cross-linking of lipase by glutaraldehyde (GA) holds much promise for immobilization.

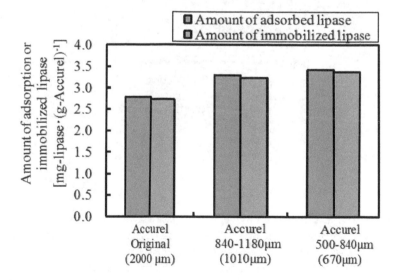

Figure 10. (a). Schematic illustration of original Accurel MP100, a polypropylene-based hydrophobic granular support. The true shape of original Accurel particles seemed to be half cylinders. (b) SEM (scanning electron microscopy) image of original Accurel particles.

The reactivity of immobilized lipase as evaluated from the oleic acid production rate strongly depended on the Accurel particle size. In particular, the 500 to 840 μm (mean diameter 670μm) particles performed significantly outstanding reactivity compared with that of 840 to 1180 μm (mean diameter 1010μm) particles and original Accurel (Fig. 11). The experimental effectiveness factor was obtained and compared with the theoretical effectiveness factor. The difference was speculated to be due to assumptions of the geometrical factor of particles and the partition equilibrium of the substrate between the carrier particle and bulk phase. Quick initiation was observed in the repeated use of immobilized lipase on the 500 to 840 μm particles. The production yield was well-preserved.

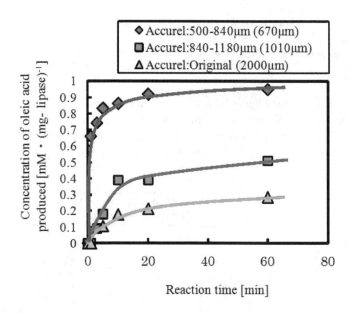

Figure 11. Comparison of reactivity of immobilized lipase for various particle sizes of Accurel.

2.3.4. Nanofiber membrane

Li, S-F. and Wu, W-T. investigated immobilized lipase activity using a nanofiber membrane [66]. The activity retention of the immobilized lipase was 87.5% of the free enzyme. Under these optimal reaction conditions, the hydrolysis conversion of soybean oil was 72% after 10min and 85% after 1.5h. In reusability, the immobilized lipase retained 65% of its initial conversion after 20 additional batch reactions. Protein loading reached 21.2mg/g material of the membrane due to the large specific surface area provided by the nanofibers. This effective enzyme immobilization method has good potential for industrial applications.

2.4. Conclusion

Soybean has been expected to be used both as a food and as a bioresource for attractive functional components. Soybean contains many proteins and much oil. Soybean oil can be hydrolyzed readily by lipase like other vegetable oils. The produced fatty acids find several applications such as in manufacturing soaps, surfactants, and detergents, and in food.

Immobilization of lipase has been investigated to improve its stability and reusability in oil hydrolysis. For practical applications, a systematic strategy is necessary to select suitable support and organic solvent. Since the novel developed method is promising, it could be used industrially for producing chemicals requiring immobilized lipases.

Nomenclature

G_{LW}: volume fraction of water in microemulsion-based organogel phase, referred from [60] (% v/v microemulsion-based organogel phase)

Log P: hydrophobicity index by Laane et al. [34]. P was defined by partition equilibrium (-)

M: molar fraction of ethanol in $SCCO_2$, referred from [33] ([mol-Ethanol]/[mol-($SCCO_2$+Ethanol)])

S: molar fraction of extracted sample in the $SCCO_2$ and ethanol binary system, referred from [33]] ([mol-extracted sample]/[mol-($SCCO_2$+Ethanol)])

W_{soln}: molar ratio of solubilized water to amphiphile, referred from [53] ([mol-H_2O_{soln}]/[mol-amphiphile])

α: the power term on the molar fraction of ethanol M, presented by Eq. (1), referred from [33]. It is summarized in Table 3 (-)

Author details

Masakazu Naya and Masanao Imai[*]

*Address all correspondence to: XLT05104@nifty.com

Course in Bioresource Utilization Sciences, Graduate School of Bioresource Sciences, Nihon University, Japan

References

[1] Izumi, T., Obata, A., Arii, M., Yamaguchi, H., & Matsuyama, A. (2007). Oral Intake of Soy Isoflavone Improves the Aged skin of Adult Women. *Journal of Nutritional Science and Vitaminology*, 53, 57-62.

[2] Lee, Y. B., Lee, H. J., Won, M. H., Hwang, I. K., Kang, T. C., Lee, J. Y., Nam, S. Y., Kim, K. S., Kim, E., Cheon, S. H., & Sohn, H. S. (2004). Soy isoflavones Improve Spatial Delayed Matching-to-Place Peformance and Reduce Cholinergic Neuron Loss in Elderly Male Rats. *The Journal of Nutrition*, 134, 1827-1831.

[3] Lee, Y. B., Lee, H. J., Kim, K. S., Lee, J. Y., Nam, S. Y., Cheon, S. H., & Sohn, H. S. (2004). Evaluation of the preventive Effect of Isoflavone Extract on Bone Loss in Ovariectomized Rats. *Bioscience Biotechnology Biochemistry.*, 68(5), 1040-1045.

[4] Suh, K. S., Koh, G., Park, C. Y., Woo, J. T., Kim, S. W., Kim, J. W., Park, I. K., & Kim, Y. S. (2003). Soybean isoflavones inhibit tumor facter-α-induced apoptosis and production of interleukin-6 and prostaglandin E_2 in osteoblastic cells. *Phytochemistry.*, 63, 209-215.

[5] Mahesha, H. G., Singh, S. A., & Rao, A. G. A. (2007). Inhibition of lipoxygenase by soy isoflavones: Evidence of isoflavones as redox inhibitors. *Archives of Biochemistry and Biophysics*, 461, 176-185.

[6] Chan, H. Y., Chan, H. Y., & Leung, L. K. (2003). A potential mechanism of soya isoflavoes against 7,12-dimetylbenz[a]anthracene tumour inhibition. *British Journal of Nutrition.*, 90, 457-465.

[7] Clubbs, E. A., & Bomser, J. A. (2007). Glycitein activates extracellular signal-regulated kinase via vascular endothelial growth factor receptor signaling in nontumorigenic (RWPE-1) prostate epithelial cells. *Journal of Nutritional Biochemistry.*, 18, 525-532.

[8] Murphy, P. A., Barua, K., & Hauck, C. C. (2002). Solvent extraction selection in the determination of isoflavones in soy foods. *Journal of Chromatography B.*, 777, 129-138.

[9] Luthria, D. L., Biswas, R., & Natarajan, S. (2007). Comparison of extraction solvents and techniques used for the assay of isoflavones from soybean. *Food Chemistry*, 105, 325-333.

[10] Rostagno, M. A., Palma, M., & Barroso, C. G. (2003). Ultrasound-assisted extraction of soy isoflavones. *Journal of Chromatography A.*, 1012, 119-128.

[11] Rostagno, M. A., Palma, M., & Barroso, C. G. (2007). Ultrasound-assisted extraction of isoflavones from beverages blended with fruit juices. *Analytica Chimica Acta*, 597, 265-272.

[12] Araújo, J. M. A., Silva, M. V., & Chaves, J. B. P. (2007). Supercritical fluid extraction of daidzein and genistein isoflavones from soybean hypocotyls after hydrolysis with endogenous β-glucosidases. *Food Chemistry.*, 105, 266-272.

[13] Kao, T. H., Chien, J. T., & Chen, B. H. (2008). Extraction yield of isoflavones from soybeans cake as affected by solvent and supercritical carbon dioxide. *Food Chemistry*, 107, 1728-1736.

[14] Rostagno, M. A., Araújo, J. M. A., & Sandi, D. (2002). Supercritical fluid extraction of isoflavones from soybean flour. *Food Chemistry*, 78, 111-117.

[15] Yu, J., Liu, Y. F., Qiu, A. Y., & Wang, X. G. (2007). Preparation of isoflavones enriched soy protein isolate from defatted soy hypocotyls by supercritical CO_2 . *LWT-Food Science and Technology*, 40, 800-806.

[16] Zuo, Y. B., Zeng, A. W., Yuan, X. G., & Yu, K. T. (2008). Extraction of soybean isoflavones from soybean meal with aqueous methanol modified supercritical carbon dioxide. *Journal of Food Engineering*, 89, 384-389.

[17] Iwai, Y., Nagano, H., Lee, G. S., Uno, M., & Arai, Y. (2006). Measurement of entrainer effect of water and ethanol on solubility in supercritical carbon dioxide by FT-IR spectroscopy. *Journal of Supercritical Fluids.*, 38, 312-318.

[18] Johannsen, M., & Brunner, G. (1994). Solubility of the xanthenes caffeine, theophylline and theobromine in supercritical carbon dioxide. *Fluid Phase Equilibria.*, 95, 215-226.

[19] Kopcak, U., & Mohamed, R. S. (2005). Caffeine solubilities in supercritical carbon dioxide/co-solvent mixtures. *Journal of Supercritical Fluids.*, 34, 209-214.

[20] Li, S., Varadarajan, G. S., & Hartland, S. (1991). Solubility of theobromine and caffeine in supercritical carbon dioxide: correlation with density-based models. *Fluid Phase Equilibria.*, 68, 263-280.

[21] Duarte, C. M. M., Crew, M., Casimiro, T., Aguiar-Ricardo, A., & Ponte, M. N. (2002). Phase equilibrium for capsaicin + water + ethanol + supercritical carbon dioxide. *Journal of Supercritical Fluids*, 22, 87-92.

[22] de la Fuente, J. C., Valderrama, J. O., Bottini, S. B., & del Valle, J. M. (2005). Measurement and modeling of solubilities of capsaicin in high-pressure CO_2 . *Journal of Supercritical Fluids*, 34, 195-201.

[23] Cygnarowicz, M. L., Maxwell, R. J., & Selder, W. D. (1990). Equilibrium solubility of β-carotene in supercritical carbon dioxide. *Fluid Phase Equilibria*, 59, 57-71.

[24] de la Fuente, J. C., Oyarzún, B., Quezada, N., & del Valle, J. M. (2006). Solubility of carotenoid pigment (lycopene and astaxanthin) in supercritical carbon dioxide. *Fluid Phase Equilibria*, 247, 90-95.

[25] Škerget, M., Knez, Ž., & Habulin, M. (1995). Solubility of β-carotene and oleic acid in dense CO_2 and data correlation by a density based model. Fluid Phase Equilibria. , 109, 131-138.

[26] Subra, P., Castellani, S., Ksibi, H., & Garrabos, Y. (1997). Contribution to the determination of the solubility of β-carotene in supercritical carbon dioxide and nitrous oxide: experimental data and modeling. *Fluid Phase Equilibria*, 131, 269-286.

[27] Berna, A., Cháfer, A., Montón, J. B., & Subirats, S. (2001). High-pressure solubility data of system ethanol (1) + catechin (2) + CO_2 (3). *Journal of Supercritical Fluids*, 20, 157-162.

[28] Cháfer, A., Berna, A., Montón, J. B., & Nuñoz, R. (2002). High-pressure solubility data of system ethanol (1) + epicatechin (2) + CO_2 (3). *Journal of Supercritical Fluids.*, 24, 103-109.

[29] Nunes, A. V. M., Matias, A. A., da Ponte, M. N., & Duarte, C. M. (2007). Quaternary Phase Equilibria for $scCO_2$ + Biophenolic Compound + Water + Ethanol. *Journal of Chemical & Engineering Data*, 52, 244-247.

[30] Wang, L. H., & Cheng, Y. Y. (2005). Solubility of Puerarin in Ethanol + Supercritical Carbon Dioxide. *Journal of Chemical Engineering Data.*, 50, 1747-1749.

[31] Huang, Z., Chiew, Y. C., Lu, W. D., & Kawi, S. (2005). Solubility of aspirin in supercritical carbon dioxide/alcohol mixture. *Fluid Phase Equilibria*, 237, 9-15.

[32] Matias, A. A., Nunes, A. V. M., Casimiro, T., & Duarte, C. M. M. (2004). Solubility of coenzyme Q10 in supercritical carbon dioxide. *J. of Supercritical Fluids.*, 28, 201-206.

[33] Nakada, M., Imai, M., & Suzuki, I. (2009). Impact of ethanol addition on the solubility of various soybean isoflavones in supercritical carbon dioxide and the effect of glycoside chain in isoflavones. *Journal of Food Engineering*, 95, 564-571.

[34] Laane, C., Boeren, S., Vos, K., & Veeger, C. (1987). Rules for optimization of biocatalysis in organic solvents. *Biotechnology and Bioengineering*, 30, 81-87.

[35] Paucar-Menacho, L. M., Amaya-Farfan, J., Berhow, M. A., Mandarino, J. M. G., Mejia, E. G., & Chang, Y. K. (2010). A high-protein soybean cultivar contains lower isoflavones and saponins but higher minerals and bioactive peptides than a low-protein cultivar. *Food Chemistry*, 120, 15-21.

[36] Comas, D. I., Wagner, J. R., & Tomas, M. C. (2006). Creaming stability of oil in water (O/W) emulsions: Influence of pH on soybean protein-lecithin interaction. *Food Hydrocolloids*, 20, 990-996.

[37] Berhow, M. A., Wagner, E. D., Vaughn, S. F., & Plewa, M. J. (2000). Characterization and antimutagenic activity of soybean saponins. *Mutation Research*, 448, 11-22.

[38] Viana, P. A., Rezende, S. T., Falkoski, D. L., Leite, T. A., Jose, I. C., Moreira, M. A., & Guimaraes, V. M. (2007). Hydrolysis of oligosaccharides in soybean products by Debaryomyces hansenii UFV-1 α-galactosidases. *Food Chemistry.*, 103, 331-337.

[39] Wang, Q., Ying, T., Jahangir, M. M., & Jiang, T. (2012). Study on removal of coloured impurity in soybean oligosaccharides extracted from sweet slurry by adsorption resins. *Journal of Food Engineering, in press.*

[40] Cao, L., Bornscheuer, U. T., & Schmid, R. D. (1999). Lipase-catalyzed solid-phase synthesis of sugar esters. Influence of immobilization on productivity and stability of the enzyme. *Journal of Molecular Catalysis B: Enzymatic*, 6, 279-285.

[41] Kiatsimkul-P, P., Sutterlin, W. R., & Suppes, G. J. (2006). Selective hydrolysis of epoxidized soybean oil by commercially available lipases: Effects of epoxy group on the enzymatic hydrolysis. *Journal of Molecular Catalysis B: Enzymatic.*, 41, 55-60.

[42] Virto, M. D., Agud, I., Montero, S., Blanco, A., Solozabal, R., Lascaray, J. M., Llama, M. J., Serra, J. L., Landeta, L. C., & Renobales, M. (1994). Hydrolysis of animal fats by immobilized Candida rugosa lipase. *Enzyme and Microbial Technology.*, 16, 61-65.

[43] Hita, E., Robles, A., Camacho, B., Gonzalez, P. A., Esteban, L., Jimenez, M. J., Munio, M. M., & Molina, E. (2009). Production of structured triacylglycerols by acidolysis catalyzed by lipases immobilized in a packed bed reactor. *Biochemical Engineering Journal*, 46, 257-264.

[44] Jimenez, M. J., Esteban, L., Robles, A., Hita, E., Gonzalez, P. A., Munio, M. M., & Molina, E. (2010). Production of triacylglycerols rich in palmitic acid at sn-2 position by lipase-catalyzed acidolysis. *Biochemical Engineering Journal*, 51, 172-179.

[45] Pilarek, M., & Szewczyk, K. W. (2007). Kinetic model of 1,3-specific triacylglycerols alcoholysis catalyzed by lipases. *Journal of Biotechnology*, 127, 736-744.

[46] Salis, A., Sanjust, E., Solinas, V., & Monduzzi, M. (2003). Characterisation of Accurel MP1004 polypropylene powder and its use as a support for lipase immobilization. *Journal of Molecular Catalysis B: Enzymatic.*, 24-25, 75-82.

[47] Watanabe, Y., Shimada, Y., Sugihara, A., & Tominaga, Y. (2002). Conversion of degummed soybean oil to biodiesel fuel with immobilized Candida antarctica lipase. *Journal of Molecular Catalysis B: Enzymatic.*, 17, 151-155.

[48] Xie, W., & Ma, N. (2010). Enzymatic transesterification of soybean oil by using immobilized lipase on magnetic nano-particles. *Biomass and Bioenergy*, 34, 890-896.

[49] Salis, A., Svensson, I., Monduzzi, M., Solinas, V., & Adlercreutz, P. (2003). The atypical lipase B from Candida antarctica is better adapted for organic media than the typical lipase from Thermomyces lanuginose. *Biochimica et Biophysica Acta.*, 1646, 145-151.

[50] Palomo, J. M., Fernandez-Lorente, G., Mateo, C., Ortiz, C., Fernandez-Lafuente, R., & Guisan, J. M. (2002). Modulation of the enantioselectivity of lipases via controlled immobilization and medium engineering: hydrolytic resolution of mandelic acid esters. *Enzyme and Microbial Technology*, 31, 775-783.

[51] Persson, M., Mladenoska, I., Wehtje, E., & Adlercreutz, P. (2002). Preparation of lipases for use in organic solvents. *Enzyme and Microbial Technology*, 31, 833-841.

[52] Naoe, K., Ohsa, T., Kawagoe, M., & Imai, M. (2001). Esterification by Rhizopus delemar lipase in organic solvent using sugar ester reverse micelles. *Biochemical Engineering Journal*, 9, 67-72.

[53] Uehara, A., Imai, M., & Suzuki, I. (2008). The most favorable condition for lipid hydrolysis by Rhizopus delemar lipase in combination with a suger-ester and alcohol W/O microemulsion system. *Colloids and Surfaces A: Physicochemical and Engineering Aspects*, 324, 79-85.

[54] Dizge, N., Aydiner, C., Imer, D. Y., Bayramoglu, M., Tanriseven, A., & Keskinler, B. (2009). Biodiesel production from sunflower, soybean, and waste cooking oils by

transesterification using lipase immobilized onto a novel microporous polymer. *Bioresource Technology.*, 100, 1983-1991.

[55] Sabbani, S., Hedenstrom, E., & Nordin, O. (2006). The enantioselectivity of Candida rugosa lipase is influenced by the particle size of the immobilising support material Accurel. *Journal of Molecular Catalysis B: Enzymatic.*, 42, 1-9.

[56] Zhou, G., Chen, Y., & Yang, S. (2009). Comparative studies on catalytic properties of immobilized Candida rugosa lipase in ordered mesoporous rod-like silica and vesicle-like silica. *Microporous and Mesoporous Materials*, 119, 223-229.

[57] Dalla, Rosa. C., Morandim, M. B., Ninow, J. L., Oliveira, D., Treichel, H., & Vladimir, Oliveira. J. (2009). Continuous lipase-catalyzed production of fatty acid ethyl esters from soybean oil in compressed fluids. *Bioresource Technology*, 100, 5818-5826.

[58] Guan, F., Peng, P., Wang, G., Yin, T., Peng, Q., Huang, J., Guan, G., & Li, Y. (2010). Combination of two lipases more efficiently catalyzes methanolysis of soybean oil for biodiesel production in aqueous medium. *Process Biochemistry*, 45, 1677-1682.

[59] Ting , W. J., Huang, C. M., Giridhar, N., & Wu, W. T. (2008). An enzymatic/acid-catalyzed hybrid process for biodiesel production from soybean oil. *Journal of the Chinese Institute of Chemical Engineering*, 39, 203-210.

[60] Nagayama, K., Yamasaki, N., & Imai, M. (2002). Fatty acid Esterification catalyzed by Candida rugosa lipase in lecithin microemulsion-based organogels. *Biochemical Engineering Journal*, 12, 231-236.

[61] Naya, M., & Imai, M. (2012). Regulation of the hydrolysis reactivity of immobilized Candida rugosa lipase with the aid of a hydrophobic porous carrier. *Asia-Pacific Journal of Chemical Engineering*, 7(S1), S157-S165.

[62] Montero, S., Blanco, A., Virto, M. D., Landeta, L. C., Agud, I., Solozabal, R., Lascaray, J. M., Renobales, M., de Llama, M. J., & Serra, J. L. (1993). Immobilization of Candida rugosa lipase and some properties of the immobilized enzyme. *Enzyme and Microbial Technol.*, 15, 239-247.

[63] Ahn, K. W., Ye, S. H., Chun, W. H., Rah, H., & Kim, S. G. (2011). Yield and component distribution of biodiesel by methanolysis of soybean oil with lipase-immobilized mesoporous silica. *Microporous and Mesoporous Materials*, 142, 37-44.

[64] Huang, D., Han, S., Han, Z., & Lin, Y. (2012). Biodiesel production catalyzed by Rhizomucor miehei lipase-displaying Pichia pastoris whole cells in an isooctane system. *Biochemical Engineering Journal*, 63, 10-14.

[65] Khare, S. K., & Nakajima, M. (2000). Immobilization of Rhizopus japonicas lipase on celite and its application for enrichment of docosahexaenoic acid in soybean oil. *Food Chemistry.*, 68, 153-157.

[66] Li, S. F., & Wu, W. T. (2009). Lipase-immobilized electrospun PAN nanofibrous membranes for soybean oil hydrolysis. *Biochemical Engineering Journal*, 45, 48-53.

[67] Li, S. F., Fan, Y. H., Hu, R. F., & Wu, W. T. (2011). Pseudomonas cepacia lipase immobilized onto the electrospun PAN nanofibrous membranes for biodiesel production from soybean oil. *Journal of Molecular Catalysis B: Enzymatic.*, 72, 40-45.

[68] Noureddini, H., Gao, X., & Philkana, R. S. (2005). Immobilized Pseudomonas cepacia lipase for biodiesel fuel production from soybean oil. *Bioresource Technology*, 96, 769-777.

[69] Ozmen, E. Y., & Yilmaz, M. (2009). Pretreatment of Candida rugosa lipase with soybean oil before immobilization on β-cyclodextrin-based polymer. *Colloids and Surfaces B: Biointerfaces.*, 69, 58-62.

[70] Rodrigues, R. C., & Záchia, Ayub. M. A. (2011). Effects of the combined use of Thermomyces lanuginosus and Rhizomucor miehei lipases for the transesterification and hydrolysis of soybean oil. *Process Biochemistry*, 46, 682-688.

[71] Wang, W., Li, T., Ning, Z., Wang, Y., Yang, B., & Yang, X. (2011). Production of extremely pure diacylglycerol from soybean oil by lipase-catalyzed glycerolysis. *Enzyme and Microbial Technology*, 49, 192-196.

[72] Xie, W., & Wang, J. (2012). Immobilized lipase on magnetic chitosan microspheres for transesterification of soybean oil. *Biomass and Bioenergy*, 36, 373-380.

[73] Gitlesen, T., & Bauer, M. (1997). Adlercreutz, P., Adsorption of lipase on polypropylene powder. *Biochimica et Biophysica Acta.*, 1345, 188-196.

[74] Al-Duri, B., & Yong, Y. P. (1997). Characterisation of the equilibrium behavior of lipase PS (from Pseudomonas) and lipolase 100L (from Humicola) onto Accurel EP100. *Journal of Molecular Catalysis B: Enzymatic.*, 3, 177-188.

[75] Stamatis, H., & Xenakis, A. (1999). Biocatalysis using microemulsion-based polymer gels containing lipase. *Journal of Molecular Catalysis B: Enzymatic*, 6, 399-406.

Chapter 10

Brazilian Soybean Varieties for Human Use

Neusa Fátima Seibel, Fernanda Périco Alves,
Marcelo Álvares de Oliveira and
Rodrigo Santos Leite

Additional information is available at the end of the chapter

1. Introduction

In the present days, the export trade in soybean and its derivatives has a major impact on the Brazilian agro-industrial system and economy. Brazil is the second largest producer, behind only the United States, and three states represent 63% of national production: Mato Grosso, Paraná and Rio Grande do Sul. The 2010/2011 crop has maintained its growth momentum, with higher volume than the previous one, with the climatic factor as primarily responsible for these results.

Soybean has a high nutritional and functional value, is source of quality protein and some essential nutrients to human diet. Due to this nutritional quality, high production, low cost and variety of derivate products, the soybean grain is an alternative for feed [10]. The benefits of soybean have increased its consumption both *in natura* and processed. Studies have shown the association between soy consumption and reduced incidence of esophageal, lung, prostate, breast and colorectal cancer, cardiovascular disease, osteoporosis, diabetes, Alzheimer's disease and menopausal symptoms. The joint action of high-quality protein and polyunsaturated and saturated fats present in soybean helps reduce LDL [37, 28].

In Brazil, the consumption of soybean and its products is still not widespread, due to the few options, exotic flavor to the Brazilian palate and presence of antinutritional factors in the grain. Some of these factors, as the protease inhibitors and lipoxygenase enzymes, can be reduced by suitable thermal processing. Coupled with this, the genetic breeding is responsible by eliminate lipoxygenase enzymes, reducing the flavor which limits the acceptability of the soy products [10, 28].

The soymilk is nutritive, lactose-free, contains no cholesterol and is highly digestible. Can be sold in liquid or powder, pasteurized or sterilized, and commonly flavored, such as juices and vitamins. The extract can also be incorporated as ingredient in breads, cakes, biscuits, chocolates and more.

The results of production, yield, chemical composition and nutritional value of soymilk depend directly on the soybean cultivar, and the quality of soymilk may also be interfered by the water proportion and initial conditions of the grains. There are 316 soybean cultivars currently available in Brazil, with different characteristics of productivity, production cycle, grain size, adaptation to regional climate and lipoxygenase presence. Some are considered commodities, and other cultivars have special purpose.

Cultivars specially developed for human consumption can contribute to the sensory quality of the extract, which directly increases the acceptability of soy as a food, since the sensory quality is decisive in the buying process. Even with important nutritional characteristics, products with undesirable sensory aspects normally lose market to other similar foods. Therefore, sensory evaluation is important to determine the consumer preference, in order to provide support for research, manufacturing, marketing and quality control in new product development [15].

2. Brazilian soybean cultivars

Soybean is currently the most important source of edible oil and high-quality plant protein for feeding both human and animals worldwide [43, 20, 40]. Originated from mid latitude regions, these species are expanding in tropical areas as a result of the development of new genotypes tolerant to the environmental adversities of these localities [9, 40]. One of the largest soybean producers of the world is Brazil, a tropical country that comprises an extensive ecological region with wide variation in the environmental conditions. In Brazil, soybean was firstly grown in the South (in mid latitude areas) and more recently next to Equator line, in the Northeastern region, owing to the development of genotypes with high productivity, well adapted to photoperiod effect and resistant to local pathogens and pests [1, 40]. Presently, in these places, soybean cultivation has great economic and social importance [40]

Water is the main factor changing soybean productivity in time and space [32, 19]. Water use by soybeans varies with climatic conditions, management practices and the life cycle of the cultivar. This crop's response to photoperiod and temperature defines the areas to which it is adapted. Water use by soybean crop increases as the crop grows and is maximal during flowering and pod-fill [19].

Most soybean cultivars respond to photoperiod as quantitative short-day plants and are adapted in a narrow band of latitudes. The soybean has a juvenile stage after emergence when it is especially sensitive to temperature and insensitive to day length [23, 19]. Cultivars with the genetically controlled long juvenile trait have wider adaptability and can be utilized over a wider range of latitudes and planting dates than cultivars without these characteristics [19].

Soybean develops well under a wide range of temperatures, although regions in which the warmest mean monthly temperature is below 20°C are considered inappropriate for soybean [7, 19]. Brown (1960) affirmed that vegetative growth is slow or nil at temperature 10ºC or less and optimum at 30ºC, decreasing thereafter. Temperatures above 40ºC are known to have adverse effects on growth rate, flower initiation and pod-set [19].

Nearly all soybean cultivars exhibit one of two possible growth habits. Cultivars with determinate growth habit have rather distinct vegetative and reproductive development periods. In the other side, indeterminate cultivars have overlapping vegetative and reproductive growth periods.

In recent years, the early planting date and harvest of soybeans, this ensures a lower use of pesticides and makes possible the cultivation of winter maize, resulted in a growth of cultivars of indeterminate habit, principal in South of Brazil. Now, they are dominating the market. Therefore, all breeding programs in Brazil have been working with the introduction of the specific characteristics on indeterminate cultivars.

Embrapa Soybeans has a specific breeding program that develops cultivars with special characteristics for human consumption. However there is still no one cultivar of indeterminate growth habit, but will be released in the near future. It is noteworthy that all cultivars and genotypes that are part of the active Germplasm Bank that give rise to these are conventional. The main cultivars released to date by this program are:

Embrapa 48 – cultivar with more than 15 years on the market. It is knew to processing soymilk with superior flavor when compared with other cultivars. However, due the market need for early cultivars, the cycle has become very long. Regarding the productivity also produces about 20% less than the current more productive cultivars.

BRS 213 – cultivar triple-null for lipoxigenase enzyme, which is responsible for a taste of the "beany flavor" in the extract. This cultivar has light hilum, but almost no more seed on the market, due to some fitossanitary problems and productivity. The cycle is also too long for the demands of today's market.

BRS 216 – cultivar with very small seeds and high protein value but the productivity is at least 30% less compared with the current cultivars. Mainly because of this very small size, a higher loss in the harvest occurs. It is indicated to produced soybean sprouts, especially because the high protein value and the small seed size.

BRS 257 – cultivar triple-null for lipoxigenase enzyme, with similar productivity with current cultivars. The soymilk and soybean flour industries are very interested in this cultivar.

BRS 258 – cultivar originated from an old Embrapa Soybean cultivar called BR 36. It also has a long cycle for the current market requirements and a lower productivity, however the soymilk and flour of this cultivar is well accepted.

BRS 267 – cultivar with very large seeds, sweet flavor and ideal for prepare soymilk and tofu. Also ideal to be consumed as a vegetable soybeans. However the cycle is long and the productivity at least 20% lower when compared with the current cultivars.

BRS 282 – cultivar originated from Embrapa 48 and was launched three years ago. This cultivar does have a cycle consistent with what the market wants today, but studies of the special characteristics of this cultivar are still scarce. The productivity is similar with current cultivars. The soymilk has excellent acceptance and is a cultivar that should be encouraged to be cultivated.

Among the cultivars released by Embrapa, there is a cultivar that did not originate in the program of special cultivars for human consumption but is suitable for this purpose, the BRS 232 cultivar. It has a size large seed and light hilum, ideal characteristics for this purpose. It is always recommended for human consumption when there isn't a special cultivar. This cultivar has well accepted soymilk and flour when compared with current cultivars.

3. Characterization of eight brazilian soybean cultivars for human use

3.1. Chemical composition

The grains of cultivars EMBRAPA 48, BRS 213, BRS 216, BRS 232, BRS 257, BRS 258, BRS 267 and BRS 282, planted in various locations in the state of Paraná, Brazil, during the 2009/10 crop were characterized, and the average results of the composition are shown in Table 1.

The calculated values were similar to those reported by other authors [6, 31]. The highest protein content was found for BRS 258 (44.37%), and differed significantly (p> 0.05) from other grains.[36] have analyzed the same variety, in organic cultivation, and reported lower levels (42.84%). [16] also reported lower contents, with values of 41.70%.

Cultivar	Moisture	Protein	Lipids	Ash	Carbohydrate
Embrapa 48	6.14 ± 0.95[a]	40.11 ± 0.58[bc]	22.45 ± 1.31[a]	4.97 ± 0.10[de]	32.47
BRS 213	5.35 ± 0.19[a]	39.50 ± 0.26[c]	21.86 ± 0.65[ab]	4.90 ± 0.30[e]	33.74
BRS 216	5.61 ± 0.23[a]	41.08 ± 0.54[bc]	19.19 ± 1.32[cd]	4.45 ± 0.15[e]	35.28
BRS 232	5.69 ± 0.07[a]	40.99 ± 0.51[bc]	20.72 ± 0.71[abcd]	5.47 ± 0.16[cd]	32.82
BRS 257	5.67 ± 1.11[a]	41.66 ± 1.38[b]	21.17 ± 0.70[abc]	6.60 ± 0.12[a]	30.57
BRS 258	6.63 ± 0.18[a]	44.37 ± 0.06[a]	18.76 ± 0.62[d]	5.86 ± 0.21[bc]	31.01
BRS 267	6.02 ± 0.16[a]	39.41 ± 1.08[c]	20.03 ± 0.39[bcd]	6.45 ± 0.30[a]	34.11
BRS 282	6.16 ± 0.38[a]	39.96 ± 0.27[bc]	20.70 ± 0.90[abcd]	6.35 ± 0.13[ab]	32.99

Table 1. Centesimal composition of eight soybeans cultivars (g.100g-¹). Means followed by same letters in columns do not differ by Tukey test (p ≤ 0.05). Means from three replicates on a dry basis. * Calculated by difference.

As reported by [11], the soybean features a unique high quality protein source. In general, the industry focus is the production of soybean meal and soybean oil. Therefore, the cultivar BRS 258, due to higher protein content, can be an interesting alternative to the industry, which seeks yield and for high protein content in soybean meal.

According with Embrapa Soja results, the average levels of protein from cultivars Embrapa 48, BRS 213, BRS 232, BRS 257, BRS 267 and BRS 282 are very similar to those found in this study [16]. An exception was found for BRS 216, which presented values of 41.08%, lower than those reported in the literature (43.06%) [17]. [25] has determined the composition of different soybeans cultivars, and found a mean value of 38% in protein. [12] reports values between 33% and 42%. In the present study, the cultivar which surpassed this variation was BRS 258. BRS 216, BRS 232, BRS 257, BRS 258 and BRS 267 were analyzed by [6], and the protein content varied between 38.47% and 39.61%.

Regarding lipids, Embrapa 48 had the highest content (22.45%), but did not differ significantly from BRS 213, BRS 232, BRS 257 and BRS 282. BRS 258 presented the lowest lipid content (18.76%) and did not differ significantly (p ≤ 0.05) from BRS 216, BRS 232, BRS 267 and BRS 282.

In general, literature reports levels between 13 and 25% for lipids in soybean [6]. According to [10], the oil content in soybeans (20%) provides enough calories, so the consumed protein is metabolized for the synthesis of new tissues, and not converted into energy, as commonly seen in diets with low caloric content. However, since industry has as main objective the production of soy oil, cultivars Embrapa 48, BRS 213, BRS 232, BRS 257 and BRS 282 are the best choice for this market.

Some authors report an inverse relation between lipids and protein in soybean [29, 41]. This relation is confirmed by the results for BRS 258, with higher protein, and consequently, lower lipids contents. The average results found in this study were lower compared to lipids and higher for the protein, when compared to that reported in the literature [16, 36].

According to [3], increasing of the planting site temperatures directly affects the oil content in the grains, increasing it. Woodrow and [31] studies about the harvest in 1999, a hot and dry year, showed smaller grains with a reduction in protein concentration, when compared to the previous crop grains (1998). However, there was an increase in lipid content of the grains. For the protein content, the temperature directly influences the composition of amino acids. At higher temperatures, the proteins are rich in methionine, desirable for human consumption.

The higher ash content was the BRS 257 (6.60%), but this did not differ significantly from cultivars BRS 267 and BRS 282. The lowest content was the BRS 216 (4.45%), with no significant differences from the levels of BRS 213 and Embrapa 48. The mineral composition of soybean has quantities that normally exceed the recommended daily dose, when consumed 100 grams of grain, with calcium as the less useful in the consuming of the whole grain [10, 37]. The highest content for total carbohydrates was found in BRS 216 (35.28%), and the lowest in grains of BRS 257 (30.57%).

It is noteworthy that the variations in results between the cultivars, and comparison with literature data using the same varieties, are normal, since the planting site, year and climatic conditions affect these values [29, 33, 31, 36].

3.2. Trypsin inhibitor of soybean grains

Trypsin inhibitor is normally present in the soybean fresh grains, and considered an antinutritional factor. The average values found in literature goes up to 18 milligrams of inhibitor per gram of soybean (HAFEZ, 1983 apud [2]). In this work, however, the average value for BRS 232 (13.82) was lower than usually reported in the literature, and did not differ statistically from BRS 216 (Table 2).

Cultivar	Trypsin Inhibitor(mg TI/g)
Embrapa 48	20.28 ± 0.35^a
BRS 213	22.97 ± 2.42^a
BRS 216	18.12 ± 1.63^{ab}
BRS 232	13.82 ± 0.73^b
BRS 257	21.02 ± 2.18^a
BRS 258	19.61 ± 0.90^a
BRS 267	23.18 ± 1.64^a
BRS 282	22.76 ± 1.92^a

Table 2. Trypsin inhibitor in soybean grains for eight different cultivars. Means followed by same letters in columns do not differ by Tukey test ($p \leq 0.05$). Means from three replicates.

Although the soybean provides high quality protein, these biochemical agents (protease inhibitors) cause a limitation in the biological utilization of the amino acids present in the grains, and may reduce protein digestibility [28, 21], by the blockade of some proteases, including human digestive enzymes. Trypsin is an enzyme secreted by the pancreas, responsible for digestion of proteins by peptide bonds break, and the presence of trypsin inhibitor causes metabolic changes in the pancreas, since the inhibitor binds with the trypsin and inhibits the digestion of proteins. With the protein concentration increasing, the pancreas is stimulated to produce more trypsin, causing pancreatic hypertrophy. Most of these proteases inhibitors are inactivated or inhibited when suitable thermal treatments are applied [11, 21, 30, 35].

3.3. Soybean Isoflavones

The Isoflavones, present in soybean with greater concentration than in the other legumes, belong to the class of phytoestrogens, and have the capacity to assist in the effects of menopause. Besides, the isoflavones are known as having anticancer properties, and antioxidant action that neutralizes free radicals, contributing to reduce LDL (bad cholesterol). The main isoflavones determined in soybean are genistein, daidzein and glycitein, which can be found in the form of aglycones (unconjugated) and glycosylated (conjugated) [4, 18].

In the present study, there was a large variation in the total isoflavones content for the studied cultivars, (Tables 3 and 4), with the highest levels in the BRS 213 (386.60 mg.100g-1) and BRS 282 (364.56 mg.100 g-1) and the lowest in BRS 258 (54.06 mg.100g-1).

The soybean grain naturally presents the isoflavones in the aglycone and glycoside form. The aglycones are absorbed directly, since they are not linked to a sugar, while the other conjugated forms require a hydrolysis for their absorption [27, 18].

The isoflavones profiles for the studied cultivars were very similar, with higher levels of the M-genistein form. However, daidzein, genistein and glycitein forms have been receiving most of attention from researchers. According to [42], genistein has the potential effect of inhibiting the growth of cancer cells at physiological concentrations, and daidzein has effect only if combined with genistein. In the present study, BRS 213 presented the highest level of genistein. Only BRS 267 and BRS 282 showed levels of glycitein, while, BRS 232 showed no levels for the isoflavones highlighted.

The acetyl form was not found in any samples, proving that the soybean did not suffer thermical treatment. According to [2], and [26], in the heat treated products the malonyl form is unstable and may be transformed into the acetyl form. [28] also points out that the processing parameters, the varieties and planting condition affect the composition and / or the isoflavones profile in soy products.

Isoflavones	EMBRAPA 48	BRS 213	BRS 216	BRS 232
G-Daidzin	34.03 ± 1.44	78.26 ± 8.82	75.64 ± 3.38	13.06 ± 0.76
G-Glycitin	8.19 ± 1.05	11.35 ± 1.61	16.84 ± 1.17	4.65 ± 0.53
G-Genistin	22.94 ± 0.59	63.71 ± 5.28	50.03 ± 1.76	10.14 ± 0.20
M-Daidzin	88.91 ± 4.83	75.12 ± 7.81	73.08 ± 3.02	33.84 ± 1.42
M-Glycitin	21.69 ± 3.35	13.94 ± 1.33	20.74 ± 1.63	12.43 ± 1.20
M-Genistin	107.45 ± 2.70	111.46 ± 9.00	89.11 ± 3.34	48.89 ± 0.97
A-Daidzin	0.00 ± 0.00	0.00 ± 0.00	0.00 ± 0.00	0.00 ± 0.00
A-Glycitin	0.00 ± 0.00	0.00 ± 0.00	0.00 ± 0.00	0.00 ± 0.00
A-Genistin	0.00 ± 0.00	0.00 ± 0.00	0.00 ± 0.00	0.00 ± 0.00
Daidzein	2.47 ± 0.21	19.03 ± 0.81	5.79 ± 0.43	0.00 ± 0.00
Glycitein	0.00 ± 0.00	0.00 ± 0.00	0.00 ± 0.00	0.00 ± 0.00
Genistein	1.87 ± 0.03	13.71 ± 0.65	3.63 ± 0.15	0.00 ± 0.00
TOTAL	287.57 ± 14.04	386.60 ± 33.66	334.86 ± 12.80	123.01 ± 2.74

Table 3. Isoflavones profile in soybean grains from the cultivars EMBRAPA 48, BRS 213, BRS 216 and BRS 232 (mg. 100g-1).

Isoflavones	BRS 257	BRS 258	BRS 267	BRS 282
G-Daidzin	39.34 ± 1.19	7.23 ± 0.32	29.82 ± 5.89	29.53 ± 2.96
G-Glycitin	10.38 ± 0.44	2.70 ± 0.32	10.55 ± 3.70	17.19 ± 1.62
G-Genistin	33.98 ± 0.94	4.09 ± 0.01	23.61 ± 1.47	35.43 ± 2.12
M-Daidzin	88.89 ± 2.35	18.31 ± 0.42	25.94 ± 6.95	69.78 ± 4.59
M-Glycitin	24.47 ± 1.41	5.97 ± 0.51	11.13 ± 4.66	31.91 ± 2.68
M-Genistin	134.59 ± 3.96	15.04 ± 0.35	35.82 ± 2.10	150.94 ± 4.97
A-Daidzin	0.00 ± 0.00	0.00 ± 0.00	0.00 ± 0.00	0.00 ± 0.00
A-Glycitin	0.00 ± 0.00	0.00 ± 0.00	0.00 ± 0.00	0.00 ± 0.00
A-Genistin	0.00 ± 0.00	0.00 ± 0.00	0.00 ± 0.00	0.00 ± 0.00
Daidzein	2.05 ± 0.25	0.43 ± 0.12	4.77 ± 0.68	8.89 ± 2.08
Glycitein	0.00 ± 0.00	0.00 ± 0.00	3.08 ± 0.66	10.96 ± 2.30
Genistein	2.59 ± 0.09	0.30 ± 0.09	4.02 ± 0.13	9.93 ± 2.57
TOTAL	329.29 ± 6.24	54.06 ± 1.05	148.74 ± 25.69	364.56 ± 12.87

Table 4. Isoflavones profile in soybean grains from the cultivars BRS 257, BRS 258, BRS 267 and BRS 282 (mg.100g^{-1}).

4. Soymilk production

The soymilk was produced at a 1:6 ratio [soybean (g): water volume (mL)], with the eight characterized cultivars (Embrapa 48, BRS 213, BRS 216, BRS 232, BRS 257, BRS 258, BRS 267 and BRS 282). Initially, the beans were submitted to soaking for five minutes at 95°C in 1:3 ratio with boiling water, and then water was discarded. After this, the soybean was submitted to heat treatment at 95°C for ten minutes, at the proportion of 1:6 with water, and seeds were ground for three minutes. The soymilk was separated from the wet okara by filtration, in which it was applied a heat treatment for two minutes under boiling.

4.1. Efficiency of the soymilk process

The yield is an important processing variable for the food industry, and should be calculated by the ratio between the mass of raw materials and final volume of extract. Thus, from the volume of processed grain, the greater the volume obtained, the better the utilization of production. According to [17], from 500 grams of grain and 4.5 liters of water, 1.5 liters of soymilk are produced. Following the same method, the extraction was performed with 250g of grains and 2.25 liters of water, splitted in 750 mL for maceration (which was discarded) and 1500 mL for grinding. The extraction yields are calculated on the weight of macerated grains, which absorbs water during this process, and the water used in grinding (Table 5).

Cultivar	Weight after soaking (g)	Soymilk(mL)	Yield (%)
Embrapa 48	385	800	42.44
BRS 213	370	820	43.85
BRS 216	390	670	35.44
BRS 232	405	720	37.79
BRS 257	365	820	43.96
BRS 258	385	760	40.32
BRS 267	390	605	32.01
BRS 282	405	840	44.09

Table 5. Yield of the process for soymilk of eight different soybean cultivars.

In this study, it was found that the different soybean cultivars resulted in different yield for the soymilk. The cultivar that showed the best results was BRS 282 (44.09%), higher than that reported by [17]. BRS 213 and BRS 257 showed very similar yields, 43.85% and 43.96% respectively. And BRS 232 (37.79%), BRS 216 (35.44%) and BRS 267 (32.01%) had the lowest yields.

4.2. Soymilk characterization

4.2.1. Freeze-dried

In accordance with the results of fresh grains, the freeze-dried soymilk with the highest protein content was the BRS 258 (42.25%) (Table 6). The lowest level was the extract of BRS 267 (35.34%), but did not differ significantly from extracts of BRS 282 and BRS 213.

Cultivar	Moisture	Protein	Lipids	Ash	Carbohydrates*
Embrapa 48	3.52 ± 0.08^d	36.25 ± 0.40^d	18.34 ± 0.14^a	8.37 ± 0.33^c	33.52
BRS 213	7.78 ± 0.16^a	36.02 ± 0.10^{de}	18.13 ± 0.22^a	9.08 ± 0.24^b	28.99
BRS 216	7.95 ± 0.26^a	33.48 ± 0.54^f	16.95 ± 0.89^{abc}	9.19 ± 0.27^b	32.43
BRS 232	7.49 ± 0.40^a	38.60 ± 0.39^c	14.99 ± 0.89^{cd}	8.69 ± 0.15^{bc}	30.23
BRS 257	7.88 ± 0.09^a	40.44 ± 0.12^b	17.73 ± 0.25^{ab}	8.52 ± 0.02^{bc}	25.43
BRS 258	4.74 ± 0.02^c	42.45 ± 0.23^a	13.57 ± 0.24^{de}	8.52 ± 0.30^{bc}	30.72
BRS 267	5.31 ± 0.27^{bc}	35.34 ± 0.22^e	12.24 ± 0.48^e	9.91 ± 0.30^{bc}	37.20
BRS 282	5.91 ± 0.16^b	35.63 ± 0.19^{de}	15.52 ± 1.20^{bcd}	10.07 ± 0.20^a	32.87

Table 6. Composition of freeze-dried soymilk from eight different soybeans cultivars (g.100g^{-1}). Means followed by same letters in columns do not differ by Tukey test (p ≤ 0.05). Means from three replicates on a dry basis. * Calculated by difference.

The higher lipid content was found in the soymilk of Embrapa 48 (18.35%), and did not differ significantly from extracts of BRS 213, BRS 216 and BRS 257. The lowest level was the soymilk of BRS 267 (12.24%), which did not differ from the extract of BRS 258. The soymilk of BRS 267 showed the highest content of carbohydrate, 37.20%.

4.2.2. *In natura (Liquid)*

The results for chemical composition of the fresh soymilk were determined indirectly, through the results of total solids (Tables 7 and 8).

Cultivar	Moisture	Solids
Embrapa 48	92.16 ± 0.06^{cd}	7.84 ± 0.06^{cd}
BRS 213	92.35 ± 0.06^{c}	7.65 ± 0.06^{d}
BRS 216	92.97 ± 0.06^{b}	7.03 ± 0.06^{e}
BRS 232	89.00 ± 0.06^{f}	11.00 ± 0.06^{a}
BRS 257	91.85 ± 0.06^{de}	8.15 ± 0.06^{bc}
BRS 258	93.71 ± 0.06^{a}	6.29 ± 0.06^{f}
BRS 267	93.68 ± 0.06^{a}	6.32 ± 0.06^{f}
BRS 282	91.67 ± 0.06^{f}	8.33 ± 0.06^{b}

Table 7. Moisture and solids of the soymilk from eight different soybean cultivars (%). Means followed by same letters in columns do not differ by Tukey test (p ≤ 0.05). Means from three replicates.

Cultivar	Protein	Lipids	Ash	Carbohydrates*
Embrapa 48	5.49 ± 0.06^{c}	2.78 ± 0.02^{ab}	1.27 ± 0.05^{cde}	6.14
BRS 213	5.08 ± 0.01^{d}	2.56 ± 0.21^{b}	1.28 ± 0.03^{c}	6.38
BRS 216	4.33 ± 0.07^{e}	2.19 ± 0.12^{c}	1.19 ± 0.04^{de}	6.35
BRS 232	7.86 ± 0.08^{a}	3.05 ± 0.18^{a}	1.77 ± 0.03^{a}	9.32
BRS 257	6.07 ± 0.02^{b}	2.66 ± 0.04^{b}	1.28 ± 0.00^{cd}	6.29
BRS 258	5.09 ± 0.03^{d}	1.63 ± 0.03^{d}	1.02 ± 0.04^{f}	4.84
BRS 267	4.23 ± 0.03^{e}	1.46 ± 0.06^{d}	1.19 ± 0.04^{e}	5.76
BRS 282	5.59 ± 0.03^{c}	2.43 ± 0.19^{bc}	1.58 ± 0.03^{b}	7.06

Table 8. Centesimal composition of soymilk from eight different soybeans cultivars (g.200mL^{-1}). Means followed by same letters in columns do not differ by Tukey test (p ≤ 0.05). Means from three replicates, wet basis. * Calculated by difference.

A simple comparison between the composition of soybeans and their respective soymilk allows observing that the solubilization rate of compounds in aqueous solutions is critical to the final results. The extract of BRS 232 showed the highest levels of soluble compounds (11.01%) obtained during the processing, being superior to others in protein levels (7.86 g. 200mL^{-1}). BRS 267 has presented high protein content in the grains (39.41%), but after processing the content was reduced and the soymilk showed the lowest level (4.23 g.200mL^{-1}), when compared to the other extracts. This reduction indicates that the cultivar has a low content of soluble proteins.

4.2.3. Trypsin Inhibitor and Isoflavones in the soymilk

The results for soymilk confirmed that the heat treatment for 15 minutes at 100°C, performed during the processing, was enough for complete inactivation of the inhibitor, with final values equal to zero. According to [35] and [28] when foods are submitted to appropriate heat treatment, the inhibitor is inactivated.

Regarding to the isoflavones in the freeze-dried soymilk, all cultivars were significantly different, and the extract of BRS 213 had the highest average (421.61 mg.100g-1). The lowest level was found in the extract of BRS 258. For the liquid soymilk, the liquor obtained from BRS 213 maintained the highest isoflavones content (64.50 mg.200mL-1). However, for the equivalent amount to one cup of drink, this did not differ from cultivar BRS 257 (62.01 mg.200mL-1). There was no significant difference between the BRS 216 (53.02 mg.200mL-1) and BRS 282 (53.65 mg.200mL-1) (Table 9).

Cultivar	Freeze-dried soymilk (mg.100g^{-1})	Liquid soymilk (mg.200mL^{-1})*
Embrapa 48	370.65 ± 0.88^b	58.12 ± 0.91^b
BRS 213	421.61 ± 2.55^a	64.50 ± 1.31^a
BRS 216	377.18 ± 3.57^b	53.02 ± 0.50^c
BRS 232	143.45 ± 3.70^e	31.55 ± 0.81^e
BRS 257	380.44 ± 3.51^b	62.01 ± 1.25^a
BRS 258	79.59 ± 0.55^f	10.01 ± 0.07^f
BRS 267	279.91 ± 8.61^d	39.58 ± 2.20^d
BRS 282	322.03 ± 4.41^c	53.65 ± 0.74^c

Table 9. Total isoflavones content of the soymilk produced from eight different cultivars. Means followed by same letters in columns do not differ by Tukey test (p ≤ 0.05). Means values from three replicates. * Equivalent to a glass of drink.

The isoflavones concentration reported in the literature, for soy beverages with original or chocolate flavor, varies between 4 and 13 mg.200mL-1 [8]. [13] reported 12.2 mg.200mL-1, and [22] found a content of 16.6 mg in 200mL. The isoflavones content and profile are also affected by the processing, environment, and the soybean varieties [8]. In the present study, the soymilk prepared with cultivars Embrapa 48, BRS 213, BRS 216, BRS 257 and BRS 282 showed values over 50 mg.200mL-1, which surpass the literature values in more than three times.

The isoflavones profile of soymilk was also determined (Tables 10 and 11), in order to verify alterations after the processing. However, it was observed that the pasteurization did not cause the appearance of acetyl form, as observed by [2] and [26]. According to them, malonyl form in products that suffered heat treatment is unstable, and can be converted in the acetyl form.

M-genistein maintained its high concentration, and after processing the glycitein form was found in all cultivars. The appearance of glycitein after soybean processing has already been reported by [24] who noted the absence of this form in raw soybean, with subsequent detection in soy-based beverage.

Isoflavones	EMBRAPA 48	BRS 213	BRS 216	BRS 232
G-Daidzin	47.16 ± 2.61	89.45 ± 1.72	86.38 ± 1.26	15.28 ± 0.41
G-Glycitin	10.99 ± 0.86	12.22 ± 1.13	19.56 ± 0.68	5.63 ± 0.16
G-Genistin	29.47 ± 0.70	63.73 ± 1.43	49.32 ± 1.11	11.99 ± 0.31
M-Daidzin	116.17 ± 0.52	91.39 ± 2.87	90.35 ± 1.47	38.12 ± 0.89
M-Glycitin	27.55 ± 0.47	16.62 ± 0.63	26.11 ± 0.65	13.50 ± 0.45
M-Genistin	127.95 ± 2.14	121.76 ± 2.62	94.01 ± 1.43	50.38 ± 0.81
A-Daidzin	0.00 ± 0.00	0.00 ± 0.00	0.00 ± 0.00	0.00 ± 0.00
A-Glycitin	0.00 ± 0.00	0.00 ± 0.00	0.00 ± 0.00	0.00 ± 0.00
A-Genistin	0.00 ± 0.00	0.00 ± 0.00	0.00 ± 0.00	0.00 ± 0.00
Daidzein	1.98 ± 0.09	9.78 ± 0.26	3.02 ± 0.26	1.13 ± 0.50
Glycitein	8.27 ± 0.89	9.77 ± 1.04	6.35 ± 0.96	7.07 ± 1.94
Genistein	1.12 ± 0.07	6.89 ± 0.13	2.03 ± 0.10	0.37 ± 0.07
TOTAL	370.65 ± 5.77	421.61 ± 8.59	377.14 ± 3.57	143.45 ± 3.69

Table 10. Isoflavones profile in the freeze-dried soymilk obtained from soybean cultivars EMBRAPA 48, BRS 213, BRS 216 and BRS 232 (mg.100g^{-1}).

Isoflavones	BRS 257	BRS 258	BRS 267	BRS 282
G-Daidzin	43.46 ± 0.41	8.33 ± 0.63	37.88 ± 1.27	38.99 ± 1.11
G-Glycitin	12.84 ± 0.84	3.35 ± 0.19	16.40 ± 0.95	18.30 ± 0.41
G-Genistin	36.41 ± 1.06	6.15 ± 0.47	32.63 ± 1.92	35.74 ± 1.01
M-Daidzin	100.23 ± 1.75	23.84 ± 0.23	60.07 ± 3.94	70.42 ± 0.89
M-Glycitin	29.21 ± 1.35	9.02 ± 0.38	25.78 ± 1.77	30.11 ± 0.58
M-Genistin	144.84 ± 6.87	18.43 ± 0.20	95.25 ± 7.09	116.37 ± 1.75
A-Daidzin	0.00 ± 0.00	0.00 ± 0.00	0.00 ± 0.00	0.00 ± 0.00
A-Glycitin	0.00 ± 0.00	0.00 ± 0.00	0.00 ± 0.00	0.00 ± 0.00
A-Genistin	0.00 ± 0.00	0.00 ± 0.00	0.00 ± 0.00	0.00 ± 0.00
Daidzein	1.62 ± 0.53	0.44 ± 0.17	1.75 ± 0.03	1.66 ± 0.05
Glycitein	9.42 ± 1.63	9.63 ± 0.31	7.92 ± 0.56	8.69 ± 0.21
Genistein	1.76 ± 0.04	0.40 ± 0.12	1.84 ± 0.06	1.76 ± 0.09
TOTAL	380.44 ± 7.64	79.59 ± 0.54	279.91 ± 17.44	322.03 ± 4.41

Table 11. Isoflavones profile in the freeze-dried soymilk obtained from soybean cultivars BRS 257, BRS 258, BRS 267 and BRS 282 (mg.100g^{-1}).

4.2.4. Sensory analysis of soymilk

Sensory analysis was performed in order to differentiate the studied cultivars, and to discuss the best features of each one in the food industry. The panel consisted of 59 judges, comprising 40% women and 60% men, aged between 16 and 54 years, and with good educational level (82.24%), ranging from Superior Incomplete (32.25%), Superior (20.96%) and Postgraduate (29.03%).

When asked about their consumption habits of soybeans "milk", 55% of the judges affirmed to consume the commercial soymilk regularly. Of these, 68% consumed with the addition of flavor, 11% consumed the original extract, and 21% affirmed to consume both (Figure 1).

Judges evaluated the samples, applying scores from 1 (dislike very much) to 10 (like very much), with 5 as an intermediary, in the scale. The mean scores given varied between 4.14 and 6.75, close to "did not like, nor dislike."

Averages were very close between the attributes and cultivars, and this turned into one of the difficulties of implementing the analysis. Judges accustomed to the consumption of commercial extract may have been hindered due to their lack of consumption habit of original extract, with no sugar added (Table 12).

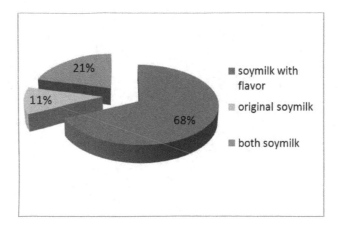

Figure 1. Type of soymilk normally consumed by the judges.

The evaluated attributes showed significant differences ($p \leq 0.05$) only for the flavor and aftertaste, with the extract of BRS 232 receiving the highest average in flavor. The highest average for aftertaste was found in BRS 213, considered the preferred for this attribute. The differences in the averages are directly linked to the composition of the grains and their extracts, and it is important to notice the relationship between the presence of lipoxygenase and the aftertaste of the extract. The soybean grain from BRS 213 has none of the lipoxygenases, which certainly contributed to achieving the highest score in the aftertaste attribute.

Cultivar	Taste	Flavor	Aftertaste	Overall appearance
Embrapa 48	4.27 ± 2.52^a	5.16 ± 2.50^b	4.14 ± 2.28^b	4.75 ± 2.39^a
BRS 213	5.24 ± 2.44^a	6.44 ± 2.05^a	5.61 ± 2.08^a	5.76 ± 2.11^a
BRS 216	4.31 ± 2.36^a	5.75 ± 2.00^{ab}	4.71 ± 2.32^{ab}	5.12 ± 2.22^a
BRS 232	5.12 ± 2.84^a	6.50 ± 2.33^a	4.73 ± 2.78^{ab}	5.82 ± 2.65^a
BRS 257	5.26 ± 2.26^a	6.05 ± 2.18^{ab}	5.14 ± 2.27^{ab}	5.90 ± 2.11^a
BRS 258	5.17 ± 2.35^a	5.87 ± 2.04^{ab}	5.37 ± 2.37^{ab}	5.78 ± 2.26^a
BRS 267	4.74 ± 2.23^a	5.80 ± 2.17^{ab}	4.84 ± 2.11^{ab}	5.34 ± 2.16^a
BRS 282	4.97 ± 2.53^a	5.98 ± 2.17^{ab}	4.98 ± 2.51^{ab}	5.51 ± 2.63^a

Table 12. Points attributed to soymilk of eight different soybean cultivars. Means followed by same letters in columns do not differ by Tukey test ($p \leq 0.05$).

The lipoxygenase enzymes (L1, L2 and L3) can be considered the primarily responsible for the undesirable taste of soybean in Brazil. The beany flavor is result of the three isoenzymes present in the grain, which catalyze the lipids oxidation. The enzyme action only begins with the breakdown and hydration of the grain, since the reaction substrate does not remain exposed in the intact grain. N-hexanal is the volatile compound produced in greater quantity, responsible for the characteristic flavor and taste [11, 29].

Although there was no significant difference between samples for the overall appearance and taste, the extract of BRS 257 deserves special attention for the highest average in both attributes. In other hand, the extract obtained from Embrapa 48 had the lowest average scores for all attributes (flavor, taste, aftertaste and overall appearance). It is interesting to note the need to produce an extract with sensory characteristics similar to the usual habits of consumption, such as the addition of flavor and aroma, allowing a better sensory evaluation.

5. Soybean products application in the food industry

Knowing the consumers profile is critical to the food industry. During the research and development of a new product, the industry focuses on knowing the market and its potential consumers, and many industries apply the sensory analysis as a tool to start or even innovate their activities. The changing of habits related to consumption of soy products has been essential for the growth of the sector [34].

[5] conducted a survey on consumer attitudes in relation to soybeans and their derivatives. They interviewed 100 individuals, 50 men and 50 women, aged 18-40 years, and mostly between 18 and 25, featuring a younger audience. When asked about soy products, tofu and "milk" were the most remembered products, and 40% of the interviewed reported never having consumed these products. A very small portion (8%) reported the consumption of soybean "milk" at least once a week. The soymilk consumption has gradually increased over

the years, by the addition of flavors capable to create a product of good flavor, which little resembles with soybean flavor.

The link between chemical composition and sensory analysis must be directly connected to yield, for the studied cultivar to become an industrial alternative. The extracts from eight soybean cultivars differed significantly in their chemical composition, and the highest protein content was found in BRS 232 (7.86 g.200mL^{-1}). In the sensorial analysis, the samples differed only in aroma and aftertaste, with the extract from BRS 232 achieving satisfactory mean. In the aroma attribute, this same cultivar had the highest average, 6.50.

The highest yield was found in BRS 282 (44.90%), followed by the cultivars Embrapa 48 (42.44%), BRS 213 (43.85%), BRS 257 (43.96%) and BRS 258 (40.32%). These values allow the use of all the studied cultivars, even those with lower yields, when considering the results of chemical composition and sensory analysis. An example is BRS 232, which showed higher levels for all compounds, but had a yield of 37.79%.

6. Final considerations

The soybean cultivars currently available in Brazil have different characteristics of productivity, production cycle, grain size, climate adaptation, lipoxygenase activity, and others. However, the Brazilian consumption of soybean as a food is still small, due to its exotic flavor to the palate, since it is an Asian grain and its development was based on the habits and customs of the orientals. These exotic flavors can be assigned to the presence of lipoxygenase enzymes, saponins and phenolic compounds, responsible for rancid or beany flavors, bitter and astringent, respectively.

A lot of products can be obtained from the soybean. However, the Brazilian food industry had to adapt them to the consumers habits, like the soymilk applied in soy beverages, which for a better acceptance is developed and commercialized with the addition of flavors or fruit juice. This is the most popular and consumed soy derivative in Brazil.

Fortunately, with more information published about the benefits of soybean consumption, its nutritional value and functional properties, this scenario is changing. After recognizing the great importance of this legume, several studies on the development of cultivars with better acceptability started, with the goal to insert soybean as an essential part of the human food. To achieve this, however, in addition to the cultivar adaptation, it is also necessary to check if this cultivar is interesting for industrialization.

The yield is a very important variable in the food industry, but cannot be considered an exclusion factor, since other important variables, such as composition, functional characteristics (isoflavones) and antinutritional compounds (trypsin inhibitor) should be observed for the choice of a soybean cultivar. Therefore, the use of cultivars specially developed for soybean based products, directed to human consumption, may contribute to improve the sensory quality and increase the soybean acceptability as a food.

Author details

Neusa Fátima Seibel[1*], Fernanda Périco Alves[2], Marcelo Álvares de Oliveira[3] and Rodrigo Santos Leite[4]

*Address all correspondence to: neusaseibel@utfpr.edu.br

1 Mestrado Profissional em Tecnologia de Alimentos. Universidade Tecnológica Federal do Paraná (UTFPR). Londrina/PR, Brasil

2 Tecnóloga em Alimentos (UTFPR). Londrina/PR, Brasil

3 Embrapa Soja. Londrina/PR, Brasil

4 Mestrando em Tecnologia de Alimentos (UTFPR). Embrapa Soja. Londrina/PR, Brasil

References

[1] Almeida, L. A., Kiihl, R. A. S., Miranda, M. A. C., & Campelo, G. J. A. (1999). Melhoramento da soja para regiões de baixas latitudes. In: Recursos genéticos e melhoramento de plantas para o nordeste brasileiro Available in http://www.cpatsa.embrapa.br/catalogo/livrorg/temas.html , 15.

[2] Anderson, R. L., & Wolf, W. J. (1995). Compositional Changes in Trypsin Inhibitors, Phytic Acid, Saponins and Isoftavones Related to Soybean Processing. *The Journal of Nutrition*, 125, 581S-588S.

[3] Barros, H. B., & Sediyama, T. (2009). Luz, umidade e temperatura. *In: SEDIYAMA, T. Tecnologias de produção e usos da soja. Londrina: Mecenas*, 17-27.

[4] Bedani, R., & Rossi, E. A. (2005, jul.-dez.) Isoflavonas; Bioquímica, fisiologia e implicações para a saúde. *Boletim CEPPA. Curitiba*, 23(2), 231-264.

[5] Behrens, BJ. H., & Silva, M. A. A. P. (2004, jul.-set..) Atitude do consumidor em relação à soja e produtos derivados. *Ciênc. Tecnol. Aliment., Campinas*, 24(3), 431-439.

[6] Benassi, V. T., Benassi, M. T., & Prudêncio, S. H. (2011). Cultivares brasileiras de soja: características para a produção de tofu e aceitação pelo mercado consumidor. *Semina: Ciências Agrárias, Londrina*, 32(1), 1901-1914.

[7] Berlato, M. A. (1981). Bioclimatologia da soja. *In: Miyasaka, S.; MEDINA, J.C. A soja no Brasil, Institute of Food Technology (ITAL), Campinas*, 175-184.

[8] Callou, K. L. A. (2009). Teor de isoflavonas e capacidade antioxidante de bebidas à base de soja. *124 f. Dissertação (Mestrado- Ciência dos Alimentos)- Faculdade de Ciências Farmacêuticas, Universidade de São Paulo, São Paulo.*

[9] Campelo, G. J. A., Kiihl, R. A. S., & Almeida, L. A. (1999). Características agronôm-
     icas e morfológicas das cultivares de soja desenvolvidas para as regiões de baixas lat-
     itudes. In: Recursos genéticos e melhoramento de plantas para o nordeste brasileiro.
     Available in http://www.cpatsa.embrapa.br/catalogo/livrorg/temas.html , 15.

[10] Carrão-Panizzi, M.C., & Mandarino, J.M.G. (1998). Soja: potencial de uso na dieta
     brasileira. In: EMBRAPA SOJA. Documento 113. Londrina: Embrapa Soja.

[11] Carrão-Panizzi, M.C., Mandarino, J.M.G., Bordingnon, J. R., & Kikuchi, A. (2000). Al-
     ternativa alimentar na dieta humana. In: A cultura da soja no Brasil. Londrina: Embrapa
     Soja,. CD-ROM.

[12] Cecchi, H.M. (2003). Fundamentos teóricos e práticos em análise de alimentos. 2 ed.. ,
     122-133.

[13] Chan, S., Ho, S. C., Kreiger, N., Darlington, G., So, K. F., & Chong, P. Y. Y. (2007).
     Dietary sources and determinants of soy isoflavone intake among midlife chinese
     women in Hong Kong. Journal of Nutrition., 137, 2451-2455.

[14] Chang, Y. K. (2001). Alimentos Funcionais e AplicaçãoTecnológica: Padaria de Saúde
     e Centro de Pesquisas em Tecnologia de Extrusão. In: EMBRAPA Soja. (Org.). Anais
     do I Simpósio Brasileiro sobre os benefícios da soja para a Saúde Humana. Londrina: Embrapa
     da soja, 41-45.

[15] Dutcosky, S. D. (2007). Análise Sensorial de alimentos. 2 ed., Curitiba: Champagnat.

[16] Embrapa soja. (2010). Cultivares de soja 2010/2011 região centro-sul. Londrina: Embra-
     pa Soja: Fundação Meridional, 60.

[17] Embrapa soja. (2003). Manual de receitas com soja. Londrina: Embrapa Soja Docu-
     mentos, 206 , 60.

[18] Esteves, E. A., & Monteiro, J. B. R. (2001, jan-abr.) Efeitos benéficos das isoflavonas de
     soja em doenças crônicas. Revista de Nutrição,Campinas, 14(1), 43-52.

[19] Farias, J. R. B. (1994). Climatic requirements. In: EMBRAPA-CNPSo. (Ed.) Tropical soy-
     bean: improvement and production. Rome: FAO, 13-17.

[20] Friedman, N. M., & Brandon, D. L. (2001). Nutritional and health benefits of soy pro-
     teins. Journal of Agricultural and Food Chemistry, 49, 1069-1086.

[21] Genovese, M. I., & Lajolo, F.M. (2006, jan./fev.) Fatores antinutricionais da soja. In-
     forme Agropecuário, Belo Horizonte, 27(230), 18-33.

[22] Genovese, M. I., & Lajolo, F. M. (2002). Isoflavones in soy based foods consumed in
     Brazil: levels, distribution and estimated intake. Journal of Agricultural and Food Chem-
     istry, 50(21), 5987-5993.

[23] Hodges, T., & FRENCH, V. (1985). Soyphen: soybean growth stages modeled from
     temperature, day length and water availability. Agronomic Journal, 77, 500-505.

[24] Jackson, J. C., Dini, J. P., Lavandier, C., Rupasinghe, H. P. V., Faulkner, H., Poysa, V., Buzzell, D., & DeGrandies, S. (2002). Effects of processing on the content and composition of isoflavones during manufacturing of soy beverage and tofu. *Process Biochemistry*, 37, 1117-1123.

[25] Kagawa, A. (1995). Standard table of food composition in Japan. *Tokyo University of Nutrition for women*, 104-105.

[26] Kurzer, M. S., & Xu, X. (1997). Dietaryphytoestrogens. *Annual Review Nutrition*, 17, 353-381.

[27] Laudanna, E. (2006, jan.-fev.) Propriedades funcionais da soja. *Informe Agropecuário, Belo Horizonte*, 27(230), 15-18.

[28] Mandarino, J. M. G. (2010). Compostos antinutricionais da soja: caracterização e propriedades funcionais. *In: COSTA, N.M.B.; ROSA, C.O.B. (ed.). Alimentos funcionais: componentes bioativos e efeitos. Rio de Janeiro: Rubio*, 177-192.

[29] Morais, A. A. C., & Silva, A. L. (1996). Complicações e resistência ao consumo. *In: MORAIS, A.A.C.; SILVA, A.L.. Soja: suas aplicações. Rio de Janeiro: Medsi*, 151-155.

[30] Morais, A. A. C., & Silva, A. L. (2000). Valor nutritivo e funcional da soja. *Rev. Bras. Nutr. Clin*, 14, 306-315.

[31] Poysa, V., & Woodrow, L. (2002). Stability of soybean seed composition and its effect on soymilk and tofu yield and quality. *Food Research International, Barking*, 35(4), 337-345.

[32] Ravelo, A. C., & Decker, W. L. (1979). Soybean weather analysis models. *In: 14th Conf. Agric. Forest Meteorol*, 72-74.

[33] Rocha, V. S. (1996). Cultura. *In: MORAIS, A.A.C.; SILVA, A.L. Soja: suas aplicações. Rio de Janeiro: Medsi*, 29-66.

[34] Sediyama, T. (2009). Tecnologias de produção e usos da soja. *Londrina: Mecenas*, 306.

[35] Silva, M. R., & Silva, M. A. P. (2000). Fatores antinutricionais: inibidores de proteases e lectinas. *Rev. Nutr.*, 13(1), 3-9.

[36] Santos, H. M. C., Oliveira, M. A., Oliveira, A. F., & Oliveira, G. B. A. (2010, jul-dez). Composição centesimal das cultivares de soja BRS 232, BRS 257 e BRS 258 cultivadas em sistema orgânico. *Revista Brasileira de Pesquisa em Alimentos, Campo Mourão*, 1(2), 07-10.

[37] Teixeira, R. C., Sediyama, H. A., & Sediyama, T. (2009). Composição, valor nutricional e propriedades funcionais. *In: SEDIYAMA, T. Tecnologias de produção e usos da soja. Londrina: Mecenas*, 247-259.

[38] Vasconcelos, I.M., Siebra, E. A., Maia, A. A. B., Moreira, R. A., Neto, A. F., Campelo, G. J. A., & Oliveira, J. T. A. (1997). Composition, toxic and antinutritional factors of

newly developed cultivars of Brazilian soybean (Glycine max). *Journal of the Science of Food and Agriculture*, 75, 419-426.

[39] Vasconcelos, VI. M., Maia, A. A. B., Siebra, E. A., Oliveira, J. T. A., Carvalho, A. F. F. U., Melo, V. M. M., Carlini, C. R., & Castelar, L. I. M. (2001). Nutritional study of two Brazilian soybean (Glycine max) cultivars differing in the contents of antinutritional and toxic proteins. *Journal of Nutritional Biochemistry*, 12, 1-8.

[40] Vasconcelos, I. M., Campello, C. C., Oliveira, J. T. A., Carvalho, A. F. U., Souza, D. O. B., & Maia, F. M. M. (2006). Brazilian soybean Glycine max (L.) Merr.cultivars adapted to low latitude regions: seed composition and content of bioactive proteins. *Rev. Bras. Bot., São Paulo*, 29(4).

[41] Wilcox, J. R., & Shibles, R. M. (2001). Interrelationships among seed quality attributes in soybean. *Crop Science*, 41, 11-14.

[42] Zava, D. T., & Duwe, G. (1997). Estrogenic and antiproliferative properties of genistein and other flavonoids in human breast cancer cells in vitro. *Nutrition and Cancer, Hillsdale*, 27(1), 31-40.

[43] Zeller, F. J. (1999). Soybean (Glycine max (L.) Merril): utilization, genetics, biotechnology. *Bodenkultur*, 50, 191-202.

Chapter 11

Food, Nutrition and Health

Eduardo Fuentes, Luis Guzmán, Gilda Carrasco,
Elba Leiva, Rodrigo Moore-Carrasco and
Iván Palomo

Additional information is available at the end of the chapter

1. Introduction

Botanically, soybean belongs to the order *Rosaceae*, family *Leguminosae* or *Papillonaceae* or *Fabaceae*, subfamily *Papilionoidae*, the genus *Glycine* and the cultivar Glycine max. It is an annual plant that measures up to 1.5 m tall, with pubescent leaves and pods; the stems are erect and rigid. In its primary and secondary roots, are located a variable number of nodes. One of the characteristics of the root system development is its sensitivity to variations in the supply and distribution of inorganic nutrients in the soil. The root system has a main root which can reach a meter deep, with an average being between 40 and 50 centimeters [1].

The soybean is a traditional oriental food, a leguminous plant native to eastern Asia, especially in China. Soybean is cultivated worldwide, the United States is the country that grows more than 50% of the world production of this important food, which has been utilized in the diet of humans around the world, due to its high content of essential amino acids (Table 1) and calcium. It is consumed as cooked beans, soy sauce, soymilk and tofu (soybean curd). Also, a vegetable oil is obtained from soybeans, rich in polyunsaturated fatty acids [2].

Soybean is an annual plant, whose seeds are the edible organ. Soybean grains are rich in protein, and also a good source of various phytochemicals such as isoflavones and lignans, molecules with antioxidant and antiplatelet activities, among other effects; also may help fight and prevent various diseases, so constitute a useful source of food. For these reasons, these compounds have been intensively studied at basic and clinical level [3].

Soybean consumption benefits, especially in several chronic diseases, have been related to its important protein content, high levels of essential fatty acids, vitamins and minerals. Consequently, the present chapter aimed at the comprehensive characterization of the antioxidant and antiplatelet activities of bioactive compounds, of soybean and its derivatives, and the extent to which soybean is a health-promoting food.

Amino acid	g/16 g Nitrogen
Isoleucine	4.54
Leucine	7.78
Lysine	6.38
Methionine	1.26
Cysteine	1.33
Phenylalanine	4.94
Tyrosine	3.14
Threonine	3.86
Tryptophan	1.28
Valine	4.80
Arginine	7.23
Histidine	2.53
Alanine	4.26
Aspartic acid	11.70
Glutamic acid	18.70
Glycine	4.18
Proline	5.49
Serine	5.12

Table 1. Amino acid composition of soybeans seeds. Source: Adapted by authors from FAO (1970) and FAO/WHO (1973).

2. Soybean: foods and bioactive compounds

Soybeans are consumed as cooked beans, which previously should be boiled for at least three hours. With these grains are prepared meals, salads and soups, which in turn, are a source of preparation of other foods. From soybean grains also is possible to obtain soy sauce, which is used especially in oriental foods, such as sushi. The soy sauce is usually made by fermenting soya grains with cracked roasted wheat, which are arranged in blocks and immersed in a cold salt water, the process takes about a year in pots mud, sometimes dried mushrooms are added as mushrooms. In Japan, it is illegal to produce or import artificial soy sauce and therefore all Japanese soy sauces are made by the traditional way [4].

Another food derived from soy is tofu, which is a widely used food in the East as well as vegetarian meals around the world. Required for preparing tofu soybeans are water and a coagulant. Initially, you get the coagulated soymilk, then is pressed and separated the liquid portion from the solid. Tofu has a firm texture similar to cheese milk; the color is cream and served in buckets. Also, from soybeans is possible to obtain vegetable oil, which is character-ized by high polyunsaturated fatty acids [5].

Soy is a source not only of proteins, vitamins and minerals, but also of many bioactive com-pounds, such as isoflavones, protease inhibitors, saponins, and phytates. The great impor-tance of these compounds is based on their biochemical activity, which results in health promotion and disease prevention, by their antioxidant, and antiplatelet activities.

Antioxidant Activity

Soybeans contain a variety of bioactive phytochemicals such as phenolic acids, flavonoids, isoflavones, saponins, phytosterols and sphingolipids; being the phenolic compounds with the highest antioxidant capacity. The key benefits of soy are related to their excellent protein content, its high content of essential fatty acids, numerous vitamins and minerals, their iso-flavones and their higher fiber content.

Polyphenolic compounds are a class of secondary metabolites biosynthesized by the vegetal kingdom [6] and involve a wide range of substances that possess one or more aromatic rings with at least one hydroxyl group. Among them, can be mentioned flavonoids, isoflavones, anthraquinones, anthocyanins, xanthones, phenols, hydroxycinnamic acids, lignin and others.

All of them act as scavengers or stabilizers of free radicals, and can produce chelation of metals, those having carboxyl groups at its end. Works have also been reported that its antioxidant action can be attributed to the inhibition of prooxidants enzymes as lipoxygenase [7].

A study by Xu (2008) in 30 samples of soybean from different regions of North Dakota, Min-nnesota (USA), found that some cultivars of black soybean had a higher antioxidant capacity measured as ORAC, FRAP and DPPH than yellow soybeans and that the phenolic acid con-tent, isoflavones and antiochians was different, suggesting that some selected cultivars can be used as producers of high quality soy, because it provides a high content of phenolic phy-tochemicals and antioxidant properties [8].

Isoflavones and equol. Of all plants, soybean contains the highest amount of different isofla-vones, a variety of phytoestrogens that have a structure similar to estrogen (figure 1). The interest in soybean isoflavones has gained importance since the 90's to today, there are a lot of evidence that these phytoestrogens possess a powerful and wide range of biological activ-ities. Isoflavones are not a steroid structure, however, has a phenolic ring than is capable of binding to the estrogen receptor (ER) and according to Makela (1995) can act as either an agonist or an antagonist [9].

The discovery of high concentrations of isoflavones in urine of adults who consume soy pro-tein, in addition to the evidence supporting its biological action, elevate the soybean to the category of functional food. The FDA in 1999 gave approval to give foods containing 6.25 g of soy protein the seal of protector of cardiovascular health, increasing significantly the sales of foods fortified with soy and isoflavone constituent.

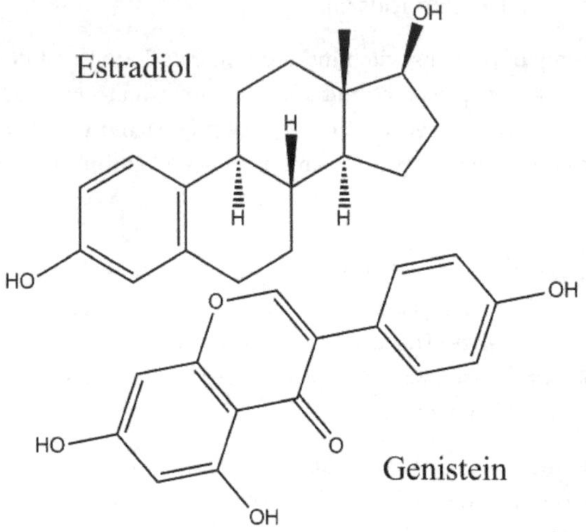

Figure 1. Comparison between the structure of the derivative of Isoflavone (genistein) and estrogen (estradiol), which shows the similarities between the two molecules.

On the other hand, traditional foods in the East, such as extracts and broth of rice and soybeans fermented with microorganisms for 21 days, were reported as antioxidants by Yen (2003) [10] and Yang (2000) [11]. The authors attributed the antioxidant power at the content of polyphenol and the presence of reductons that only occur during the fermentation process. The antioxidant supplements or foods containing antioxidants may be used to reduce the oxidative damage related to age and diseases such as artherosclerosis, diabetes, cancer, cirrhosis, among other [12].

For 10 to 15 years, has been a strong interest in the use of products of botanical origin for the protection, whitening and skin aging. According Baunmann (2009) the mechanism of action of botanical products has been known given the use of advanced technologies applied to research, this is how there are several reports showing that soy components play an important role in the extracellular matrix of the dermis [13].

Moreover, it is accepted that isoflavones act through different mechanisms such as modulation of cell growth and proliferation, extracellular matrix synthesis, inhibition of inflammation and oxidative stress. The isoflavones reduced renal injury by decreasing the concentration of lipoproteins in plasma and acting as an antioxidant reducing the lipid peroxidation [14].

The Equol, whose structure corresponds to 7-hydroxy-3-(4'-hydroxyphenyl)-chroman, is a nonesteroidal estrogen, which was discovered in the early 80's in the urine of adults who consumed soy foods [15]. It has been shown to be a metabolite of daidzein, one of the major isoflavones present in foods containing soy, which is formed after hydrolysis of the isoflavone glycoside [16] at intestinal level and subsequent bacterial biotransformation in the colon [17], leaving an intermediary called dihydro-equol [18-20].

Equol does not originate in plants, but is the product of degradation of the isoflavone glycoside in the intestine [21], situation that was confirmed in infants of 4 months who were fed with formulations containing soy [22, 23]. All mammals can biotransform isoflavone glycoside permanently, except the man who for reasons still unexplained, only 20-35% of adults produces equol after eating foods made from soybeans or that have been enriched with pure isoflavones [17, 24, 25].

Several studies have suggested that those mammals who are equol producers show a greater response to isoflavone-enriched diets, leading to the conclusion that equol is a more potent isoflavone than genistein and is the only one that has a chiral carbon at position 3 of the furan ring, making two enantiomeric forms, S and R that differ significantly in their conformational structure [26].

Gopaul (2012) investigated the effect of equol on gene expression of proteins in the skin, using a cellular model of human dermis and found that equol significantly increased gene expression of collagen, elastin (ELN), and tissue factor inhibitor of metalloproteinases and decreased metalloproteinases (MMPs), causing positive changes in the skin's antioxidant and anti-aging genes. The same occurred in cultured human fibroblasts (hMFC), in which equol significantly increased type I collagen (COL1A1), while 5a-dihydrotestosterone (5a-DHT) significantly decreased cell viability. These findings suggest that equol has great potential for topical applications to the skin, for the treatment and prevention of aging of the skin by increasing the extracellular matrix components [27].

Has been found that Equol have affinity for the estrogen receptor beta, which is abundant in keratinocytes of the epidermis and dermal fibroblasts [28-30]. On the other hand, equol is a selective androgen modulator and has the ability to bind to 5a-dihydrotestosterone (5a-DHT) and inhibit its potent action on the skin [31]. In this sense, we can quote the opposite effect that have androgens and estrogens, the former producing an injury to the skin by increasing MMPs, while the latter have a positive effect on the aging of the skin by increasing collagen, elastin and decreasing MMPs [13, 32-35].

In turn, Muñoz (2009) studied the inhibitory effect of soy isoflavones and the metabolite equol derived from daidzein, an agonist that has biological effects attributed to an antagonism of the thromboxane A2 receptor (TXA2R), which helps explain the beneficial effects of dietary isoflavones in the prevention of thrombotic events [36].

Recent works by Ronis (2012) studied the effect of mice fed with extracts of soy protein or isoflavones finding that can reduced the metabolic syndrome in rats via activation of peroxisome proliferation activated receptor (PPAR), liver X receptor (LXR) and decreased signaling protein binding to the sterol regulatory element binding proteins (SREBP) [37].

Anthocyanins. The anthocyanins are known to have antioxidant effects and play an important role in preventing various degenerative diseases. Structurally, this are a suitable chemical structure to act as antioxidants, since they can donate hydrogens or electrons to free radicals or catch and move them in its aromatic structure. There are about 300 anthocyanins in nature, with different glycosidic substitutions in the basic structure of the ion 2-phenyl-benzopirilio or flavilio [38].

Paik (2012) examined the effect of anthocyanin extracts from the cover of black soybean in an animal model of retinal degeneration (RD), the leading cause of death of the photorecep-tor cells that lead to blindness and noted that extracts of anthocyanins may protect retinal neurons from damage induced by degenerative agents such as N-methyl-N-nitrosourea (MNU) at a dose of (50mg/kg), which acts as a methylating agent that causes DNA damage to the photoreceptors [39].

In general, bioactive compounds from soybean are many, but still exists wide variety of in-formation of the beneficial effects and also adverse effects of isoflavones and anthocyanins, so likewise it is necessary to study more thoroughly this compound and its relation to chronic diseases through scientific studies with larger number of patients and longer study periods, in order to clarify all the diffuse concepts, labile or poorly sustained, so as to give the isoflavones, a clear place in the diet therapy.

Antiplatelet Activity

There has been much recent interest in the cardiovascular benefits of dietary soybean on potential anti-thrombogenic and anti-atherogenic effects [40]. Extracts containing isofla-vones and soy saponins inhibit the platelet aggregation *ex vivo* induced by ADP and colla-gen in diabetic rats [41]. Moreover, black soybean extracts inhibited platelet aggregation induced by collagen *in vitro* and *ex vivo*, and attenuates the release of serotonin and P-selec-tin expression [42].

The effects of soybean products on platelet aggregation were initially described for genistein [43]. In these reports, genistein was able to inhibit platelet activation induced by collagen and thromboxane A2 analog (TXA2), but not by thrombin. Genistein (10 mg/kg) in mouse significantly prolonged the thrombotic occlusion time and significantly inhibited *ex vivo* and *in vitro* (30 µM) platelet aggregation induced by collagen [44]. Genistein is a well-known in-hibitor of protein tyrosine kinases, however, on platelet functions *in vitro* genistein inhibits activation of phospholipase C in stimulated platelets, apparently independent of its effects on tyrosine kinases. These results suggest that dietary supplementation of soy may prevent the progression of thrombosis and atherosclerosis [45]. Daidzein, another soy flavonoid that lacks tyrosine kinase inhibitory activity also inhibited the response to collagen and TXA2, suggesting that these flavonoids inhibit platelet aggregation by competition for the throm-boxane A2 receptor (TXA2R) rather than through tyrosine kinase inhibition. Genistein and daidzein have effect on platelets, macrophages and endothelial cells: inhibited collagen-in-duced platelet aggregation in a dose-dependent manner and in macrophage cell line activat-ed with interferon γ, plus lipopolysaccharide inhibit tumoral necrosis factor α (TNF-α) secretion, dose-dependently. Both isoflavones also dose-dependently decreased monocyte chemoattractant protein-1 secretion induced by TNF-α in human umbilical vein endothelial cells [40].

Equol is more active than soy isoflavone itself to compete for binding to TXA2R in human platelets (in the range of micromoles / L), so that inhibits the platelet aggregation and secretion induced by U46619 [36]. Under equilibrium conditions, the following order of the relative affinity in inhibiting [(3)H]-SQ29585 binding was: equol > genistein > daidzein > glycitein > genistin > daidzin > glycitin [36]. Guerrero and colleagues suggested that this competitive binding was due to structural features of these flavonoids such as the presence of a double bond in C2-C3 and a keto group in C4 [36].

From a extraction of soy sauce, two kinds of components with anti-platelet activity were isolated and structurally identified: 1-methyl-1,2,3,4-tetrahydro-β-carboline (MTBC) and 1-methyl-β-carboline (MBC). MTBC shows IC$_{50}$ ranging from 2.3 to 65.8 μg/mL for aggregation response induced by epinephrine, platelet-activating factor (PAF), collagen, ADP and thrombin [46]. Membrane fluidity regulates the platelet function and various membrane-fluidizing agents are known to inhibit platelet aggregation [47]. Certain β-carbolines influence the fluidity of model membranes. The alteration of membrane fluidity may be involved in the antiplatelet effects of MTBC and MBC [48].

Soybean protein inhibits platelet aggregation induced by thrombin, collagen and ADP, and prolongs the clotting time [49]. Also was observed that most fractions obtained of soy protein hydrolysates (gel filtration chromatography, reverse-phase HPLC and cation exchange HPLC) inhibited rat platelet aggregation induced by ADP, which suggests that most peptides have some degree of antiplatelet effect. From the inhibitory fractions, two new peptides were identified, SSGE and DEE, and at concentrations of 480 and 460 μM, respectively, inhibited in 50% platelet aggregation [50].

The diet may be the most important factor influencing the risk of cardiovascular disease. Soybean derivatives can be denominated as functional ingredients as they contain bioactive compounds that inhibit platelet aggregation, which gives a preventive effect on thrombus formation [51].

3. Functional Food

Soybean is a very rich source of essential nutrients and one of the most versatile foodstuffs. It possesses good quality protein and highly digestible (92–100%) and contains all the essential amino acids. Soybean-protein products also contain a high concentration of isoflavones (1 g/kg) [52]. Therefore, consumption of soy-based foods has been associated with multiple health benefits [53, 54]. Among a variety of soybeans, black soybean is known to display diverse biological activities superior to those of yellow and green soybeans, such as antioxidant, anti-inflammatory and anticancer activities [42].

Soy food intake has been shown to have beneficial effects on cardiovascular disease risk factors. Data directly linking soy food intake to clinical outcomes of cardiovascular disease, evidence that soy food consumption may reduce the risk of coronary heart disease in women and may be protective against the development of subarachnoid hemorrhage [54, 55]. Based on that 1% reduction in cholesterol values is associated with an approximate 2-3% reduction in the risk of coronary heart disease, it can be assumed that a daily intake of 20-50 grams of isolated soy protein could result in a 20- 30% reduction in heart disease risk [56, 57].

Several components associated with soy protein have been implicated in lowering cholesterol: trypsin inhibitors, phytic acid, saponins, isoflavins, fiber and proteins [58]. Apparently, there is a synergy among the components of intact soy protein, which provides the maximum hypocholesterolemic benefit. A variety of clinical trials have demonstrated that consuming 25 to 50 g/daily of soy protein is both safe and effective in reducing LDL cholesterol by ≈4% to 8% [58]. Therefore, maturation of SREBP and induction of SRE-regulated genes produce an increase in surface LDL receptor expression that increases the clearance of plasma cholesterol, thus decreasing plasma cholesterol levels [59]. However, other results present direct evidence for the existence of LDL receptor and plasma lipoprotein-independent pathways by which dietary soy protein isolate inhibits atherosclerosis [60, 61].

The addition of soy protein in diet or replacing animal protein in the diet with soy, lowers blood cholesterol. Moreover, defatted soy flour is a widely used in these applications as a partial replacement for nonfat dry milk [62-64]. Soy protein can increase protein content and its used in compounded foods (breads, crackers, doughnuts, and cakes) for their functional properties, including water and fat absorption, emulsification, aeration (whipping), heat setting, and for increasing total protein content and improving the essential amino acids profile [65].

The low breast cancer mortality rates in Asian countries and the putative anti-estrogenic effects of isoflavones have fueled speculation that soyfood intake reduces breast cancer risk [66]. Soy sauce promotes digestion, because the consumption of a cup of clear soup containing soy sauce enhances gastric juice secretion in humans. The feeding of a diet containing 10% soy sauce to male mice for 13 months also reduces the frequency and multiplicity of spontaneous liver tumors [67]. Over the past decades, enormous research efforts have been made to identify bioactive components in soy [68]. The Health effects of soy dietary are variable depending on individuals' metabolism and in particular to their ability to convert daidzein to equol that seems to be restricted to approximately 1/3 of the population. Equol production has been indeed linked to a decreased on arterial stiffness and antiatherosclerotic effects via nitric oxide production [69]. Despite being a biotransformation product of daidzein, the equol at low dosage can prevent skeletal muscle cell damage induced by H_2O_2 [70] and possesses anticancer activity via apoptosis induction in mammary gland tumors of rats [71].

Hydroponic cultivation improved the nutritional quality of soybean seeds with regard to fats and dietary fiber. This suggest that specific cultivars should be selected to obtain the desired nutritional features of the soybean raw material [72]. Irrigation enhanced the isoflavone content of both early- and late-planted soybeans as much as 2.5-fold. Accumulation of individual isoflavones, daidzein and genistein, are also elevated by irrigation [53].

4. Digestion and Absorption

A number of factors including the amount consumed, chemical speciation, interactions with other ingredients, physiological state (e.g., gender, ethnicity, age, health status) and intestinal microflora influence the absorption of dietary isoflavonoids by the gastrointestinal tract [73]. Additionally, the absorption and disposition of isoflavones (daidzein and genistein) appears to be independent of age, menopausal status and probiotic or prebiotic consumption [74, 75].

After absorption, isoflavones are reconjugated predominantly to glucuronic acid and to a lesser extent to sulphuric acid. Only a small portion of the free aglycone has been detected in blood, demonstrating that the rate of conjugation is high [76]. The isoflavone aglycones are absorbed faster and in greater amounts than glucosides in humans, dependent on the initial concentrations. Thus, products rich on aglycone can be more effective in the prevention of chronic diseases [77]. Concentrations of isoflavones and their gut microflora metabolites in the plasma, urine, and feces are significantly higher in the subjects who consume a high-soy diet than in those who consume a low-soy diet [78].

Bioavailability of isoflavone glycosides (daidzein and genistein) as pure compounds or in a soyfood matrix (soymilk) requires initial hydrolysis of the sugar moiety by intestinal β-glucosidases for the uptake to the peripheral circulation [16]. Twenty-four hours after dosing, both plasma and urine isoflavone concentrations were nearly null [79]. The genistein compound is absorbed from the lumen partly unhydrolyzed and transported directly (by an unknown transporter or diffusion) to the vascular side [80]. Conjugates of daidzein are more bioavailable than those of genistein. Thus, after oral administration of soy extract in rats providing 74 micromol of genistein and 77 micromol of daidzein / kg (as Conjugates), were found that plasma concentration of daidzein was maximal at 2 h and it was almost double that of genistein. Since about 50% of the genistein dose is excreted as 4-ethyl phenol (the main end product from genistein) [81]. The end product of the biotransformation of the phytoestrogen daidzein, is the equol, that is not produced in all healthy adults in response to dietary challenge with soy or daidzein [21]. However, plasma genistein concentrations are consistently higher than daidzein when equal amounts of the two isoflavones are administered, and this is accounted for by the more extensive distribution of daidzein (236 L) compared with genistein (161 L). In addition to the conjugated state, the chemical structures of isoflavones play a major role in its pharmacokinetics with marked qualitative and quantitative differences depending on the type of supplement ingested [82-84].

5. Health

Functional food may act as an adjunctive therapy/alternative treatment of different pathologies, and scientific studies are appearing more frequently demonstrating that this hypothesis is, indeed, a reality. Soybean containing isoflavone and protein is considered a functional food item [85].

Epidemiological studies suggest that soybean consumption is associated, at least in part, with lower incidences of a number of chronic diseases. The lower rates of several chronic diseases in Asia, including type 2 diabetes, certain types of cancer and cardiovascular diseases, between others, have been partly attributed to consumption of large quantities of soy foods (figure 2) [86].

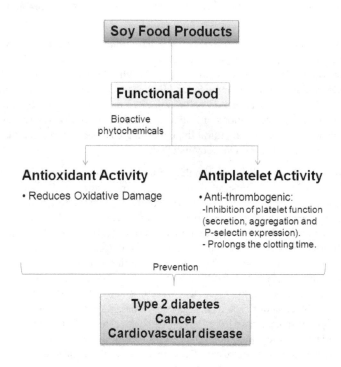

Figure 2. Biological activities from soy food products and its effect on health.

a) *Soybean and type 2 diabetes.* Type 2 diabetes mellitus is a multifactorial metabolic disorder disease, which results from defects in both insulin secretion and insulin action. Insulin stimulates uptake, utilization and storage of glucose in cells throughout the body by inducing multiple signaling pathways in the tissues that express the transmembrane insulin receptor [87], especially in skeletal muscle that accounts for 75% of whole-body insulin-stimulated glucose uptake. The reduced responsiveness of cells to insulin is due to defective intracellular signaling processes [88, 89]. Millions of people have been diagnosed with type 2 diabetes, and many more are unaware they are at high risk [90]. Obesity is the major risk factor for diabetes and accounts for ≈70% of the variance in the prevalence of this common disease [91]. Dry beans and soyfoods offer benefits in the prevention of diabetes and in the clinical management of established diabetes, soybeans, in particular, have a low glycemic index.

They are rich in phytates, soluble fiber, and tannins, all of which correlate inversely with carbohydrate digestion and glycemic response [62, 92].

Soybean and its natural bioactive products have been studied as an antidiabetic potential. Studies have been conducted to examine the therapeutic effect of different bioactive compounds such as aglycin, phenolic-rich extract and glyceollins, all derivates of soybean. Soybean peptides have been widely used as a natural health food and supplement. It should be good for preventing obesity and diabetes because long-term feeding of soy peptide induced weight loss in obese mice [93]. In healthy and diabetic animal models, soybean peptides decreased blood glucose by increasing insulin sensitivity and improving glucose tolerance [89, 94].

Aglycin, a natural bioactive peptide isolated from soybean, structurally, it has a high stability with six cysteines embedded in three disulfide bonds. It is also resistant to digestion by trypsin, pepsin, Glu-C and bovine rumen fluid and has an antidiabetic potential [95, 96], at this respect, Lu J (2012), studied the effect of aglycin administration on the glucose homeostasis. For this, the diabetes was induced in BALB/c mice fed with a high-fat diet and a single intraperitoneal injection of streptozotocin. With onset of diabetes, the mice were administered daily with aglycin (50 mg/kg/d) for 4 weeks, blood glucose was monitored once a week [89]. The administration of aglycin restored insulin-signaling transduction by maintaining the insulin receptor (IR) and the insulin receptor substrate 1 (IRS1) expression at both the mRNA and protein levels, as well as elevating the expression of p-IR, p-IRS1, p-Akt and membrane GLUT4 protein. The results hence demonstrate that oral administration of aglycin can potentially attenuate or prevent hyperglycemia by increasing insulin receptor signaling pathway in the skeletal muscle of streptozotocin/high-fat-diet-induced diabetic mice [89].

Complex polysaccharides are hydrolyzed by α-amylase to oligosaccharides that are further hydrolyzed to liberate glucose by intestinal α-glucosidase before being absorbed into the intestinal epithelium and entering blood circulation. Therefore, α-amylase and α-glucosidase inhibitors may help to reduce postprandial hyperglycemia by inhibiting the enzymatic hydrolysis of carbohydrates, and hence may delay the absorption of glucose [97]. Therefore, effort has been directed in finding a natural and safer α-amylase and α-glucosidase inhibitors. Phytochemicals such as phenolics with strong antioxidant properties has been reported to be good inhibitors of these enzymes [98]. The phenolic compounds of soybean have also been studied, Ademiluyi (2011) [99] assessed the inhibitory activities of different phenolic-rich extracts from soybean on key-enzyme linked to type 2 diabetes (α-amylase and α-glucosidase) [99]. Their results shown that the different phenolic-rich extracts used inhibited α-amylase, α-glucosidase activities in a dose dependent pattern and the free phenolic extract had higher α-glucosidase inhibitory activity when compared to that of α-amylase; this property confer an advantage on soybean phenolic-rich extracts over commercial antidiabetic drugs with little or no side effect [99].

In turn, the flavonoid family of phytochemicals, particularly those derived from soy, has received attention regarding their estrogenic activity as well as their effects on human health and disease. In addition to these flavonoids other phytochemicals, including phytostilbene, enterolactone, and lignans, possess endocrine activity; the health benefits of soy-based foods may, therefore, be dependent upon the amounts of the various hormonally active phytochemicals within these foods, especial attention have recieved the isoflavonoid phytoalexin compounds, glyceollins, in soy plants grown under stressed conditions [100]. A glyceollin-containing fermented soybeans was assessed by Park S (2012), where diabetic mice, induced by intraperitoneal injections of streptozotocin (20 mg/kg bw), were administered a high fat diet with no soybeans (control), 10% unfermented soybeans and 10% fermented soybeans containing glyceollins (FSG), respectively, for 8 weeks; among the diabetic mice, FSG-treated mice exhibited the lowest peak for blood glucose levels with an elevation of serum insulin levels during the first part of oral glucose tolerance testing. FSG also made blood glucose levels drop quickly after the peak and it decreased blood glucose levels more than the control during insulin tolerance testing [101]. The enhancement of glucose homeostasis was comparable to the effect induced by rosiglitazone, a commercial peroxisome proliferator-activated receptor-γ agonist, but it did not match the level of glucose homeostasis in the non-diabetic mice [101].

In vitro studies suggest that isoflavones have antidiabetic properties such as the inhibition of the intestinal brush border uptake of glucose, α-glucosidase inhibitor actions, and tyrosine kinase inhibitory properties [102]. Animal studies have indicated that soy protein or isoflavones improve glycemic control, lower insulin requirement, and increase insulin sensitivity [96, 103]. Several observational studies have also suggested that soy intake was associated with improved glycemic control or lowered risk of diabetes [104, 105].

Nevertheless, data from human clinical trials that evaluated the possible beneficial effects of isoflavone-rich soy products on glycemic control and insulin sensitivity have generated mixed findings. Some studies showed that soy products and foods significantly improve glycemic control [106-108], whereas others observed no significant effect [109-111]. This inconsistency could be due to a number of possible reasons. A wide variety of soy products, such as traditional soy foods, isolated soy protein, soy extracts, or purified isoflavones, and a variety of controls have been used [112]. Varied amounts and compositions of protein and isoflavones in soy products and the menopausal status of participants, study duration, baseline health status of participants, intervention adherence, and degree to which dietary intake is controlled may have contributed to variations in studies [113].

b) Soybean and cancer. For a long time has been described the effect of soy on human health, this legume present since ancient times in the Oriental diet is now present all over the world. The eastern population has low rates of various cancers, among these; breast cancer is undoubtedly a classic example.

Liener and Seto in 1955, described an *in vivo* effect in rats inoculated with Walker tumor. Daily injection of a soy extract at 25 mg / kg in these animals results in a decrease in tumor size and weight at the end of the period. This effect was observed only in animals that were treated from the time of inoculation of the tumor and not in those treated after tumor establishment [114]. From Liener investigations have been reported countless works that characterize the effect of soy on tumor growth.

Since 1996, it has been reported the use of extracts rich in soy isoflavones in human clinical studies. Has recently been published the results of a clinical study of soy isoflavones used in women at high risk of breast cancer, the results at 6 months of intervention is that soy isoflavones decrease epithelial proliferation in women with risk, modulating the expression of a large number of genes involved in carcinogenesis [115]. One of the markers associated with risk of breast cancer is the IGF-1 (Insulin-like growth factor 1), this growth factor is involved in various processes that stand between growth and cell development. The increase in plasma concentration is associated with up to seven times greater risk of developing this malignancy. However, high intake of soy in American women produces a slight increase in the concentration of IGF-1, these data are extremely complex to interpret. Eating a diet high in soy and algae produce a decrease of up to 40% of the levels of IGF-1 [116].

There is a wide range of cancers in which have been found some degree of association with soy consumption. Recently, a meta-analysis has shown that high-soy intake determined a low risk of lung cancer [117]. In this context, researchers have demonstrated *in vivo*, in nude mice (immune compromised) that were intravenously injected human tumor cell line A549, these cells generate tumor nodules in the lungs of these animals. By analyzing histologically the tissues of mice that received a treatment of 1 mg / day for 30 days orally soybeans, it was determined that the tumor cells were more sensitive to radiation therapy in addition to reducing vascular damage, inflammation and fibrosis caused by radiation on healthy tissue [118].

The relationship between breast cancer and the presence of estrogen receptor in these tumor cells is a fact and is related primarily to the aggressiveness of the tumor and its size. The non-steroidal phytoestrogens chemical compounds present in plants and especially in soy and structurally resemble human estrogen, may play a central role in the effect on the risk of breast cancer in postmenopausal women. This effect, albeit has been reported, is still not entirely clear, while some studies are very conclusive as that conducted by Zaineddin et al in 2012, recently, over 3000 cases and over 5000 controls in a case-control study in German women, the results did show a relationship between the consumption of phytoestrogens including soy and reduced the relative risk of breast cancer in postmenopausal women [119].

Genistein a predominant isoflavonoid in soybeans has long shown a beneficial effect on the prevention and treatment of some cancers. Has been studied this soy isoflavone and the mechanisms that may be associated with its antitumor effect. Among the mechanisms described is the inhibition of nuclear factor NFkappa B a key molecule in tumor cell survival, in 2011 a group of researchers reported that genistein inhibited the proliferation and induced apoptosis in the BGC-823 cell line, a cell line of human gastric cancer, in a dose and time dependent. The mechanism by which this isoflavonoid is capable of producing this effect is by decreasing cyclooxygenase 2 (COX-2), through inhibition of the transcriptional activation of NFkB [120]. Recently another research group reported that genistein is capable of reducing the growth of the cell line of breast cancer MDA-MB -231 inhibit NFkB transcriptional activity through a mechanism dependent signaling pathway Notch-1 [121], These researchers found that genistein negatively regulates the expression of cyclin B1, Bcl-2 and Bcl-xl. Genistein has proven to be a potentiating agent of other compounds with anticarcinogenic effect, a recent example is the reported effect on the cell line A549 lung cancer, where genistein potentiates the apoptotic effect of Trichostatin A, the mechanism involved in this effect would be enhancing the positive regulation of the expression of mRNA encoding the tumor necrosis factor receptor 1(TNFR-1), which play a role of death receptor and therefore may at least partly explain this phenomenon [122]. One line of research to elucidate the mechanisms associated with the anti tumor effect of genistein is the ability of this molecule to inhibit the progression of tumor stem cells, in this context, Zhang et al 2012 have found that genistein has an effect until now not reported to prevent carcinogenesis in a model of early colon cancer using as markers of this process the signaling pathway WNT / beta-catenin, some of the genes that are under the control of this pathway are: cyclin D1 and c-myc, being genistein and an extract of soy protein capable of inhibiting various genes involved in the WNT pathway including the mentioned Wnt5a, sFRP1, sFRP2 and Sfrp5 [123]. It has also been shown that genistein could also exert its effect by regulating the immune system, is how Iranian researchers showed that genistein is capable of protecting carcinogenesis in a mouse model of cervical cancer by an immune modulatory mechanism [124].

In summary, it appears that soy consumption is a protective agent in carcinogenesis of various cancers. The mechanisms involved in this process are still a mystery although there are efforts to discover them. One of the molecules characterized in this protective effect is genistein, a soy isoflavone, to which today have described multiple benefits and is now the therapeutic target for the generation of other molecules with structural similarity to her and that enhance its effects *in vivo*.

c) Soybean and cardiovascular disease. Another major concern in today's world is cardiovascular disease, in which the nutritional properties of soybean proteins are well known. Within the past 25 years, numerous studies have reported inverse associations between soy protein intake and plasma cholesterol concentrations; this association is particularly evident in hyper-cholesterolemic men and women [125-127].

Several studies comparing isoflavone-rich soy diets with isoflavone-free soy diets have been performed in experimental animals and humans. Soy consumption could reduce the cardiovascular disease risk factors through its beneficial components, including complex carbohydrates, unsaturated fatty acids, vegetable protein, soluble fiber, oligosaccharides, vitamins, minerals, inositol-derived substances such as lipintol and pinitol, and phytoestrogens, particularly the isoflavones genistein, diadzein, and glycitein [128, 129].

Different studies have been carried out to assess the influence of soy-protein on serum concentrations of total cholesterol, LDL cholesterol, triacylglycerol, and apoB-100. Beneficial results have also been seen among subjects with different types of diseases [130, 131]. Beneficial effects of soy consumption on blood lipids were the most consistently reported findings by Azadbakht (2007) [108]. In turn, Anderson (1995), showed significant reductions in total cholesterol (9%), LDL cholesterol (13%), and triacylglycerol's (11%) with the consumption, on average, of 47 g soy-protein/daily [125].

For its part, Jenkins (2002) studied the effects of high- and low-isoflavone soy-protein foods on both lipid and nonlipid risk factors for coronary artery disease. They found that compared with the control diet, however, both soy diets resulted in significantly lower total cholesterol, estimated CAD risk, and ratios of total to HDL cholesterol, LDL to HDL cholesterol, and apolipoprotein B to A-I [132]. No significant sex differences were observed, except for systolic blood pressure, which in men was significantly lower after the soy diets than after the control diet. On the basis of blood lipid and blood pressure changes, the calculated CAD risk was significantly lower with the soy diets, by $10.1 \pm 2.7\%$ [132].

Other group reported that monkeys fed isoflavone-rich soy-protein-isolate diets had significantly better serum lipid values (lower total cholesterol and higher HDL-cholesterol concentrations) than monkeys fed isoflavone-poor soy-protein-isolate diets. Whereas the administration of the antiestrogen tamoxifen is accompanied by an increase in serum triacylglycerol concentrations, soy-protein administration is associated with a decrease in serum triacylglycerol concentrations [133].

Other cardiac benefits of soy intake, independent of cholesterol reduction, have been identified and investigated. Clarkson (1994) [134] used monkeys with experimental atherosclerosis as a model to examine the effects of estrogen administration on vascular dilatation *in vivo*. Using a similar animal model they also showed that an isoflavone-rich soy-protein-isolate diet has a favorable effect on dilatation of coronary arteries similar to that of estrogen administration [62, 135].

The amount of soy protein that should be recommended for use to achieve "therapeutic effects" is unknown. Also, further research is required to determine the safety of isoflavones in pharmaceutical doses. Animal studies suggest that small amounts of isoflavones have favorable effects on lipoprotein oxidation and cholesterol reduction. Much more work is required to determine the minimum amount needed to have a specific beneficial health effect.

6. Conclusion

Epidemiological studies suggest that soybean consumption is associated, at least in part, with lower incidences of a number of chronic diseases, including type 2 diabetes, certain types of cancer and cardiovascular diseases, between others, mainly due antioxidant and antiplatelet activities.

Soybean derivatives can be denominated as functional ingredients as they contain bioactive compounds that inhibit platelet aggregation, which gives a preventive effect on thrombus formation. Also soybeans contain a variety of bioactive phytochemicals such as phenolic acids, flavonoids, isoflavones, saponins, phytosterols and sphingolipids; being the phenolic compounds with the highest antioxidant capacity. The key benefits of soy are related to their excellent protein content, its high content of essential fatty acids, numerous vitamins and minerals, their isoflavones and their higher fiber content. The experimental evidences related soy protein more than soy isoflavones as responsible by effects observed. At the present is not possible to discard another component present in soy as responsible by the effects.

Functional food may act as an adjunctive therapy/alternative treatment of different pathologies, and scientific studies are appearing more frequently demonstrating that this hypothesis is, indeed, a reality.

These evidences were considered by FDA, that published claims that recommended soy
protein extract as an alternative to reduce blood cholesterol concentrations and prevent cancer, diabetes and increase protection cardiovascular.

Acknowledgements

The authors acknowledge the support by Centro de Estudios en Alimentos Procesados
(CEAP), Conicyt-Regional, Gore Maule, R09I2001, Talca, Chile.

Author details

Eduardo Fuentes[1,3], Luis Guzmán[1], Gilda Carrasco[2,3], Elba Leiva[1],
Rodrigo Moore-Carrasco[1,3] and Iván Palomo[1,3*]

*Address all correspondence to: ipalomo@utalca.cl

1 Departamento de Bioquímica Clínica e Inmunología, Facultad de Ciencias de la Salud,
Universidad de Talca, Chile

2 Departamento de Horticultura, Facultad de Ciencias Agrarias, Universidad de Talca, Chile

3 Centro de Estudios en Alimentos Procesados (CEAP), Conicyt-Regional, Gore Maule,
R09I2001, Talca, Chile

References

[1] Forde, B., & Lorenzo, H. (2001). The nutritional control of root development. *Plant
    and Soil*, 232, 51-68.

[2] Liu, X., Jin, H., Wang, G., & Herbert, S. (2008). Soybean yield physiology and devel-
    opment of high-yielding practices in Northeast China. *Field Crops Research*, 105,
    157-171.

[3] Setchell, K. D. (1998). Phytoestrogens the biochemistry, physiology, and implications
    for human health of soy isoflavones. *Am J Clin Nutr*, 68, 1333S -1346S .

[4] Heaney, R. P., Weaver, C. M., & Fitzsimmons, M. L. (1991). Soybean phytate content:
    effect on calcium absorption. *Am J Clin Nutr*, 53, 745-747.

[5]   Liu, Z. S., & Chang, S. K. (2004). Effect of soy milk characteristics and cooking conditions on coagulant requirements for making filled tofu. *J Agric Food Chem*, 52, 3405-3411.

[6]   Wood, J. E., Senthilmohan, S. T., & Peskin, A. V. (2002). Antioxidant activity of procyanidin-containing plant extracts at different pHs. *Food Chemistry*, 77, 155-161.

[7]   Decker, E. A. (1995). The role of phenolics, conjugated linoleic acid, carnosine, and pyrroloquinoline quinone as nonessential dietary antioxidants. *Nutr Rev*, 53, 49-58.

[8]   Xu, B., & Chang, S. K. (2008). Characterization of phenolic substances and antioxidant properties of food soybeans grown in the North Dakota-Minnesota region. *J Agric Food Chem*, 56, 9102-9113.

[9]   Makela, S., Poutanen, M., Lehtimaki, J., Kostian, M. L., et al. (1995). Estrogen-specific 17 beta-hydroxysteroid oxidoreductase type 1 (E.C. 1.1.1.62) as a possible target for the action of phytoestrogens. *Proc Soc Exp Biol Med*, 208, 51-59.

[10] Yen, G. C., Chang, Y. C., & Su, S. W. (2003). Antioxidant activity and active compounds of rice koji fermented with Aspergillus candidus. *Food Chemistry*, 83, 49-54.

[11] Yang, J. H., Mau, J. L., Ko, P. T., & Huang, L. C. (2000). Antioxidant properties of fermented soybean broth. *Food Chemistry*, 71, 249-254.

[12] Siddhuraju, P., Mohan, P. S., & Becker, K. (2002). Studies on the antioxidant activity of Indian Laburnum (Cassia fistula L.): a preliminary assessment of crude extracts from stem bark, leaves, flowers and fruit pulp. *Food Chemistry*, 79, 61-67.

[13] Baunmann, L.S. (2009). Cosmetic Dermatology: Principles and Practice, Revised 2nd ed. *McGraw-Hill, Columbus, Ohio.*

[14] Ranich, T., Bhathena, S. J., & Velasquez, M. T. (2001). Protective effects of dietary phytoestrogens in chronic renal disease. *J Ren Nutr*, 11, 183-193.

[15] Axelson, M., Kirk, D. N., Farrant, R. D., Cooley, G., et al. (1982). The identification of the weak oestrogen equol [7-hydroxy-3-(4'-hydroxyphenyl)chroman] in human urine. *Biochem J*, 201, 353-357.

[16] Setchell, K. D., Brown, N. M., Zimmer-Nechemias, L., Brashear, W. T., et al. (2002). Evidence for lack of absorption of soy isoflavone glycosides in humans, supporting the crucial role of intestinal metabolism for bioavailability. *Am J Clin Nutr*, 76, 447-453.

[17] Setchell, K. D., Borriello, S. P., Hulme, P., Kirk, D. N., & Axelson, M. (1984). Nonsteroidal estrogens of dietary origin: possible roles in hormone-dependent disease. *Am J Clin Nutr*, 40, 569-578.

[18] Kelly, G. E., Nelson, C., Waring, M. A., Joannou, G. E., & Reeder, A. Y. (1993). Metabolites of dietary (soya) isoflavones in human urine. *Clin Chim Acta*, 223, 9-22.

[19] Heinonen, S., Wahala, K., & Adlercreutz, H. (1999). Identification of isoflavone metabolites dihydrodaidzein, dihydrogenistein, 6'-OH-O-dma, and cis-4-OH-equol in

human urine by gas chromatography-mass spectroscopy using authentic reference compounds. *Anal Biochem, 274,* 211-219.

[20] Atkinson, C., Berman, S., Humbert, O., & Lampe, J. W. (2004). In vitro incubation of human feces with daidzein and antibiotics suggests interindividual differences in the bacteria responsible for equol production. *J Nutr, 134,* 596-599.

[21] Setchell, K. D., Brown, N. M., & Lydeking-Olsen, E. (2002). The clinical importance of the metabolite equol-a clue to the effectiveness of soy and its isoflavones. *J Nutr, 132,* 3577-3584.

[22] Setchell, K. D., Zimmer-Nechemias, L., Cai, J., & Heubi, J. E. (1998). Isoflavone content of infant formulas and the metabolic fate of these phytoestrogens in early life. *Am J Clin Nutr, 68,* 1453S -1461S .

[23] Setchell, K. D., Zimmer-Nechemias, L., Cai, J., & Heubi, J. E. (1997). Exposure of infants to phyto-oestrogens from soy-based infant formula. *Lancet, 350,* 23-27.

[24] Lampe, J. W., Karr, S. C., Hutchins, A. M., & Slavin, J. L. (1998). Urinary equol excretion with a soy challenge: influence of habitual diet. *Proc Soc Exp Biol Med, 217,* 335-339.

[25] Rowland, I. R., Wiseman, H., Sanders, T. A., Adlercreutz, H., & Bowey, E. A. (2000). Interindividual variation in metabolism of soy isoflavones and lignans: influence of habitual diet on equol production by the gut microflora. *Nutr Cancer, 36,* 27-32.

[26] Setchell, K. D, Clerici, C, Lephart, E. D., Cole, S. J., et al. (2005). S-equol, a potent ligand for estrogen receptor beta, is the exclusive enantiomeric form of the soy isoflavone metabolite produced by human intestinal bacterial flora. *Am J Clin Nutr, 81,* 1072-1079.

[27] Gopaul, R., Knaggs, H. E., & Lephart, E. D. (2012). Biochemical investigation and gene analysis of equol: a plant and soy-derived isoflavonoid with antiaging and antioxidant properties with potential human skin applications. *Biofactors, 38,* 44-52.

[28] Kuiper, G. G., Lemmen, J. G., Carlsson, B., Corton, J. C., et al. (1998). Interaction of estrogenic chemicals and phytoestrogens with estrogen receptor beta. *Endocrinology, 139,* 4252-4263.

[29] Kuiper, G. G., Carlsson, B., Grandien, K., Enmark, E., et al. (1997). Comparison of the ligand binding specificity and transcript tissue distribution of estrogen receptors alpha and beta. *Endocrinology, 138,* 863-870.

[30] Thornton, M. J., Taylor, A. H., Mulligan, K., Al-Azzawi, F., et al. (2003). The distribution of estrogen receptor beta is distinct to that of estrogen receptor alpha and the androgen receptor in human skin and the pilosebaceous unit. *J. Investig Dermatol Symp Proc, 8,* 100-103.

[31] Lund, T. D., Munson, D. J., Haldy, M. E., Setchell, K. D., et al. (2004). Equol is a novel anti-androgen that inhibits prostate growth and hormone feedback. *Biol Reprod*, 70, 1188-1195.

[32] Makrantonaki, E., & Zouboulis, C. C. (2009). Androgens and ageing of the skin. *Curr Opin Endocrinol Diabetes Obes*, 16, 240-245.

[33] Nitsch, S. M., Wittmann, F., Angele, P., Wichmann, M. W., et al. (2004). Physiological levels of 5 alpha-dihydrotestosterone depress wound immune function and impair wound healing following trauma-hemorrhage. *Arch Surg*, 139, 157-163.

[34] Stevenson, S., & Thornton, J. (2007). Effect of estrogens on skin aging and the potential role of SERMs. *Clin Interv Aging*, 2, 283-297.

[35] Verdier-Sevrain, S., Bonte, F., & Gilchrest, B. (2006). Biology of estrogens in skin: implications for skin aging. *Exp Dermatol*, 15, 83-94.

[36] Munoz, Y., Garrido, A., & Valladares, L. (2009). Equol is more active than soy isoflavone itself to compete for binding to thromboxane A(2) receptor in human platelets. *Thromb Res*, 123, 740-744.

[37] Ronis, M. J., Chen, Y., Badeaux, J., & Badger, T. M. (2009). Dietary soy protein isolate attenuates metabolic syndrome in rats via effects on PPAR, LXR, and SREBP signaling. *J Nutr*, 139, 1431-1438.

[38] Harborne, J. B, & Williams, C. A. (2000). Advances in flavonoid research since 1992. *Phytochemistry*, 55, 481-504.

[39] Paik, S. S., Jeong, E., Jung, S. W., Ha, T. J., et al. (2012). Anthocyanins from the seed coat of black soybean reduce retinal degeneration induced by N-methyl-N-nitrosourea. *Exp Eye Res*, 97, 55-62.

[40] Gottstein, N., Ewins, B. A., Eccleston, C., Hubbard, G. P., et al. (2003). Effect of genistein and daidzein on platelet aggregation and monocyte and endothelial function. *Br J Nutr*, 89, 607-616.

[41] Yin, X. Z., Quan, J. S., Takemichi, K., & Makoto, T. (2004). Anti-atherosclerotic effect of soybean isofalvones and soyasaponins in diabetic rats. *Zhonghua Yu Fang Yi Xue Za Zhi*, 38, 26-28.

[42] Kim, K., Lim, K. M., Kim, C. W., Shin, H. J., et al. (2011). Black soybean extract can attenuate thrombosis through inhibition of collagen-induced platelet activation. *J Extract Black*, 22, 964-970.

[43] Mc Nicol, A. (1993). The effects of genistein on platelet function are due to thromboxane receptor antagonism rather than inhibition of tyrosine kinase. *ProstaglandinsLeukot Essent Fatty Acids*, 48, 379-384.

[44] Kondo, K., Suzuki, Y., Ikeda, Y., & Umemura, K. (2002). Genistein, an isoflavone included in soy, inhibits thrombotic vessel occlusion in the mouse femoral artery and in vitro platelet aggregation. *Eur J Pharmacol*, 455, 53-57.

[45] Atsushi, O. (2004). Anti-platelets Effects of Genistein, an Isoflavonoid from Soybean. *Soy Protein Research*, 7, 145-148.

[46] Tsuchiya, H., Sato, M., & Watanabe, I. (1999). Antiplatelet activity of soy sauce as functional seasoning. *J Agric Food Chem*, 47, 4167-4174.

[47] Kitagawa, S., Orinaka, M., & Hirata, H. (1993). Depth-dependent change in membrane fluidity by phenolic compounds in bovine platelets and its relationship with their effects on aggregation and adenylate cyclase activity. *Biochim Biophys Acta*, 1179, 277-282.

[48] Peura, P., Mackenzie, P., Koivusaari, U., & Lang, M. (1982). Increased fluidity of a model membrane caused by tetrahydro-beta-carbolines. *Mol Pharmacol*, 22, 721-724.

[49] Silva, M., Santana, L., Silva-Lucca, R., Lima, A., et al. (2011). Immobilized Cratylia mollis lectin: An affinity matrix to purify a soybean (Glycine max) seed protein with in vitro platelet antiaggregation and anticoagulant activities. *Process Process*, 46, 74-80.

[50] Lee, K., & Kim, S. (2005). SSGE and DEE, new peptides isolated from a soy protein hydrolysate that inhibit platelet aggregation. *Food Chemistry*, 90, 389-393.

[51] Tsuchiya, H., Sato, M., & Watanabe, I. (1999). Antiplatelet activity of soy sauce as functional seasoning. *J Agric Food Chem*, 47, 4167-4174.

[52] Setchell, K. D., & Cassidy, A. (1999). Dietary isoflavones: biological effects and relevance to human health. *J Nutr*, 129, 758S -767S .

[53] Bennett, J. O., Yu, O., Heatherly, L. G., & Krishnan, H. B. (2004). Accumulation of genistein and daidzein, soybean isoflavones implicated in promoting human health, is significantly elevated by irrigation. *J Agric Food Chem*, 52, 7574-7579.

[54] Okamoto, K., & Horisawa, R. (2006). Soy products and risk of an aneurysmal rupture subarachnoid hemorrhage in Japan. *Eur J Cardiovasc Prev Rehabil*, 13, 284-287.

[55] Zhang, X., Shu, X. O., Gao, Y. T., Yang, G., et al. (2003). Soy food consumption is associated with lower risk of coronary heart disease in Chinese women. *J Nutr*, 133, 2874-2878.

[56] Potter, S. M., Bakhit, R. M., Essex-Sorlie, D. L., Weingartner, K. E., et al. (1993). Depression of plasma cholesterol in men by consumption of baked products containing soy protein. *Am J Clin Nutr*, 58, 501-506.

[57] Bakhit, R. M., Klein, B. P., Essex-Sorlie, D., Ham, J. O., et al. (1994). Intake of 25 g of soybean protein with or without soybean fiber alters plasma lipids in men with elevated cholesterol concentrations. *J Nutr*, 124, 213-222.

[58] Erdman, J. W., Jr. (2000). AHA Science Advisory: Soy protein and cardiovascular disease: A statement for healthcare professionals from the Nutrition Committee of the AHA. *Circulation*, 102, 2555-2559.

[59] Mullen, E., Brown, R. M., Osborne, T. F., & Shay, N. F. (2004). Soy isoflavones affect sterol regulatory element binding proteins (SREBPs) and SREBP-regulated genes in HepG2 cells. *J Nutr*, 134, 2942-2947.

[60] Adams, M. R., Golden, D. L., Anthony, M. S., Register, T. C., & Williams, J. K. (2002). The inhibitory effect of soy protein isolate on atherosclerosis in mice does not require the presence of LDL receptors or alteration of plasma lipoproteins. *J Nutr*, 132, 43-49.

[61] Adams, M. R., Register, T. C., Golden, D. L., Anthony, M. S., et al. (2002). The athero-protective effect of dietary soy isoflavones in apolipoprotein E-/- mice requires the presence of estrogen receptor-alpha. *Arterioscler Thromb Vasc Biol*, 22, 1859-1864.

[62] Anderson, J. W., Smith, B. M., & Washnock, C. S. (1999). Cardiovascular and renal benefits of dry bean and soybean intake. *Am J Clin Nutr*, 70, 464S -474S .

[63] Iritani, N., Sugimoto, T., Fukuda, H., Komiya, M., & Ikeda, H. (1997). Dietary soy-bean protein increases insulin receptor gene expression in Wistar fatty rats when dietary polyunsaturated fatty acid level is low. *J Nutr*, 127, 1077-1083.

[64] Zhan, S., & Ho, S. C. (2005). Meta-analysis of the effects of soy protein containing iso-flavones on the lipid profile. *Am J Clin Nutr*, 81, 397-408.

[65] Lusas, E. W., & Riaz, M. N. (1995). Soy protein products: processing and use. J Nutr ., 125, 573S-580S.

[66] Messina, M. J. (1999). Legumes and soybeans: overview of their nutritional profiles and health effects. *Am J Clin Nutr*, 70, 439S -450S .

[67] Kataoka, S. (2005). Functional effects of Japanese style fermented soy sauce (shoyu) and its components. *J Biosci Bioeng*, 100, 227-234.

[68] Kang, J., Badger, T. M., Ronis, M. J., & Wu, X. Non-isoflavone phytochemicals in soy and their health effects. *J Agric Food Chem*, 58, 8119-8133.

[69] Gil-Izquierdo, A., Penalvo, J. L., Gil, J. I., Medina, S., et al. Soy isoflavones and cardi-ovascular disease epidemiological, clinical and-omics perspectives. *Curr Pharm Bio-technol*, 13, 624-631.

[70] Wei, X., Wu, J., Ni, Y., Lu, L., & Zhao, R. (2011). Antioxidant effect of a phytoestrogen equol on cultured muscle cells of embryonic broilers. *In Vitro Cell Dev Biol Anim*, 47, 735-741.

[71] Choi, E., & Kim, G. (2011). Anticancer mechanism of equol in dimethylbenz(a)anthra-cene-treated animals. 7 12 . *Int J Oncol*, 39, 747-754.

[72] Palermo, M., Paradiso, R., De Pascale, S., & Fogliano, V. Hydroponic cultivation im-proves the nutritional quality of soybean and its products. *J Agric Food Chem*, 60, 250-255.

[73] Hendrich, S., & Fisher, K. (2001). What do we need to know about active ingredients in dietary supplements? Summary of workshop discussion. *J Nutr*, 131, 1387S -8S .

[74] Setchell, K. D., Brown, N. M., Desai, P. B., Zimmer-Nechimias, L., et al. (2003). Bioa-
      vailability, disposition, and dose-response effects of soy isoflavones when consumed
      by healthy women at physiologically typical dietary intakes. *J Nutr*, 133, 1027-1035.

[75] Larkin, T. A., Astheimer, L. B., & Price, W. E. (2009). Dietary combination of soy with
      a probiotic or prebiotic food significantly reduces total and LDL cholesterol in mildly
      hypercholesterolaemic subjects. *Eur J Clin Nutr*, 63, 238-245.

[76] Rowland, I., Faughnan, M., Hoey, L., Wahala, K., et al. (2003). Bioavailability of phy-
      to-oestrogens. *Br J Nutr, Suppl 1, S,* 89, 45-58.

[77] Izumi, T., Piskula, M. K., Osawa, S., Obata, A., et al. (2000). Soy isoflavone aglycones
      are absorbed faster and in higher amounts than their glucosides in humans. *J Nutr*,
      130, 1695-1699.

[78] Wiseman, H., Casey, K., Bowey, E. A., Duffy, R., et al. (2004). Influence of 10 wk of
      soy consumption on plasma concentrations and excretion of isoflavonoids and on
      gut microflora metabolism in healthy adults. *Am J Clin Nutr*, 80, 692-699.

[79] Xu, X., Wang, H. J., Murphy, P. A., Cook, L., & Hendrich, S. (1994). Daidzein is a
      more bioavailable soymilk isoflavone than is genistein in adult women. *J Nutr*, 124,
      825-832.

[80] Andlauer, W., Kolb, J., & Furst, P. (2000). Isoflavones from tofu are absorbed and me-
      tabolized in the isolated rat small intestine. *J Nutr*, 130, 3021-3027.

[81] King, R. A. (1998). Daidzein conjugates are more bioavailable than genistein conju-
      gates in rats. *Am J Clin Nutr*, 68, 1496S -1499S .

[82] Setchell, K. D., Brown, N. M., Desai, P., Zimmer-Nechemias, L., et al. (2001). Bioavail-
      ability of pure isoflavones in healthy humans and analysis of commercial soy isofla-
      vone supplements. *J Nutr*, 131, 1362S -75S .

[83] Piazza, C., Privitera, M. G., Melilli, B., Incognito, T., et al. (2007). Influence of inulin
      on plasma isoflavone concentrations in healthy postmenopausal women. *Am J Clin
      Nutr*, 86, 775-780.

[84] Benvenuti, C., & Setnikar, I. (2011). Effect of Lactobacillus sporogenes on oral isofla-
      vones bioavailability: single dose pharmacokinetic study in menopausal women.
      Arzneimittelforschung, 61, 605-609.

[85] Miguez, A. C., Francisco, J. C., Barberato, S. H., Simeoni, R., et al. (2012). The func-
      tional effect of soybean extract and isolated isoflavone on myocardial infarction and
      ventricular dysfunction: The soybean extract on myocardial infarction. *J Nutr Bio-
      chem*.

[86] Palomo, I., Guzmán, L., Leiva, E., Mujica, V., et al. (2011). Soybean and Health. *Hany
      El-Shemy: InTech*.

[87] Ruiz-Alcaraz, A. J., Liu, H. K., Cuthbertson, D. J., Mc Manus, E. J., et al. (2005). A novel regulation of IRS1 (insulin receptor substrate-1) expression following short term insulin administration. *Biochem J*, 392, 345-352.

[88] Bjornholm, M., & Zierath, J. R. (2005). Insulin signal transduction in human skeletal muscle: identifying the defects in Type II diabetes. *Biochem Soc Trans*, 33, 354-357.

[89] Lu, J., Zeng, Y., Hou, W., Zhang, S., et al. (2012). The soybean peptide aglycin regulates glucose homeostasis in type 2 diabetic mice via IR/IRS1 pathway. *J Nutr Biochem*.

[90] Mizokami-Stout, K., Cree-Green, M., & Nadeau, K. J. (2012). Insulin resistance in type 2 diabetic youth. *Curr Opin Endocrinol Diabetes Obes*, 19, 255-262.

[91] Everhart, J. E., Pettitt, D. J., Bennett, P. H., & Knowler, W. C. (1992). Duration of obesity increases the incidence of NIDDM. *Diabetes*, 41, 235-240.

[92] Liener, I. E. (1994). Implications of antinutritional components in soybean foods. *Crit Rev Food Sci Nutr*, 34, 31-67.

[93] Ishihara, K., Oyaizu, S., Fukuchi, Y., Mizunoya, W., et al. (2003). A soybean peptide isolate diet promotes postprandial carbohydrate oxidation and energy expenditure in type II diabetic mice. *J Nutr*, 133, 752-757.

[94] Davis, J., Higginbotham, A., O'Connor, T., Moustaid-Moussa, N., et al. (2007). Soy protein and isoflavones influence adiposity and development of metabolic syndrome in the obese male ZDF rat. *Ann Nutr Metab*, 51, 42-52.

[95] Dun, X. P., Wang, J. H., Chen, L., Lu, J., et al. (2007). Activity of the plant peptide aglycin in mammalian systems. *FEBS J*, 274, 751-759.

[96] Lu, M. P., Wang, R., Song, X., Chibbar, R., et al. (2008). Dietary soy isoflavones increase insulin secretion and prevent the development of diabetic cataracts in streptozotocin-induced diabetic rats. *Nutr Res*, 28, 464-471.

[97] Saito, N., Sakai, H., Suzuki, S., Sekihara, H., & Yajima, Y. (1998). Effect of an alpha-glucosidase inhibitor (voglibose), in combination with sulphonylureas, on glycaemic control in type 2 diabetes patients. *J Int Med Res*, 26, 219-232.

[98] Kwon, Y. I., Vattem, D. A., & Shetty, K. (2006). Evaluation of clonal herbs of Lamiaceae species for management of diabetes and hypertension. *Asia Pac J Clin Nutr*, 15, 107-118.

[99] Ademiluyi, A. O., & Oboh, G. (2011). Soybean phenolic-rich extracts inhibit key-enzymes linked to type 2 diabetes (alpha-amylase and alpha-glucosidase) and hypertension (angiotensin I converting enzyme) in vitro. *Exp Toxicol Pathol*.

[100] Burow, M. E., Boue, S. M., Collins-Burow, B. M., Melnik, L. I., et al. (2001). Phytochemical glyceollins, isolated from soy, mediate antihormonal effects through estrogen receptor alpha and beta. *J Clin Endocrinol Metab*, 86, 1750-1758.

[101] Park, S., Kim, da. S., Kim, J. H., Kim, J. S., & Kim, H. J. (2012). Glyceollin-containing fermented soybeans improve glucose homeostasis in diabetic mice. *Nutrition, 28,* 204-211.

[102] Vedavanam, K., Srijayanta, S., O'Reilly, J., Raman, A., & Wiseman, H. (1999). Antioxidant action and potential antidiabetic properties of an isoflavonoid-containing soyabean phytochemical extract (SPE). *Phytother Res,* 13, 601-608.

[103] Ascencio, C., Torres, N., Isoard-Acosta, F., Gomez-Perez, F. J., et al. (2004). Soy protein affects serum insulin and hepatic SREBP-1 mRNA and reduces fatty liver in rats. *J Nutr,* 134, 522-529.

[104] Villegas, R., Gao, Y. T., Yang, G., Li, H. L., et al. (2008). Legume and soy food intake and the incidence of type 2 diabetes in the Shanghai Women's Health Study. *Am J Clin Nutr,* 87, 162-167.

[105] Nanri, A., Mizoue, T., Takahashi, Y., Kirii, K., et al. (2010). Soy product and isoflavone intakes are associated with a lower risk of type 2 diabetes in overweight Japanese women. *J Nutr,* 140, 580-586.

[106] Aubertin-Leheudre, M., Lord, C., Khalil, A., & Dionne, I. J. (2008). Isoflavones and clinical cardiovascular risk factors in obese postmenopausal women: a randomized double-blind placebo-controlled trial. *J Womens Health (Larchmt),* 17, 1363-1369.

[107] Azadbakht, L., Atabak, S., & Esmaillzadeh, A. (2008). Soy protein intake, cardiorenal indices, and C-reactive protein in type 2 diabetes with nephropathy a longitudinal randomized clinical trial. *Diabetes Care,* 31, 648-654.

[108] Azadbakht, L., Kimiagar, M., Mehrabi, Y., Esmaillzadeh, A., et al. (2007). Soy inclusion in the diet improves features of the metabolic syndrome: a randomized crossover study in postmenopausal women. *Am J Clin Nutr,* 85, 735-741.

[109] Blakesmith, S. J., Lyons-Wall, P. M., George, C., Joannou, G. E., et al. (2003). Effects of supplementation with purified red clover (Trifolium pratense) isoflavones on plasma lipids and insulin resistance in healthy premenopausal women. *Br J Nutr,* 89, 467-474.

[110] Gonzalez, S., Jayagopal, V., Kilpatrick, E. S., Chapman, T., & Atkin, S. L. (2007). Effects of isoflavone dietary supplementation on cardiovascular risk factors in type 2 diabetes. *Diabetes Care,* 30, 1871-1873.

[111] Liao, F. H., Shieh, M. J., Yang, S. C., Lin, S. H., & Chien, Y. W. (2007). Effectiveness of a soy-based compared with a traditional low-calorie diet on weight loss and lipid levels in overweight adults. *Nutrition,* 23, 551-556.

[112] Erdman, J. W., Jr , , Badger, T. M., Lampe, J. W., Setchell, K. D., & Messina, M. (2004). Not all soy products are created equal: caution needed in interpretation of research results. *J Nutr,* 134, 1229S -1233S .

[113] Liu, Z. M., Chen, Y. M., & Ho, S. C. (2011). Effects of soy intake on glycemic control: a meta-analysis of randomized controlled trials. *Am J Clin Nutr,* 93, 1092-1101.

[114] Liener, I. E., & Seto, T. A. (1955). Nonspecific effect of soybean hemagglutinin on tumor growth. *Cancer research*, 15, 407-409.

[115] Khan, S. A., Chatterton, R. T., Michel, N., Bryk, M., et al. (2012). Soy isoflavone supplementation for breast cancer risk reduction: a randomized phase II trial. *Cancer
Prev Res (Phila)*, 5, 309-319.

[116] Teas, J., Irhimeh, M. R., Druker, S., Hurley, T. G., et al. (2011). Serum IGF-1 concentrations change with soy and seaweed supplements in healthy postmenopausal
American women. *Nutr Cancer*, 63, 743-748.

[117] Yang, W. S., Va, P., Wong, M. Y., Zhang, H. L., & Xiang, Y. B. (2011). Soy intake is
associated with lower lung cancer risk: results from a meta-analysis of epidemiologic
studies. *Am J Clin Nutr*, 94, 1575-1583.

[118] Hillman, G. G., Singh-Gupta, V., Runyan, L., Yunker, C. K., et al. (2011). Soy isoflavones radiosensitize lung cancer while mitigating normal tissue injury. *Radiotherapy
and oncology journal of the European Society for Therapeutic Radiology and Oncology*, 101,
329-336.

[119] Zaineddin, A. K., Buck, K., Vrieling, A., Heinz, J., et al. (2012). The association between dietary lignans, phytoestrogen-rich foods, and fiber intake and postmenopausal breast cancer risk: a german case-control study. *Nutr Cancer*, 64, 652-665.

[120] Li, Y. S., Wu, L. P., Li, K. H., Liu, Y. P., et al. (2011). Involvement of nuclear factor
kappaB (NF-kappaB) in the downregulation of cyclooxygenase-2 (COX-2) by genistein in gastric cancer cells. *The Journal of international medical research*, 39, 2141-2150.

[121] Pan, H., Zhou, W., He, W., Liu, X., et al. (2012). Genistein inhibits MDA-MB-231 tri-
ple-negative breast cancer cell growth by inhibiting NF-kappaB activity via the
Notch-1 pathway. *International journal of molecular medicine*, 30, 337-343.

[122] Wu, T. C., Yang, Y. C., Huang, P. R., Wen, Y. D., & Yeh, S. L. (2012). Genistein enhances the effect of trichostatin A on inhibition of A549 cell growth by increasing expression of TNF receptor-1. *Toxicol Appl Pharmacol*, 262, 247-254.

[123] Zhang, Y., Li, Q., Zhou, D., & Chen, H. (2012). Genistein, a soya isoflavone, prevents
azoxymethane-induced up-regulation of WNT/beta-caten in signalling and reduces
colon pre-neoplasia in rats. *Br J Nutr*, 1-10.

[124] Ghaemi, A., Soleimanjahi, H., Razeghi, S., Gorji, A., et al. (2012). Genistein induces a
protective immunomodulatory effect in a mouse model of cervical cancer. *Iranian
journal of immunology : IJI*, 9, 119-127.

[125] Anderson, J. W., Johnstone, B. M., & Cook-Newell, M. E. (1995). Meta-analysis of the
effects of soy protein intake on serum lipids. *N Engl J Med*, 333, 276-282.

[126] Carroll, K. K. (1991). Review of clinical studies on cholesterol-lowering response to
soy protein. *J Am Diet Assoc*, 91, 820-827.

[127] Wangen, K. E., Duncan, A. M., Xu, X., & Kurzer, M. S. (2001). Soy isoflavones im-
       prove plasma lipids in normocholesterolemic and mildly hypercholesterolemic post-
       menopausal women. *Am J Clin Nutr*, 73, 225-231.

[128] Kim, J. I., Kim, J. C., Kang, M. J., Lee, M. S., et al. (2005). Effects of pinitol isolated
       from soybeans on glycaemic control and cardiovascular risk factors in Korean pa-
       tients with type II diabetes mellitus: a randomized controlled study. *Eur J Clin Nutr*,
       59, 456-458.

[129] Crisafulli, A., Altavilla, D., Marini, H., Bitto, A., et al. (2005). Effects of the phytoes-
       trogen genistein on cardiovascular risk factors in postmenopausal women. *Meno-
       pause*, 12, 186-192.

[130] Hoie, L. H., Graubaum, H. J., Harde, A., Gruenwald, J., & Wernecke, K. D. (2005).
       Lipid-lowering effect of 2 dosages of a soy protein supplement in hypercholesterole-
       mia. *Adv Ther*, 22, 175-186.

[131] Hermansen, K., Hansen, B., Jacobsen, R., Clausen, P., et al. (2005). Effects of soy sup-
       plementation on blood lipids and arterial function in hypercholesterolaemic subjects.
       Eur J Clin Nutr, 59, 843-850.

[132] Jenkins, D. J., Kendall, C. W., Jackson, C. J., Connelly, P. W., et al. (2002). Effects of
       high- and low-isoflavone soyfoods on blood lipids, oxidized LDL, homocysteine, and
       blood pressure in hyperlipidemic men and women. *Am J Clin Nutr*, 76, 365-372.

[133] Anthony, M. S., Clarkson, T. B., Hughes, C. L. Jr, Morgan, T. M., & Burke, G. L.
       (1996). Soybean isoflavones improve cardiovascular risk factors without affecting the
       reproductive system of peripubertal rhesus monkeys. *J Nutr*, 126, 43-50.

[134] Clarkson, T. B., Anthony, M. S., & Klein, K. P. (1994). Effects of estrogen treatment on
       arterial wall structure and function. *Drugs*, 2, 42-51.

[135] Williams, J. K., & Clarkson, T. B. (1998). Dietary soy isoflavones inhibit in-vivo con-
       strictor responses of coronary arteries to collagen-induced platelet activation. *Coron
       Artery Dis*, 9, 759-764.

Chapter 12

Potential Use of Soybean Flour (*Glycine max*) in Food Fortification

O. E. Adelakun, K. G. Duodu, E. Buys and
B. F. Olanipekun

Additional information is available at the end of the chapter

1. Introduction

Protein-energy malnutrition (PEM) results from prolonged deprivation of essential amino acids and total nitrogen and/or energy substrates. Growth which is a continuous process results from the complex interaction between inheritance and environment. Protein-energy malnutrition (PEM) also results from food insufficiency as well as from poor social and economic conditions [6, 20]. Malnutrition originates from a cellular imbalance between nutrient/energy supply and the body's demand to ensure growth and maintenance [22]. Dietary energy and protein deficiencies usually occur together, although one sometimes predominates the other and if severe enough, may lead to the clinical syndrome of kwashiorkor (predominant protein deficiency) or marasmus (mainly energy deficiency).

The origin of PEM can be primary, when it is the result of inadequate food intake, or secondary, when it is the result of other diseases that lead to low food ingestion, inadequate nutrient absorption or utilization, increased nutritional requirements and/or increased nutrient losses [13]. In as much as protein-energy malnutrition applies to a group of related disorders that develop in children and adults whose consumption of protein and energy is insufficient to satisfy the bodies nutritional needs, about one third of children worldwide suffer from malnutrition [8]. This continues to be an important problem mostly in developing countries. One way to curb the global menace of PEM is through food fortification of plant origin. Food fortification is broadly aimed to allow all people to obtain from their diet all the energy, macro- and micronutrients they need to enjoy a healthy and productive life [3].

Legumes are one of the world's most important sources of food supply especially in the developing countries in terms of food, energy as well as nutrients. It has been recognized as an important source of protein and in some cases oil. As a legume, soybeans are an important global crop that provides oil and protein for users. It is the richest sources of protein among the plant foods. Features that motivate researchers to explore its utility in a wide range include the good quality and functionality of its proteins, surplus availability and low cost. Soybeans in the form of full fat flour, concentrate, isolate and texturised proteins have been used in a wide range of food products. These attributes made soybeans to be considered as an ideal food for meeting the protein needs of the population.

The objective of this chapter is to briefly describe the quality attributes of soybean and the potential use of its flour in food fortification.

2. Soybeans

Soybeans *(Glycine max)* belonging to the family *leguminosae* constitute one of the oldest cultivated crops of the tropics and sub-tropical regions, and one of the world's most important sources of protein and oil. Soybeans are probably the most important oil seed legume which has its origin in Eastern Asia, mainly China. The cultivar *Glycine max* is thought to be derived from *Glycine ussuriensis* and *Glycine tomentosa* which grow wild in China, and can be found in great quantities in Asian countries such as Japan and Indonesia [16].

The seeds vary in shape and colour depending on the cultivar. In shape, they can be spherical to flatten while the colour varies from white, yellow and brown to black. Also, the chemical composition of each variety of soybeans differs from each other. Due to the long and tedious processing technique of soybeans, Japan which is one of the largest suppliers of soybean has developed highly advanced processing technologies in the processing and manufacturing of highly acceptable and palatable soya products [14]. As a result of high protein content in soybean, it can be used as a substitute for expensive meat and meat products [5].

Soybean is particularly very unique for different reason and hence classify as a valuable and economical agricultural commodity. In the first instance, it possesses agronomic characteristics with its ability to adapt to a wide range of soil and climate; and its nitrogen fixing ability. This makes it to be a good rotational crop for use with high nitrogen – consuming crops such as corn and rice. Secondly, soybean unique chemical composition on an average dry matter basis is about 40% of protein and 20% of oil. This composition makes it to rank highest in terms of protein content among all food crops and second in terms of oil content after peanut (48%) among all food legumes. Furthermore, soybean is a very nutritious food crop.

3. Nutritional content of soybean

Soybean *(Glycine max)* first emerged as a domesticated crop in the eastern half of North China around the 11th century B.C of Zhou Dynasty. It is easy to grow and has adaptability to a wide range of soils and climate. Because it contains high amount of protein and oil, the soybean was considered one of the five sacred grains along with rice, wheat, barley and millet [4]. The protein and oil component of soybean are not only high in terms of quantity but also in quality. For instance, soy oil has a highest proportion of unsaturated fatty acids such as linoleic and linolenic acid making it a healthy oil to use. Soybeans are known to be typical of such crops that contain all three of the macro-nutrients required particularly for human nutrition. They also contain protein which provides all the essential amino acids in the amounts needed for human health. Most of the essential amino acid present in soybean is available in an amount that is close to those required by animals and humans. The protein – digestibility – correlated amino acid score is close to 1, a rating that is the same for animal proteins such as an egg white and casein. The profile of various nutrients in raw matured seeds of soybeans as highlighted in USDA Nutrient database is as indicated in Table 1 below. Additionally, soybean contains phytochemicals which have been shown to offer unique health benefit. Soybean also has versatile end uses which include human food, animal feed and industrial materials [10].

4. Utilization of soybeans

Soybean can be processed to give soy milk, a valuable protein supplement in infant feeding, soycurds and cheese [23]. It is also used to produce soysauce used extensively in cooking and as a sauce. Soybeans are also used for making candies and ice cream and soybean flour which could be mixed with wheat flour to produce a wide variety of baked goods such as bread and biscuits [16]. Soybean oil is used for edible purposes, particularly as a cooking, and salad oil and, for manufacture of margarine [16]. The oil can also be used industrially in the processing of paints, soap, oil, cloth and printing inks. The meal and soybean proteins are used in the manufacture of synthetic fibre (artificial wool) adhesives and textile [14]. Soybeans could be made into such products as tempeh, miso and natto which may include other sub-products [16].

Soybean protein fibre has been reportedly produced from bean dregs that are produced when extracting oil [14]. From these, globular protein is extracted, made into a spinning solution of a consistent concentration with the addition of a functional auxiliary, and spun into yarn by the wet method [21]. Effect of fermentation on soybean has the tendency of altering the features of the arising dregs when oil is so extracted from soybean. This thus has tendency of either skewing up or otherwise the various arrays of benefits known to accrue from the development of soybean fiber blends with other fibers [21]. [7] has also reported on the chemical composition and total digestible nutrients (TDN) of fermented soybean paste residue. This is usually exploited for the utilization of such residues in livestock rations. Furthermore, [9] had suggested the likelihood of the use of fermented soybean paste residue for livestock feed in the near future as a form of turning waste to wealth and thus serving as anchor for many other accruing benefits.

Nutrient	value per 100g
Energy	1,866kJ (446kcal)
Carbohydrates	30.16g
Sugars	7.33g
Dietary fiber	9.3g
Fat	19.94g
Saturated	2.884g
Monounsaturated	4.404g
Polyunsaturated	11.255g
Protein	36.49g
Tryptophan	0.591g
Threonine	1.766g
Isoleucine	1.971g
Leucine	3.309g
Lysine	2.706g
Methionine	0.547g
Phenylalanine	2.122g
Tyrosine	1.539g
Valine	2.029g
Arginine	3.153g
Histidine	1.097g
Alanine	1.915g
Aspartic acid	5.112g
Glutamic acid	7.874g
Glycine	1.880g
Proline	2.379g
Serine	2.357g
Water	8.54g
Vitamin A equiv.	1μg (0%)
Vitamin B6	0.377mg (29%)
Vitamin B12	0μg (0%)
Vitamin C	6.0 mg (10%)
Vitamin K	47 μg (45%)
Calcium	277mg (28%)
Iron	15.70 mg (126%)
Magnesium	280 mg (76%)
Phosphorus	704 mg (101%)
Potasium	1797 mg (38%)
Sodium	2 mg (0%)
Zinc	4.89 mg (49%)

Table 1. Source: USDA Nutrient database, Percentage relative to US recommendations for adultsNutritional content of raw mature seeds of soybean

5. Antinutritional factors in soybeans

Flatulence is characterized by stomach cramps, nausea, diarrhoea, intestinal and gastric discomfort resulting from the production of large amounts of gas in the gastrointestinal tract. Although all the causative factors in flatulence formation are still unknown, it has been suggested and generally accepted that low-molecular weight oligosaccharides, primarily raffinose, stachyose and verbascose present in most legume seeds are linked to flatulence [2]. Researchers have suggested practical procedures for the reduction of flatulence in cooked and processed soybeans. These include fermentation, removal of seed coat prior to cooking or processing, soaking in water, germination and cooking with a mixture of sodium carbonate and bicarbonate [11]. Oligosaccharides can also be reduced through fermentation thereby leading to elimination of flatulence.

6. Soybean in food fortification

In many developing nations, cereal based foods are widely utilized as food and as dietary staples for adults and weaning foods for infants including Africa where it accounts for up to 77% of total caloric consumption. The major cereal based foods in these regions are derived mainly from maize, sorghum, millet, rice, or wheat. Oilseeds are the largest single source for production of protein concentrates. Of these, soybeans, in terms of tonnage produced, are the most important source. Properly defatted soybean flour will contain 50% or more of protein. By removing soluble carbohydrates and minerals, concentrates containing up to 70% protein can be prepared, and dispersible isolates containing 90% or more of protein are being made. The isolates are of interest as highly concentrated fortification media and also as bases for a variety of high-protein beverages, desserts, and similar products. Soybean concentrates have the virtue of low cost and good nutritive value. Fortification of staple cereals with soybean can help improves their nutritive value and may aid in alleviating malnutrition in developing countries [17].

[12] determined the effect of germination and drying on the functional and nutritional properties of common red bean flour and evaluated the effect of incorporating different levels of cowpea and a constant level of soybean into red bean flour on the functional properties of the composite. Incorporation of soybean and cowpea flour into germinated bean flour at levels of 10 and 30%, respectively, produced a composite with higher functional properties. [15] studied the effect of fermentation with *Rhizopus oligosporus* on some physico-chemical properties of starch extracts from soybean flour. Their study revealed that the length of fermentation with *R. oligosporus* within the period of 0–72 h on soybean (G. max) affected many physico-chemical properties of starch extract as well as pasting characteristics of extracted starch. The physico-chemical attributes of starch 'extract' from fermented soybean flour revealed that water binding capacity, water absorption capacity, swelling capacity and solubility power decreased slightly with increases in fermentation period. Also, while the starch yield and amylose content decreased and amylopectin contents increased with fermentation

period. The pasting characteristics of starch 'extract' from the fermented soybean flour revealed the potential use of the flours in weaning food formulation due to reduce viscosity trend as fermentation period increased.

[1] also worked on the effect of soybean substitution on some physical, compositional and sensory properties of kokoro (a local maize snack). Kokoro which is a finger-shaped snack made from maize is widely acceptable and consumed by children and adults, especially in the southwestern part of Nigeria in the Oyo and Ogun states. Soybean (*Glycine max*), which has high quality protein with high contents of sulphur-containing amino acids, is a good supplement in maize products. Because Snacks provide an avenue for introducing plant proteins such as soybeans to people who normally resist trying any unfamiliar food, kokoro was prepared from maize–soybean flour mixtures in ratios of 100:0, 90:10, 80:20, 70:30, 60:40 and 50:50. The physical, compositional, and sensory characteristics of kokoro were evaluated. Protein and fat contents increased, while carbohydrate content decreased as the soy flour proportion of the flour mixture used in the kokoro was increased.

The bulk density and water-holding capacity increased with increasing proportion of soybean flour, while the swelling capacity was found to decrease. High soy-substitution significantly reduced the sensory acceptance of kokoro. Sensory evaluation indicated that maize:soybean

flour mixture ratios of 100:0 and 90:0 were the most acceptable to the panellists. Several studies have been conducted to improve the protein quality of food products by fortification with plant proteins such as soybean, which is less expensive. Amino acid fortification was suggested by [18]. [19] also determined the acceptability of fermented maize meal fortified with defatted soy-flour in traditional Ghanaian foods.

7. Conclusion

Soybean flour has huge potentials of being used to enrich foods in order to provide adequate nutrients for individuals not meeting daily needs. Based on the available information on the nutrients profile of soybean including the amino profiles, human consumption of soybean flour can be promoted because of its positive effect on nutritional enhancement on different fortified food products. However, more efforts need to be directed at addressing associated technological issue which is flatulence to further increase effective utilization of the food product in food fortification.

Author details

O. E. Adelakun[1*], K. G. Duodu[1], E. Buys[1] and B. F. Olanipekun[2]

*Address all correspondence to: olu_yemisi2000@yahoo.com

1 Department of Food Science, University of Pretoria, Pretoria 0002, South Africa

2 Department of Food Science and Engineering, Ladoke Akintola University of Technology, Nigeria

References

[1] Adelakun, O. E., Adejuyitan, J. A., Olajide, J. O., & Alabi, B. K. (2005). Effect of Soybean Substitution on some Physical, Compositional and Sensory Properties of Kokoro (A Local Maize Snack). *European Food Research and Technology.*, 220, 79-82.

[2] Akpapunan, M. A., & Markakis, P. (1979). Oligosaccharides of 13 Cultivars of Cowpea (Vigna Sinensis). *J. Food Science*, 44, 317-318.

[3] Allan, L., Benoist, B., Dary, O., & Hurrell, R. (2006). Guidelines on food fortification with micronutrients. *A Report: Word Health Organization and Food and Agriculture Organization of the United States.*

[4] Ang, C. Y. W., Liu, K., & Huang, Y. W. (1999). Asian Foods Science and Technology. *Technomic Publishing Company Inc*, 139.

[5] Charles, A., & Guy, L. (1999). Food Biochemistry. *Aspen Publishers Inc. Gaitheasburg, Maryland*, 39-41.

[6] Dulger, H., Arik, M., Sekeroglu, M. R., Tarakcioglu, M., Noyan, T., & Cesur, Y. (2002). Proinflammatory cytokines in Turkish children with protein energy malnutrition. *Mediators Inflamm*, 11, 363-5.

[7] FFTC. (2002). Food and Fertilizer Technology. *An International Center for Farmers in the Asia Pacific Region. Taiwan.*

[8] Gray, V. B., Cossman, J. S., & Powers, E. L. (2006). Stunted growth is associated with physical indicators of malnutrition but not food insecurity among rural school children in Honduras. *Nutr Res*, 26(11), 549-55.

[9] Jong-Kyu, Ha., Kim, S. W., & Kim, W. Y. (1996). Use of Agro-Industrial. By-products a Animal Feeds in Korea. *A publication of FFTC. An International Centre for Farmers in the Asia Pacific Region.*

[10] Liu, K. (2000). Expanding soybean food utilization. *Food Technol*, 54(7), 46-47.

[11] Ndubuaku, V. O., Nwaegbule, A. C., & Nnanyeluyo, D. O. (1989). Flatulence and Other Discomfort Associated with Consumption of Cowpea (Vigna uguiculata). *Appetite*, 13, 171-181.

[12] Njintang, N. Y., Mbofung, C. M. F., & Waldron, K. W. (2001). In Vitro Protein Digestibility and Physicochemical Properties of Dry Red Bean (Phaseolus vulgaris) Flour:

Effect of Processing and Incorporation of Soybean and Cowpea Flour. *J. Agric. Food Chem.*, 49, 2465-2471.

[13] Nnakwe, N. (1995). The effect and causes of protein-energy malnutrition in Nigerian children. *Nutrition Research*, 15(6), 785-794.

[14] NSRL. (2002). A Publication of China National Soybean Research Laoratory on Soybean Procesing:. *From Field to Consumer*, http://www/NSRL.

[15] Olanipekun, B. F., Otunola, E. T., Adelakun, O. E., & Oyelade, O. J. (2009). Effect of fermentation with Rhizopus oligosporus on some physico-chemical properties of starch extracts from soybean flour. *Food and Chemical Toxicology*, 1401-1405.

[16] Onwueme, I. C., & Sinha, T. D. (1999). Field crop production in tropical Africa. *England, Michael Health Ltd.*, 402-414.

[17] Parman, G. K. (1968). Fortification of Cereals and Cereal Products with Proteins and Amino Acids. *J. Agric Food Chem.*, 16.

[18] Potter, N. N. (1978). Food Science. *3rd Edition, Ilhiala, New York*, 356.

[19] Plahar, W. A., & Leung, H. K. (1983). Composition of Ghanian fermented maize meal and the effect of soya fortification on sensory properties. *J Sci Food Agric*, 34, 407-411.

[20] Sereebutra, P., Solomons, N., Aliyu, M. H., & Jolly, P. E. (2006). Sociodemographic and environmental predictors of childhood stunting in rural Guatemala. *Nutr Res*, 26, 65-70.

[21] Senshoku, K. S. (2002). Physical Characteristics and Processing method of Chinese Soybean Fiber. *Home Technology Jianghe Tianrongbi. Fiber Co. Ltd. (China)*.

[22] Shils, M. E., Olson, J. A., & Moshe, S. (1999). Modern Nutrition in Health and Disease. *Philadelphia Lippincott Williams & Wilkins*.

[23] Tunde-Akintunde, T. Y. (2000). Predictive Models for Evaluating the Effect of Processing Parameters or Yield and Quality of Some Soybean (Glycine max (L) Merri) Products. *Ph.D Thesis- Universityof Ibadan, Nigeria*.

Permissions

The contributors of this book come from diverse backgrounds, making this book a truly international effort. This book will bring forth new frontiers with its revolutionizing research information and detailed analysis of the nascent developments around the world.

We would like to thank Hany A. El-Shemy, for lending his expertise to make the book truly unique. He has played a crucial role in the development of this book. Without his invaluable contribution this book wouldn't have been possible. He has made vital efforts to compile up to date information on the varied aspects of this subject to make this book a valuable addition to the collection of many professionals and students.

This book was conceptualized with the vision of imparting up-to-date information and advanced data in this field. To ensure the same, a matchless editorial board was set up. Every individual on the board went through rigorous rounds of assessment to prove their worth. After which they invested a large part of their time researching and compiling the most relevant data for our readers. Conferences and sessions were held from time to time between the editorial board and the contributing authors to present the data in the most comprehensible form. The editorial team has worked tirelessly to provide valuable and valid information to help people across the globe.

Every chapter published in this book has been scrutinized by our experts. Their significance has been extensively debated. The topics covered herein carry significant findings which will fuel the growth of the discipline. They may even be implemented as practical applications or may be referred to as a beginning point for another development. Chapters in this book were first published by InTech; hereby published with permission under the Creative Commons Attribution License or equivalent.

The editorial board has been involved in producing this book since its inception. They have spent rigorous hours researching and exploring the diverse topics which have resulted in the successful publishing of this book. They have passed on their knowledge of decades through this book. To expedite this challenging task, the publisher supported the team at every step. A small team of assistant editors was also appointed to further simplify the editing procedure and attain best results for the readers.

Our editorial team has been hand-picked from every corner of the world. Their multi-ethnicity adds dynamic inputs to the discussions which result in innovative outcomes. These outcomes are then further discussed with the researchers and contributors who give their valuable feedback and opinion regarding the same. The feedback is then collaborated with the researches and they are edited in a comprehensive manner to aid the understanding of the subject.

Apart from the editorial board, the designing team has also invested a significant amount of their time in understanding the subject and creating the most relevant covers. They scrutinized every image to scout for the most suitable representation of the subject and create an appropriate cover for the book.

The publishing team has been involved in this book since its early stages. They were actively engaged in every process, be it collecting the data, connecting with the contributors or procuring relevant information. The team has been an ardent support to the editorial, designing and production team. Their endless efforts to recruit the best for this project, has resulted in the accomplishment of this book. They are a veteran in the field of academics and their pool of knowledge is as vast as their experience in printing. Their expertise and guidance has proved useful at every step. Their uncompromising quality standards have made this book an exceptional effort. Their encouragement from time to time has been an inspiration for everyone.

The publisher and the editorial board hope that this book will prove to be a valuable piece of knowledge for researchers, students, practitioners and scholars across the globe.

List of Contributors

Samuel N. Nahashon
Department of Agricultural Sciences, Tennessee State University, Nashville, TN, USA

Agnes K. Kilonzo-Nthenge
Department of Family and Consumer Sciences, Tennessee State University, Nashville, TN, USA

D. Pettersson and K. Pontoppidan
Department of Feed Applications, Novozymes A/S, Denmark

Fred A. Kummerow
University of Illinois, USA

A. Cabrera-Orozco, C. Jiménez-Martínez and G. Dávila-Ortiz
Graduates in Food, National School of Biological Sciences. National Polytechnic Institute, Mexico

Sherif M. Hassan
College of Agricultural and Food Sciences, King Faisal University, Kingdom of Saudi Arabia

Eugene A. Borodin
Amur State Medical Academy, Blagoveshchensk, Russian Federation

Igor E. Pamirsky
Institute of Geology and Natural Management of the Far Easteren Branch of Russia Academy of Sciences, Russian Federation

Mikhail A. Shtarberg
Amur State Medical Academy, Blagoveshchensk, Russian Federation

Vladimir A. Dorovskikh
Amur State Medical Academy, Blagoveshchensk, Russian Federation

Alexander V. Korotkikh
Amur State Medical Academy, Blagoveshchensk, Russian Federation

Chie Tarumizu
Asian Nutrition and Food Culture Research Center, Jumonji University, Saitama, Japan

Shigeru Yamamoto
Asian Nutrition and Food Culture Research Center, Jumonji University, Saitama, Japan

Kiyoharu Takamatsu
Fuji Oil Co., Ltd Osaka, Japan

Kenjiro Koga
Faculty of Pharmaceutical Sciences, Hokuriku University, Japan

Masakazu Naya and Masanao Imai
Course in Bioresource Utilization Sciences, Graduate School of Bio resource Sciences, Nihon University, Japan

Neusa Fátima Seibel
Mestrado Profissional em Tecnologia de Alimentos. Universidade Tecnológica Federal do Paraná (UTFPR). Londrina/PR, Brazil

Fernanda Périco Alves
Tecnóloga em Alimentos (UTFPR), Londrina/PR, Brazil

Marcelo Álvares de Oliveira
Embrapa Soja. Londrina/PR, Brazil

Rodrigo Santos Leite
Mestrando em Tecnologia de Alimentos (UTFPR). Embrapa Soja. Londrina/PR, Brazil

Eduardo Fuentes
Departamento de Bioquímica Clínica e Inmunología, Facultad de Ciencias de la Salud, Universidad de Talca, Chile
Centro de Estudios en Alimentos Procesados (CEAP), Conicyt-Regional, Gore Maule, R09I2001, Talca, Chile

Luis Guzmán
Departamento de Bioquímica Clínica e Inmunología, Facultad de Ciencias de la Salud, Universidad de Talca, Chile

Gilda Carrasco
Departamento de Horticultura, Facultad de Ciencias Agrarias, Universidad de Talca, Chile
Centro de Estudios en Alimentos Procesados (CEAP), Conicyt-Regional, Gore Maule, R09I2001, Talca, Chile

Elba Leiva
Departamento de Bioquímica Clínica e Inmunología, Facultad de Ciencias de la Salud, Universidad de Talca, Chile

Rodrigo Moore-Carrasco
Departamento de Bioquímica Clínica e Inmunología, Facultad de Ciencias de la Salud, Universidad de Talca, Chile
Centro de Estudios en Alimentos Procesados (CEAP), Conicyt-Regional, Gore Maule

Iván Palomo
Departamento de Bioquímica Clínica e Inmunología, Facultad de Ciencias de la Salud, Universidad de Talca, Chile
Centro de Estudios en Alimentos Procesados (CEAP), Conicyt-Regional, Gore Maule, R09I2001, Talca, Chile

O. E. Adelakun
Department of Food Science, University of Pretoria, Pretoria 0002, South Africa

K. G. Duodu
Department of Food Science, University of Pretoria, Pretoria 0002, South Africa

E. Buys
Department of Food Science, University of Pretoria, Pretoria 0002, South Africa

B. F. Olanipekun
Department of Food Science and Engineering, Ladoke Akintola University of Technology, Nigeria